CONCEPTS IN QUANTUM MECHANICS

PURE AND APPLIED PHYSICS
A SERIES OF MONOGRAPHS AND TEXTBOOKS

CONSULTING EDITORS

H. S. W. MASSEY

University College, London, England

KEITH A. BRUECKNER

*University of California, San Diego
La Jolla, California*

CONCEPTS IN QUANTUM MECHANICS

F. A. Kaempffer
Department of Physics
The University of British Columbia
Vancouver, Canada

1965

ACADEMIC PRESS • New York and London

ACADEMIC PRESS INC.
111 Fifth Avenue, New York, New York 10003

United Kingdom Edition published by
ACADEMIC PRESS INC. (LONDON) LTD.
Berkeley Square House, London W.1

LIBRARY OF CONGRESS CATALOG CARD NUMBER: 64-24660

PRINTED IN THE UNITED STATES OF AMERICA

Preface

Quantum mechanics is no longer a revolutionary theory. In the more than 35 years since its inception it has become an established branch of physics.

Students of quantum mechanics are saved trouble if they are not led through all the historical pitfalls, and instead acquainted from the very beginning with concepts, such as spin, that cannot be grasped except by quantum mechanical means.

The first aim of this work has thus been to break away from the traditional presentation of the subject matter, by casting aside as much as possible reliance on arguments based on the correspondence with classical physics.

With this in mind, certain terminological usages, such as calling a ψ-function a "wave function" or referring to the particle occupation number representation as "second quantization," have been eliminated deliberately, because teaching experience has shown that such terms are the source of misunderstanding or of misleading thinking habits that are difficult to eradicate once they are established.

The second aim has been to emphasize topics of contemporary interest to physicists engaged in experimental research.

There are excellent texts, for example the classics by Schiff or by Landau and Lifshitz, which treat standard problems, such as the energy levels of the hydrogenic electron or the eigenstates of the angular momentum operators, and no need was felt to compete with these treatments. Accordingly, standard problems have been included only if they were needed for illustration of basic concepts.

On the other hand, topics such as time reversal invariance, or superselection rules, or the interaction picture, are in most texts treated as highbrow and couched in group theoretical or field mechanical language inaccessible to many experimental physicists, although in fact these topics are elementary and should be treated as such.

In this sense the present work is an attempt to present advanced quantum mechanics from an elementary point of view.

v

This need not disqualify it to serve as an introductory text. The content of this work has actually been given, in the author's opinion successfully, together with assigned problems and numerous references as an introductory two-year course for graduate students at The University of British Columbia.

Habits die hard, however, and the author realizes that in fact this work may find its most common use as a supplementary text to the more standard treatments.

Vancouver, Canada   F. A. KAEMPFFER

Table of Contents

*Sections preceded by an asterisk may be omitted on first reading.

solely on analysis of experimental situations realized by
selective measurements, with compound spin orientation
experiments serving as illustration. One is led in this context
to consider in

the quasi-particle concept properties of macroscopic objects with the same formalism originally designed for description of so-called elementary particles. The division into fermions and bosons is defined. The exclusion principle is stated and used to describe in

quantum number. The principle of reciprocity is proved and various conditions are examined under which this leads to validity of the principle of detailed balance.

CONCEPTS IN QUANTUM MECHANICS

Prelude

Quantum mechanics purports to be a description of physical reality which deliberately eliminates from theory all features not demanded by experiment. That there should be any need at all for special efforts to accomplish this obviously sensible aim is partly due to a peculiar feature of human language.

Language has been largely fashioned after macroscopic models. For example, if the word "particle" is used as a subject to which various physical properties are attributed, one notices the subversive effect of language when one tries to completely avoid the surreptitious use of some mental image of a "particle" between measurements. However, to refrain from using inappropriate mental models is just the kind of intellectual asceticism demanded by quantum mechanics.

Development of quantum mechanics was forced by the recognition of two distinct experimental aspects of physical reality which cannot be accommodated within the framework of classical mechanics.

(1) Beginning with the turn of this century it transpired that some dynamical quantities, which in classical mechanics are always accorded a continuous range of values, may in fact assume only certain discrete values. Any physical quantity A which assumes only discrete values $a_1, \ldots, a_n, \ldots$ is said to be "quantized." Examples are the energy values of an electron bound to an atomic nucleus, and the intrinsic angular momentum or "spin" of an elementary "particle."

Although temporary amends were made by imposition of more or less mysterious "quantum rules" on classical mechanics, these artifices became untenable when one finally had to face in atomic physics yet another fundamental aspect of reality, namely

(2) the impossibility of simultaneous exact determination of the totality of physical attributes of an object. Observation of atomic phenomena and their logical analysis, in particular, led to the discovery that

(a) the interaction between the object of a measurement and the measuring apparatus cannot be indefinitely weakened, and

1

(*b*) the disturbance produced by the interaction object-apparatus is only statistically predictable and cannot, therefore, be compensated.

Thus measurement of one attribute of an object can produce uncontrollable changes in the value previously assigned to another attribute of the same object. It is therefore meaningless to simultaneously assign numerical values to *all* attributes of an object.

These facts governing the atomic domain are inconsistent with the classical theory of measurement which is based on the belief that the interaction object-apparatus, if it cannot be made negligibly small, can at least be taken into account precisely and can thus, in principle, be compensated.

Two physical attributes $A^{(1)}$ and $A^{(2)}$ will be called "compatible" if measurement of one does not affect the value assigned to the other by a preceding measurement. Examples of compatible attributes are: The absolute value J and one component, J_3 say, of the angular momentum of an object. The three components of the momentum of an object. The energy and any one other conserved quantity in a closed system.

Two physical attributes A and B which are not compatible are called "incompatible." Examples of incompatible attributes are: any two components of angular momentum of an object; momentum and parity of an object, provided the momentum does not have the value zero; strangeness and conjugality of an elementary particle, provided the strangeness is not zero.

Every physical system will now be assumed to possess a *complete* set of compatible physical attributes $A^{(1)} \ldots A^{(k)}$, so that any two of these attributes are compatible and that no other attributes exist which are compatible with every member of the set. This assumption is not trivial, because at present there exists no experimental criterion which would allow one to determine whether a compatible set is complete. For example, it was thought for a long time that the complete set of attributes of an electron consisted of momentum $\mathbf{p}$, mass m, charge e, spin s, s_3. Recently yet another attribute of the electron has emerged, the lepton number L, which is compatible with all members of the set $\mathbf{p}$, m, e, s, s_3.

It should be stressed again that the number of physical attributes in one complete set A is, in general, much smaller than the number of *all possible* physical attributes of an object. One can usually find other complete sets B, C, ... which are mutually incompatible. In the example above s_3 can be replaced by a component of spin other than s_3, resulting in a different complete set.

A "complete measurement" on an object means a set of observations enabling one to ascribe definite values $a_i^{(k)}$ to a maximum number of compatible attributes $A^{(k)}$. Since any experiment designed to find the value b of another attribute B will now affect one or more of the previously established values $a_i^{(k)}$ in an uncontrollable way, any complete measurement realizes the optimum state of knowledge about a given object.

If the set of compatible attributes $A^{(k)}$ of an object is known to have the values $a_i^{(k)}$, the object is said to be in a "pure quantum state" or simply "state" characterized by the quantum numbers $a_i^{(k)}$. One is thus led to search for a description of the state of an object, containing all the information represented by the set of numbers $a_i^{(k)}$, and containing *only* that information.

Pure States

The basic mathematical concepts needed for the description of the state of an object are most simply developed for idealized physical systems in which any physical attribute A can assume only a finite number of discrete values or quantum numbers $a_1, ..., a_n$.

In reality many physical attributes are, of course, even in atomic physics, capable of assuming a continuous range of values. To grasp such continuous sets of quantum numbers requires some mathematical sophistication, however, which might tend to obscure the basic simplicity of the quantum mechanical formalism, and for this reason they will be excluded from consideration temporarily, to be taken up later in Section 8.

A very simple state is the one depending on a dichotomic attribute, i.e. an attribute which can have two values only. Examples of dichotomic attributes are: the electron spin component in direction of an applied magnetic field; the electric charge that may be carried by a nucleon; the parity of a set of pions in the center of mass frame; the number of fermions in a given quantum state.

As a starting point consider an experimental situation in which a spin $\frac{1}{2}$ associated with a magnetic moment is known to be aligned in a given direction, which may be sufficiently characterized by two polar angles ϑ, φ. Such a spin state can always be prepared by performing a Stern-Gerlach type of experiment with the external field in direction ϑ, φ and letting only the appropriate component of the split beam emerge from the apparatus. This state will be represented by a two-dimensional complex unit vector or "state vector" denoted

$$(1.1) \qquad |a_1\rangle = \begin{pmatrix} \alpha_1 \\ \beta_1 \end{pmatrix}$$

where the two complex numbers α_1 and β_1 are functions of the direction ϑ, φ of the spin. The label "a_1" shall represent the quantum number "spin up in direction ϑ, φ." The requirement that $|a_1\rangle$ be a unit vector, meaning that the scalar product of $|a_1\rangle$ with its hermitean conjugate, denoted

$$(1.2) \qquad \langle a_1| = \widehat{\alpha_1^* \quad \beta_1^*}$$

5

be unity, namely

(1.3) $\langle a_1 | a_1 \rangle = \overbrace{\alpha_1^* \quad \beta_1^*} \begin{pmatrix} \alpha_1 \\ \beta_1 \end{pmatrix} = |\alpha_1|^2 + |\beta_1|^2 = 1$

qualifies the quantities $|\alpha_1|^2$ and $|\beta_1|^2$ for a probability interpretation which will be arranged to suit the following experimental fact.

If on an atomic beam with electronic spins $\frac{1}{2}$ aligned in direction ϑ, φ and represented by the state vector $|a_1\rangle$ another Stern-Gerlach experiment is performed with external field in parallel with the z-axis, this observation of the z component of spin interferes with the state $|a_1\rangle$ such that some spins are aligned parallel to the z-axis and some antiparallel to it. The outcome of this experiment is, in principle, only statistically predictable. Thus, there are required two numbers, adding up to unity, one of which is the probability for finding the spin aligned parallel to the z-axis and the other the probability for finding the spin aligned antiparallel to the z-axis. This suggests arranging the dependence of α_1 and β_1 on ϑ and φ such that $|\alpha_1|^2$ and $|\beta_1|^2$ can be identified as these probabilities, respectively. With this convention the states $|a_+\rangle = \binom{1}{0}$ and $|a_-\rangle = \binom{0}{1}$ represent situations in which the spin in z direction has with certainty the values $+1$ and -1, respectively. [In the following, components of a state vector $|a_i\rangle$ will generally be denoted $(a_i)_k$. With this convention $\alpha_1 = (a_1)_1$ and $\beta_1 = (a_1)_2$.]

In generalization of this probability interpretation it will now be assumed that the projection of a state

$$|a\rangle = \begin{pmatrix} (a)_1 \\ (a)_2 \\ \vdots \end{pmatrix} \qquad \text{on another state} \qquad |b\rangle = \begin{pmatrix} (b)_1 \\ (b)_2 \\ \vdots \end{pmatrix}$$

represents the amplitude for finding upon measurement the quantum number b if the object is known to be in the state $|a\rangle$, where the projection is defined as

(1.4) $\langle b | a \rangle = (b)_1^* (a)_1 + (b)_2^* (a)_2 + \ldots = \langle a | b \rangle^*.$

Thus, $|\langle b | a \rangle|^2$ *shall be interpreted as the probability for observing the quantum number* b, *if the object is known to be in a state characterized by the quantum number* a. For example, if $|a_+\rangle = \binom{1}{0}$ is the state in which the spin is with certainty parallel to the z-axis, and $|a_1\rangle = \binom{\alpha_1}{\beta_1}$ is the state in which the spin is parallel to the direction ϑ, φ, then

$$\langle a_+ | a_1 \rangle = 1 . \alpha_1 + 0 . \beta_1 = \alpha_1,$$

and $|\alpha_1|^2$ is the probability for finding the spin $+1$ in z direction if the object is in the spin state $|a_1\rangle$.

This basic postulate of quantum mechanics enables one to infer from the dichotomic nature of electronic spin existence of another unit vector

$$(1.5) \qquad |a_2\rangle = \begin{pmatrix} \alpha_2 \\ \beta_2 \end{pmatrix} = \begin{pmatrix} (a_2)_1 \\ (a_2)_2 \end{pmatrix}; \qquad |\alpha_2|^2 + |\beta_2|^2 = 1$$

representing the state of spin antiparallel to the direction ϑ, φ, which is orthogonal to the vector $|a_1\rangle$,

$$(1.6) \qquad \langle a_2|a_1\rangle = 0,$$

because if the object is known to be in the spin state $|a_1\rangle$, the probability for finding the spin in opposite direction is zero. The two possible spin states that can be found upon observation of spin in a given direction ϑ, φ are thus represented by a system of two orthogonal complex vectors,

$$(1.7) \qquad \langle a_i|a_k\rangle = \delta_{ik} \qquad (i, k = 1, 2).$$

Since observation of the spin in a given direction ϑ, φ is a complete measurement and realizes the optimum knowledge attainable about electronic spin, the system of basis vectors $|a_1\rangle$, $|a_2\rangle$ should be complete, i.e. satisfy the closure relation

$$(1.8) \qquad \sum_i |a_i\rangle\langle a_i| = I,$$

the identity symbol I being defined by $\langle b|I|a\rangle = \langle b|a\rangle$ for any vectors $|b\rangle$, $|a\rangle$. Indeed, if one considers measurement of spin in some other direction ϑ', φ', symbolized by two orthonormal vectors $|b_1\rangle$, $|b_2\rangle$, the basic interpretation postulate of quantum mechanics requires that these vectors can be written as linear superpositions

$$(1.9) \quad |b_k\rangle = |a_1\rangle c_{1k} + |a_2\rangle c_{2k} = |a_1\rangle\langle a_1|b_k\rangle + |a_2\rangle\langle a_2|b_k\rangle.$$

From this follows by projection and after use of (1.7)

$$(1.10) \qquad \langle b_j|b_k\rangle = \sum_i \langle b_j|a_i\rangle\langle a_i|b_k\rangle$$

which is consistent only if relation (1.8) holds.

The probability interpretation, the orthonormality condition, and the closure relation are not affected if any state vector is multiplied by a number of modulus unity. Thus, *the same physical state* $|a\rangle$ *is described by all vectors* $e^{i\alpha}|a\rangle$, *where* α *is any real number*.

The concepts developed here for the special case of the spin $\frac{1}{2}$ are easily adapted to any state depending on a dichotomic variable. Thus the state vector

$$(1.11) \qquad |a\rangle = \begin{pmatrix} \alpha \\ \beta \end{pmatrix}; \qquad |\alpha|^2 + |\beta|^2 = 1$$

will describe the charge state of a nucleon, if $|\alpha|^2$ is interpreted as the probability for finding the nucleon as a proton, $|\beta|^2$ as the probability for finding it as a neutron. Similarly, the same symbol can describe the occupation state of a fermion quantum state, if $|\alpha|^2$ and $|\beta|^2$ are interpreted as the probabilities for finding that state empty or full, respectively.

Extension of these notions to states depending on attributes that may have more than two values proceeds without difficulty. A simple example is the charge state of a pion, which requires for its description a state vector having three components

$$(1.12) \qquad |a\rangle = \begin{pmatrix} \alpha \\ \beta \\ \gamma \end{pmatrix}; \qquad |\alpha|^2 + |\beta|^2 + |\gamma|^2 = 1,$$

because according to experiment a pion may have positive, zero, or negative electric charge. In expression (1.12) α, β, γ are taken as the probability amplitudes for finding the pion positive, neutral, negative, respectively.

Additional compatible attributes of an object can always be accommodated by extending the dimensionality of the state vector space. Formally, this is done by forming the direct product of any two state vectors, namely

$$(1.13) \qquad |a_i^{(1)} a_k^{(2)}\rangle = |a_i^{(1)}\rangle \times |a_k^{(2)}\rangle = \begin{pmatrix} [a_i^{(1)}]_1 \\ [a_i^{(1)}]_2 \\ \vdots \end{pmatrix} \times \begin{pmatrix} [a_k^{(2)}]_1 \\ [a_k^{(2)}]_2 \\ \vdots \end{pmatrix}$$

$$= \begin{pmatrix} [a_i^{(1)}]_1 [a_k^{(2)}]_1 \\ [a_i^{(1)}]_1 [a_k^{(2)}]_2 \\ \vdots \\ [a_i^{(1)}]_2 [a_k^{(2)}]_1 \\ [a_i^{(1)}]_2 [a_k^{(2)}]_2 \\ \vdots \end{pmatrix}$$

where $|a_i^{(1)}\rangle$ is the state in which the attribute $A^{(1)}$ has the value $a_i^{(1)}$, and $|a_k^{(2)}\rangle$ the state in which $A^{(2)}$ has the value $a_k^{(2)}$. For example, a nucleon possesses two dichotomic attributes, spin and charge. The combined spin and charge space is spanned by four unit vectors, which in a self-explanatory notation may be written $|p\uparrow\rangle$; $|p\downarrow\rangle$; $|n\uparrow\rangle$; $|n\downarrow\rangle$ and represented by the set of direct products

$$|p\uparrow\rangle = \begin{pmatrix} 1 \\ 0 \end{pmatrix}_{charge} \times \begin{pmatrix} 1 \\ 0 \end{pmatrix}_{spin} = \begin{pmatrix} 1 \\ 0 \\ 0 \\ 0 \end{pmatrix}; \qquad |p\downarrow\rangle = \begin{pmatrix} 1 \\ 0 \end{pmatrix}_c \times \begin{pmatrix} 0 \\ 1 \end{pmatrix}_s = \begin{pmatrix} 0 \\ 1 \\ 0 \\ 0 \end{pmatrix};$$

(1.14)

$$|n\uparrow\rangle = \begin{pmatrix}0\\1\end{pmatrix}_c \times \begin{pmatrix}1\\0\end{pmatrix}_s = \begin{pmatrix}0\\0\\1\\0\end{pmatrix}; \qquad |n\downarrow\rangle = \begin{pmatrix}0\\1\end{pmatrix}_c \times \begin{pmatrix}0\\1\end{pmatrix}_s = \begin{pmatrix}0\\0\\0\\1\end{pmatrix}.$$

NOTES

Gerlach and Stern [1] were first to observe the splitting into two components of an atomic beam in an inhomogeneous magnetic field. This was later interpreted as being caused solely by the magnetic moment associated with the spin $\frac{1}{2}$ of the remaining valence electron.

Fermi [2] gives a lucid account of the use of vectors depending on a dichotomic variable, and uses them for the description of the charge state of systems of nucleons in particular.

Dirac [3] invented the bracket notation used throughout this work.

REFERENCES

[1] W. Gerlach and O. Stern, *Z. Physik.* **8**, 110 (1922).

[2] E. Fermi, Lectures on pions and nucleons, *Nuovo Cimento Suppl.* **2**, 18 (1955).

[3] P. A. M. Dirac, "Quantum Mechanics," 4th ed. Oxford Univ. Press, London and New York, 1958.

Observables

The measurement of an attribute A on an object in a state $|b\rangle$ will, in general, interfere with the state $|b\rangle$ such that after observation of A the object will be found in one of the states $|a_i\rangle$ which affix the value a_i of A to the object. Which of the states $|a_1\rangle$, $|a_2\rangle$, ... will result from the measurement, i.e. which of the possible values a_1, a_2, ... will actually be observed in an individual measurement of A, can be predicted only statistically.

Although it is thus not, in general, possible to associate a specific value a_i of A with a given state $|b\rangle$, an average value $\bar{A}$ of A will emerge if the observation of A is carried out on an ensemble of objects all in the same pure state $|b\rangle$.

For example, consider again the spin state $|a_1\rangle = \left(\begin{smallmatrix}\alpha_1\\\beta_1\end{smallmatrix}\right)$, describing a spin $\boldsymbol{\sigma} = 2\mathbf{s}(|\mathbf{s}| = \frac{1}{2})$ having with certainty the value $+1$ in direction ϑ, φ, and suppose an observation of the spin component in z direction, σ_3, is carried out. The result of observation, performed by means of a Stern-Gerlach experiment with the magnetic field in z direction, will be the establishment of either state $|a_+\rangle = \left(\begin{smallmatrix}1\\0\end{smallmatrix}\right)$ or state $|a_-\rangle = \left(\begin{smallmatrix}0\\1\end{smallmatrix}\right)$, i.e. σ_3 will be found to have either the value $+1$ or the value -1, with probabilities $|\alpha_1|^2$ and $|\beta_1|^2$, respectively. However, if the average value of the spin transforms under rotations of the coordinate system as a vector, one should expect $\bar{\sigma}_3$ to be simply the projection of the unit vector in (ϑ, φ) direction on the z-axis, namely

$$(2.1) \qquad \bar{\sigma}_3 = \cos\vartheta.$$

Similarly, the average values of σ_1 and σ_2 in the spin state $|a_1\rangle$ should come out to be

$$(2.2) \qquad \bar{\sigma}_1 = \sin\vartheta\cos\varphi$$

$$(2.3) \qquad \bar{\sigma}_2 = \sin\vartheta\sin\varphi.$$

The formalism which reflects the interference of observation with states and at the same time provides a means for computing the average value of an observable A in any given state $|b\rangle$ consists of representing an observable by a hermitean linear operator such that the eigenvectors

11

of this operator are the states $|a_i\rangle$, in which A has definite values a_i, and such that the corresponding eigenvalues are the numbers a_i themselves,

$$(2.4) \qquad A|a_i\rangle = a_i|a_i\rangle.$$

Such a representation is possible because the eigenvalues of a hermitean linear operator form an orthogonal system that can be normalized. This representation of the observable A as a linear operator satisfying (2.4) is also convenient, because A will transform a given state $|b\rangle$, in general, into another state $|c\rangle$

$$(2.5) \qquad |c\rangle = A|b\rangle$$

such that the projection of $|c\rangle$ on $|b\rangle$ is the average value of A in the state $|b\rangle$,

$$(2.6) \qquad \bar{A} = \langle b|c\rangle = \langle b|A|b\rangle.$$

To see this, expand the state $|b\rangle$ in terms of the basis vectors $|a_i\rangle$

$$(2.7) \qquad |b\rangle = |a_1\rangle\langle a_1|b\rangle + |a_2\rangle\langle a_2|b\rangle + \ldots$$

where $\langle a_i|b\rangle$ is the probability amplitude for finding the value a_i of A in state $|b\rangle$. As a consequence of (2.4) one has

$$(2.8) \qquad A|b\rangle = a_1|a_1\rangle\langle a_1|b\rangle + a_2|a_2\rangle\langle a_2|b\rangle + \ldots$$

and thus

$$(2.9) \qquad \langle b|A|b\rangle = a_1|\langle a_1|b\rangle|^2 + a_2|\langle a_2|b\rangle|^2 + \ldots$$

which is, by definition, the average value of A, because each value a_i contributes to A with the corresponding probability $|\langle a_i|b\rangle|^2$.

Equation (2.5) can be written explicitly in matrix and vector notation

$$(2.10) \qquad \begin{pmatrix} (c)_1 \\ (c)_2 \\ \vdots \end{pmatrix} = \begin{pmatrix} A_{11} & A_{12} & \cdots \\ A_{21} & A_{22} & \\ \vdots & & \end{pmatrix} \begin{pmatrix} (b)_1 \\ (b)_2 \\ \vdots \end{pmatrix}.$$

The requirement that the average value of a physical observable be a real number restricts the representation of observables by operators to hermitean operators, i.e. A must satisfy the condition

$$(2.11) \qquad A^+ = A$$

where the hermitean adjoint A^+ is defined by the equation

$$(2.12) \qquad \langle A^+c|b\rangle = \langle c|A|b\rangle.$$

Applying the matrix notation (2.10) to this definition, one verifies the operation of hermitean adjunction to be equivalent to transposing the matrix A and taking the complex conjugate of its elements,

$$(2.13) \qquad A_{ik}^{+} = A_{ki}^{*}.$$

The reality of $\bar{A}$ for hermitean operators follows from

$$(2.14) \quad \bar{A} = \langle b|A|b \rangle = \langle A^{+}b|b \rangle = \langle b|A^{+}|b \rangle^{*} = \langle b|A|b \rangle^{*} = \bar{A}^{*}.$$

It is sometimes useful to keep in mind that, although any observable must be represented by a hermitean operator, the converse need not be true, i.e. a hermitean operator need not necessarily represent an observable.

From the foregoing, it follows that the observable spin in direction ϑ, φ must be represented by a hermitean operator, which may be written as a 2×2 matrix

$$(2.15) \qquad \boldsymbol{\sigma}_{\vartheta,\varphi} = \begin{pmatrix} A_{11} & A_{12} \\ A_{12}^{*} & A_{22} \end{pmatrix}; \qquad A_{11}, A_{22} \text{ real};$$

so that $|a_1\rangle = \binom{\alpha_1}{\beta_1}$ and $|a_2\rangle = \binom{\alpha_2}{\beta_2}$ are the two eigenstates of $\boldsymbol{\sigma}_{\vartheta,\varphi}$ with eigenvalues $+1$ and -1, respectively,

$$(2.16) \qquad \boldsymbol{\sigma}_{\vartheta,\varphi}|a_1\rangle = +|a_1\rangle; \qquad \boldsymbol{\sigma}_{\vartheta,\varphi}|a_2\rangle = -|a_2\rangle$$

or, in components,

$$(2.17) \qquad \begin{array}{ll} A_{11}\alpha_1 + A_{12}\beta_1 = \alpha_1 & A_{11}\alpha_2 + A_{12}\beta_2 = -\alpha_2 \\ A_{12}^{*}\alpha_1 + A_{22}\beta_1 = \beta_1 & A_{12}^{*}\alpha_2 + A_{22}\beta_2 = -\beta_2 \end{array}$$

The necessary and sufficient conditions that these equations for α_i, β_i have solutions are that the coefficients have vanishing determinants, namely

$$(2.18) \qquad \begin{array}{l} A_{11}A_{22} - |A_{12}|^2 + 1 - A_{11} - A_{22} = 0; \\ A_{11}A_{22} - |A_{12}|^2 + 1 + A_{11} + A_{22} = 0 \end{array}$$

These equations are equivalent to the restrictions

$$(2.19) \qquad A_{11} + A_{22} = 0; \qquad A_{11}A_{22} - |A_{12}|^2 + 1 = 0$$

so that $\boldsymbol{\sigma}_{\vartheta,\varphi}$ can be represented in the form

$$(2.20) \quad \boldsymbol{\sigma}_{\vartheta,\varphi} = \begin{pmatrix} A_{11} & A_{12} \\ A_{12}^{*} & -A_{11} \end{pmatrix}; \qquad A_{11}^2 + |A_{12}|^2 = +1; \qquad A_{11} \text{ real}.$$

To obtain the dependence of α_i, β_i and A_{11}, A_{12} on the polar angles ϑ, φ, consider as a starting point the z component of spin σ_3. The corresponding operator must be a special case of (2.20), obtained by putting $\vartheta = 0$, and will be written

$$(2.21) \quad \sigma_3 = \begin{pmatrix} Z_{11} & Z_{12} \\ Z_{12}^* & -Z_{11} \end{pmatrix}; \qquad Z_{11}^2 + |Z_{12}|^2 = 1; \qquad Z_{11} \text{ real.}$$

The matrix elements Z_{11}, Z_{12} must be chosen such that the two states

$$(2.22) \qquad\qquad |a_+\rangle = \begin{pmatrix} 1 \\ 0 \end{pmatrix}, \qquad |a_-\rangle = \begin{pmatrix} 0 \\ 1 \end{pmatrix}$$

are the eigenstates of σ_3 with eigenvalues $+1$ and -1, respectively,

$$(2.23) \qquad\qquad \sigma_3|a_+\rangle = +|a_+\rangle; \qquad \sigma_3|a_-\rangle = -|a_-\rangle$$

or, in components,

$$(2.24) \qquad Z_{11} = 1, \quad Z_{12}^* = 0; \qquad Z_{12} = 0, \quad -Z_{11} = -1.$$

Thus σ_3 is found to be represented by the matrix

$$(2.25) \qquad\qquad\qquad \sigma_3 = \begin{pmatrix} 1 & 0 \\ 0 & -1 \end{pmatrix}.$$

If one demands now in accordance with (2.1) that the average value of σ_3 in the state $|a_1\rangle = \begin{pmatrix} \alpha_1 \\ \beta_1 \end{pmatrix}$ be $\cos\vartheta$, one obtains for α_1 and β_1 the condition

$$(2.26) \quad \bar{\sigma}_3 = \langle a_1|\sigma_3|a_1\rangle = \overbrace{\alpha_1^* \quad \beta_1^*}\begin{pmatrix} 1 & 0 \\ 0 & -1 \end{pmatrix}\begin{pmatrix} \alpha_1 \\ \beta_1 \end{pmatrix} = |\alpha_1|^2 - |\beta_1|^2 = \cos\vartheta$$

which together with the normalization condition

$$(2.27) \qquad\qquad\qquad |\alpha_1|^2 + |\beta_1|^2 = 1$$

yields the solution

$$(2.28) \qquad \alpha_1 = e^{i\psi}\cos(\vartheta/2); \qquad \beta_1 = e^{i(\chi+\psi)}\sin(\vartheta/2)$$

with arbitrary phases ψ and χ. The phase of β_1 has been written $(\chi + \psi)$ because any state vector is determined only up to a common phase of all its components in any case, and by convention the phase ψ will now be put equal to zero, so that $|a_+\rangle$ becomes the special case of $|a_1\rangle$ for $\vartheta = 0$. Thus

$$(2.29) \qquad\qquad\qquad |a_1\rangle = \begin{pmatrix} \cos(\vartheta/2) \\ e^{i\chi}\sin(\vartheta/2) \end{pmatrix}.$$

The remaining unknown phase χ will now be determined by utilizing conditions (2.2) and (2.3). Denoting the operators representing σ_1 and σ_2 by

$$(2.30) \qquad \sigma_1 = \begin{pmatrix} X_{11} & X_{12} \\ X_{12}^* & -X_{11} \end{pmatrix};$$

$$\sigma_2 = \begin{pmatrix} Y_{11} & Y_{12} \\ Y_{12}^* & -Y_{11} \end{pmatrix}; \qquad \begin{array}{l} X_{11}^2 + |X_{12}|^2 = 1; \\ Y_{11}^2 + |Y_{12}|^2 = 1; \end{array} \qquad \begin{array}{l} X_{11} \quad \text{real} \\ Y_{11} \quad \text{real} \end{array}$$

equations (2.2) and (2.3) read explicitly, if the average is taken for the state $|a_1\rangle$ as given in (2.29), using some elementary trigonometry,

$$(2.31) \qquad \begin{aligned} \bar{\sigma}_1 &= \langle a_1 | \sigma_1 | a_1 \rangle \\ &= X_{11} \cos\vartheta + \tfrac{1}{2}(X_{12}\, e^{i\chi} + X_{12}^*\, e^{-i\chi}) \sin\vartheta = \sin\vartheta \cos\varphi \\ \bar{\sigma}_2 &= \langle a_1 | \sigma_2 | a_1 \rangle \\ &= Y_{11} \cos\vartheta + \tfrac{1}{2}(Y_{12}\, e^{i\chi} + Y_{12}^*\, e^{-i\chi}) \sin\vartheta = \sin\vartheta \sin\varphi. \end{aligned}$$

Since $\sin\vartheta$ and $\cos\vartheta$ are linearly independent these equations require

$$(2.32) \qquad\qquad X_{11} = Y_{11} = 0$$

so that according to (2.30) X_{12} and Y_{12} must have modulus unity,

$$(2.33) \qquad\qquad X_{12} = e^{i\xi}; \qquad Y_{12} = e^{i\eta}.$$

Substitution into (2.31) gives the equations

$$(2.34) \qquad \begin{aligned} \tfrac{1}{2}[e^{i(\chi+\xi)} + e^{-i(\chi+\xi)}] &= \cos\varphi \\ \tfrac{1}{2}[e^{i(\chi+\eta)} + e^{-i(\chi+\eta)}] &= \sin\varphi. \end{aligned}$$

From the first of these equations follows

$$(2.35) \qquad\qquad \chi + \xi = \varphi$$

which, substituted into the second equation (2.34) gives

$$(2.36) \qquad \tfrac{1}{2}[e^{i(\varphi+\eta-\xi)} + e^{-i(\varphi+\eta-\xi)}] = \sin\varphi.$$

This can be true only if

$$(2.37) \qquad\qquad \eta - \xi = -\pi/2.$$

Thus one of the phases χ, η, ξ remains undetermined, one chooses *by convention*

$$(2.38) \qquad\qquad \xi = 0$$

so that

$$(2.39) \qquad\qquad \eta = -\pi/2; \qquad \chi = \varphi$$

and therefore

$$(2.40) \qquad\qquad X_{12} = 1; \qquad Y_{12} = -i.$$

Applying the same line of reasoning to the state $|a_2\rangle$, one thus finally obtains the following representations in terms of the polar angles ϑ, φ

$$(2.41) \qquad |a_1\rangle = \begin{pmatrix} \cos(\vartheta/2) \\ \sin(\vartheta/2)\, e^{i\varphi} \end{pmatrix}$$

$$(2.42) \qquad |a_2\rangle = \begin{pmatrix} -\sin(\vartheta/2)\, e^{-i\varphi} \\ \cos(\vartheta/2) \end{pmatrix}$$

$$(2.43) \quad \sigma_1 = \begin{pmatrix} 0 & 1 \\ 1 & 0 \end{pmatrix}; \qquad \sigma_2 = \begin{pmatrix} 0 & -i \\ i & 0 \end{pmatrix}; \qquad \sigma_3 = \begin{pmatrix} 1 & 0 \\ 0 & -1 \end{pmatrix}.$$

The so-called Pauli matrices (2.43) are special cases of the general spin operator which, with help of (2.42) and (2.43), is found from Eqs. (2.17) and (2.20)

$$(2.44) \qquad \boldsymbol{\sigma}_{\vartheta,\,\varphi} = \begin{pmatrix} \cos\vartheta & \sin\vartheta\, e^{-i\varphi} \\ \sin\vartheta\, e^{i\varphi} & -\cos\vartheta \end{pmatrix}.$$

It is interesting to note that the operator $\boldsymbol{\sigma}_{\vartheta,\,\varphi}$ and its "cartesian components" σ_1, σ_2, σ_3 satisfy the same relation that holds between an ordinary vector pointing in direction ϑ, φ and its cartesian components,

$$(2.45) \qquad \boldsymbol{\sigma}_{\vartheta,\,\varphi} = \sigma_1 \sin\vartheta \cos\varphi + \sigma_2 \sin\vartheta \sin\varphi + \sigma_3 \cos\vartheta.$$

The Pauli matrices (2.34) have the algebraic properties

$$(2.46) \quad \sigma_i \sigma_k + \sigma_k \sigma_i = 2\delta_{ik} I; \qquad \sigma_1 \sigma_2 - \sigma_2 \sigma_1 = 2i\sigma_3 \quad \text{(cyclically)}.$$

Introducing $s_i = \tfrac{1}{2}\sigma_i$ one finds the commutation relations (C.R.s)

$$(2.47) \qquad s_1 s_2 - s_2 s_1 = i s_3 \qquad \text{(cyclically)}.$$

Such C.R.s will be recognized in Section 12 as a general property of any three operators that represent an angular momentum in quantum mechanics.

The optimum information obtainable about an object is, in general, contained in any state vector that is a linear superposition of the simultaneous eigenstates $|a_i^{(1)} a_k^{(2)} \ldots\rangle$ of the set of compatible observables $A^{(1)}$, $A^{(2)}$, ... The requirement that $|a_i^{(1)} a_k^{(2)} \ldots\rangle$ form a complete set can be met only if *the operators representing compatible observables* $A^{(1)} \ldots A^{(n)}$ *commute*,

$$(2.48) \qquad A^{(n)} A^{(m)} - A^{(m)} A^{(n)} = 0.$$

To see this, suppose $|a_i^{(1)} a_k^{(2)}\rangle$ is a simultaneous eigenstate of $A^{(1)}$ and $A^{(2)}$ with eigenvalues $a_i^{(1)}$ and $a_k^{(2)}$, respectively, so that

$$(2.49) \quad [A^{(1)} A^{(2)} - A^{(2)} A^{(1)}] |a_i^{(1)} a_k^{(2)}\rangle = (a_i^{(1)} a_k^{(2)} - a_k^{(2)} a_i^{(1)}) |a_i^{(1)} a_k^{(2)}\rangle = 0.$$

If now $|a_i^{(1)} a_k^{(2)}\rangle$ are required to form a complete set, then any state $|b\rangle$ can be expanded

$$(2.50) \qquad |b\rangle = \sum_i \sum_k c_{ik} |a_i^{(1)} a_k^{(2)}\rangle; \qquad c_{ik} = \langle a_i^{(1)} a_k^{(2)} | b\rangle$$

so that because of (2.49)

$$(2.51) \qquad [A^{(1)} A^{(2)} - A^{(2)} A^{(1)}] |b\rangle = 0$$

for all vectors $|b\rangle$, which can be true only if

$$(2.52) \qquad A^{(1)} A^{(2)} - A^{(2)} A^{(1)} = 0.$$

The general case (2.48) follows from this by induction.

One may, of course, count two or more compatible observables as a single observable which upon observation yields two or more numbers. This freedom is reflected in the possibility of representing, in the space spanned by the direct products, $|a_i^{(1)}\rangle \times |a_k^{(2)}\rangle$ the operator

$$A^{(1)} A^{(2)} = A^{(2)} A^{(1)}$$

by the direct product

$$(2.53) \qquad A = A^{(1)} \times A^{(2)}$$

which is a diagonal matrix if $A^{(1)}$ and $A^{(2)}$ are diagonal. This follows by straightforward computation from the following definitions.

The direct product of two matrices A and B

$$(2.54) \qquad C = A \times B$$

is meant to have the matrix elements

$$(2.55) \qquad C_{ij,\,nm} = A_{in} B_{jm}.$$

The labeling of the rows in C is done in the sequence $ij = 11, 12, \ldots, 1n, 21, 22, \ldots, 2n, \ldots$, similarly for the columns. Thus the direct product of the matrices

$$(2.56) \qquad A = \begin{pmatrix} A_{11} & A_{12} \\ A_{21} & A_{22} \end{pmatrix} \qquad B = \begin{pmatrix} B_{11} & B_{12} \\ B_{21} & B_{22} \end{pmatrix}$$

is

$$(2.57) \qquad C = \begin{pmatrix} A_{11} B_{11} & A_{11} B_{12} & A_{12} B_{11} & A_{12} B_{12} \\ A_{11} B_{21} & A_{11} B_{22} & A_{12} B_{21} & A_{12} B_{22} \\ A_{21} B_{11} & A_{21} B_{12} & A_{22} B_{11} & A_{22} B_{12} \\ A_{21} B_{21} & A_{21} B_{22} & A_{22} B_{21} & A_{22} B_{22} \end{pmatrix}.$$

The following statements are easily verified.

(I) If A and B are square matrices, then C is also square.

(II) If $AA' = A''$ and $BB' = B''$ and if $A \times B = C$ and $A' \times B' = C'$ then also $CC' = A'' \times B''$ or $(A \times B)(A' \times B') = AA' \times BB'$.

(III) The direct product of two diagonal matrices is again a diagonal matrix.

(IV) The direct product of two unit matrices is again a unit matrix.

NOTES

Pauli [1] invented the description of dichotomic spin in terms of the three operators σ_1, σ_2, σ_3.

The definition of the direct product used in this work is identical with the one given by Wigner [2].

REFERENCES

[1] W. Pauli, *Z. Physik* **43**, 601 (1927).
[2] E. Wigner, "Group Theory and Its Application to the Quantum Mechanics of Atomic Spectra." Academic Press, New York, 1959.

Transformations in State Vector Space That Leave the Physical Content of Quantum Mechanics Invariant

Summing up the results of the two preceding sections one may say there are two quantities which represent the physical content of quantum mechanical states.

(I) The absolute value of the projection of one state vector $|a\rangle$ on another state vector $|b\rangle$, $|\langle b|a\rangle|$. It is identified with the probability amplitude for finding the value b of the observable B if the observable A is known to have the value a.

(II) The average value of an observable A in a state $|b\rangle$, $\bar{A} = \langle b|A|b\rangle$. The observable A is represented by a linear hermitean operator $A^+ = A$, whose eigenvectors $|a_i\rangle$, defined by $A|a_i\rangle = a_i|a_i\rangle$, form an orthonormal basis, $\langle a_i|a_k\rangle = \delta_{ik}$, $\sum_i |a_i\rangle \langle a_i| = I$, spanning the state vector space.

Obviously, these two experimentally accessible quantities are not affected by any transformation

(3.1)
$$|a\rangle \to |a'\rangle; \qquad |b\rangle \to |b'\rangle; \qquad A \to A'$$

such that

(3.2)
$$|\langle b'|a'\rangle| = |\langle b|a\rangle|$$

and

(3.3)
$$\bar{A}' = \bar{A}.$$

One possibility of effecting such a transformation consists of a change in the basis vectors which corresponds, geometrically speaking, to a rotation of the coordinate frame in state vector space. This linear operation U and its inverse U^{-1} are defined by

(3.4)
$$|a\rangle = U|a'\rangle; \qquad U^{-1}|a\rangle = |a'\rangle; \qquad U(\alpha|a\rangle + \beta|b\rangle) = \alpha U|a\rangle + \beta U|b\rangle$$

with the understanding that (3.3) should hold, and that (3.2) be satisfied through the stronger condition

(3.5)
$$\langle b'|a'\rangle = \langle b|a\rangle$$

19

which requires U to be a unitary operator,

(3.6) $$UU^+ = U^+U = I \quad \text{or} \quad U^{-1} = U^+.$$

Condition (3.3) entails that A' be obtained from A by the similarity transformation

(3.7) $$A' = U^+AU.$$

The transformed operator A' retains the hermitean property of A.

Another possibility of effecting a transformation (3.1) which satisfies (3.2) consists of performing the *nonlinear* operation Θ defined by

(3.8) $$\Theta(\alpha|a\rangle + \beta|b\rangle) = \alpha^* \Theta|a\rangle + \beta^* \Theta|b\rangle$$

such that

(3.9) $$\langle\Theta a|\Theta b\rangle = \langle a|b\rangle^* = \langle b|a\rangle.$$

Operators of this kind are called antiunitary operators.

The importance of this invariance property of quantum mechanical description of reality resides in the possibility of representing symmetry operations in physical space by suitably chosen unitary or antiunitary operations in the abstract space spanned by the state vectors. The operation of inversion of spatial coordinates, for example, can be represented by a unitary operator, whereas the operation of reversal of motion, formally equivalent to a reversal of the time axis, requires representation by an antiunitary operator, as will be shown later in Sections 14 and 15.

Of immediate usefulness is the possibility, opened by the invariance under unitary transformations, of choosing as basis a set of vectors such that the matrices representing observables appear in the simplest possible form. In particular, it is always possible to diagonalize a hermitean matrix A by a unitary transformation. The diagonal elements of the thus transformed A are then the eigenvalues of A, and the columns of the transformation matrix U are made up out of the untransformed eigenvectors of the matrix A.

To prove this, write $A|a_i\rangle = a_i|a_i\rangle$ in components

(3.10) $$\sum_n A_{mn}(a_i)_n = a_i(a_i)_m.$$

Having then introduced the transformation U by

(3.11) $$|a_i\rangle = U|a_i'\rangle; \quad U^+|a_i\rangle = |a_i'\rangle$$

one deduces immediately from

(3.12) $$a_i|a_i'\rangle = a_i U^+|a_i\rangle = U^+A|a_i\rangle = U^+AUU^+|a_i\rangle = A'|a_i'\rangle$$

that the eigenvalues of A' are the same as those of A, namely a_i, the eigenfunctions being now $|a_i'\rangle = U^+|a_i\rangle$. By taking

(3.13) $U_{nm} = (a_m)_n$ i.e.

$$U = \begin{pmatrix} (a_1)_1 & (a_2)_1 & \cdots \\ (a_1)_2 & (a_2)_2 & \\ \vdots & & \end{pmatrix}; \qquad U^+ = \begin{pmatrix} (a_1)_1^* & (a_1)_2^* & \cdots \\ (a_2)_1^* & (a_2)_2^* & \\ \vdots & & \end{pmatrix}$$

the eigenvectors reduce to the simple form

(3.14) $$|a_i'\rangle = U^+|a_i\rangle = \begin{pmatrix} \langle a_1|a_i\rangle \\ \langle a_2|a_i\rangle \\ \vdots \end{pmatrix} = \begin{pmatrix} \delta_{1i} \\ \delta_{2i} \\ \vdots \end{pmatrix}$$

i.e. $$|a_1'\rangle = \begin{pmatrix} 1 \\ 0 \\ \vdots \end{pmatrix}; \qquad |a_2'\rangle = \begin{pmatrix} 0 \\ 1 \\ \vdots \end{pmatrix} \qquad \text{etc.}$$

The demonstration that A' is diagonalized by (3.13), with a_i as diagonal elements, is now straightforward. One has

(3.15) $$(AU)_{mi} = \sum_n A_{mn} U_{ni} = \sum_n A_{mn}(a_i)_n = a_i(a_i)_m$$

and therefore, because of the orthogonality of eigenvectors belonging to different eigenvalues,

(3.16) $$(U^+AU)_{ki} = \sum_m (a_k)_m^* a_i(a_i)_m = a_i \delta_{ik}$$

i.e. $$A' = U^+AU = \begin{pmatrix} a_1 & & & 0 \\ & a_2 & & \\ & & \cdot & \\ & & & \cdot \\ 0 & & & \cdot \end{pmatrix}.$$

It is often useful to write a unitary operator in the form

(3.17) $$U = e^{iS}$$

where the generator of the transformation, S, is a hermitean operator,

(3.18) $$S^+ = S$$

and the symbol e^{iS} stands for the expansion

(3.19) $$e^{iS} = I + iS + (i^2/2!)S^2 + \ldots$$

As an example, consider the unitary matrix whose columns are the eigenvectors (2.41), (2.42) of the general spin matrix (2.44),

$$(3.20) \qquad U = \begin{pmatrix} \cos(\vartheta/2) & -\sin(\vartheta/2)\,e^{-i\varphi} \\ \sin(\vartheta/2)\,e^{i\varphi} & \cos(\vartheta/2) \end{pmatrix};$$

$$U^+ = \begin{pmatrix} \cos(\vartheta/2) & \sin(\vartheta/2)\,e^{-i\varphi} \\ -\sin(\vartheta/2)\,e^{i\varphi} & \cos(\vartheta/2) \end{pmatrix}.$$

By straightforward computation one verifies unitarity

$$(3.21) \qquad U U^+ = U^+ U = I = \begin{pmatrix} 1 & 0 \\ 0 & 1 \end{pmatrix}$$

and the ensuing diagonalization of the general spin operator (2.44)

$$(3.22) \qquad \boldsymbol{\sigma}' = U^+ \boldsymbol{\sigma} U = \begin{pmatrix} 1 & 0 \\ 0 & -1 \end{pmatrix}$$

with diagonal elements $+1$, -1, as expected.

From the interpretation of ϑ, φ as polar angles in physical space, it follows that the unitary operator (3.20) represents in abstract spin space a rotation of coordinates in physical space such that the z'-axis points in a direction described by the angles ϑ, φ in the untransformed coordinate system.

The representation of spatial rotations in spin space is seen to be double valued, i.e. rotation through an angle $\vartheta = 2\pi$ does not regenerate the original states $|a_1\rangle$ and $|a_2\rangle$, one finds $-|a_1\rangle$ and $-|a_2\rangle$ instead, and it requires a rotation through $\vartheta = 4\pi$ to recover $|a_1\rangle$ and $|a_2\rangle$. This does not lead to any inconsistencies with observation, however, because $|a_i\rangle$ and $-|a_i\rangle$ represent the same physical state.

It will be shown in Section 12 that the rotation by angle ϕ around an axis with direction cosines α, β, γ is represented by

$$(3.23) \qquad U = \exp i\phi(\alpha s_1 + \beta s_2 + \gamma s_3) \qquad \text{where} \qquad s_i = \sigma_i/2.$$

This may be written, using the expansion (3.19) and the C.R.s of the s_i,

$$(3.24) \qquad U = \cos(\phi/2) + 2i(\alpha s_1 + \beta s_2 + \gamma s_3)\sin(\phi/2).$$

It must be kept in mind that

$$\exp[i\phi(\alpha s_1 + \beta s_2 + \gamma s_3)] \neq \exp(i\phi\alpha s_1)\exp(i\phi\beta s_2)\exp(i\phi\gamma s_3).$$

NOTES

Unitary operators have been widely employed ever since their significance in quantum mechanics was completely elucidated by Dirac [1] and Jordan [2].

Antiunitary operators, on the other hand, although their importance for quantum mechanics was pointed out a long time ago by Wigner [3], did not acquire equal status with unitary operators until recently, when time-reversal symmetry has come to the forefront of experimental research in particle physics.

REFERENCES

[1] P. A. M. Dirac, *Proc. Roy. Soc.* **A113**, 621 (1927).
[2] P. Jordan, *Z. Physik* **40**, 809 (1927).
[3] E. Wigner, *Nachr. Akad. Wiss. Goettingen Math.-Phys. Kl.* p. 546 (1932).

The Density Matrix

An alternative way of apprehending the state of an object, free from the arbitrariness of phase which afflicts the representation of states by state vectors, consists of describing the state by a hermitean matrix $M(b)$, whose elements are in terms of the components $(b)_i$ of the corresponding state vector $|b\rangle$

$$(4.1) \qquad M_{ik}(b) = (b)_i (b)_k^* \qquad \text{or} \qquad M = |b\rangle\langle b|.$$

The normalization condition $\langle b|b \rangle = 1$ is then equivalent to the condition

$$(4.2) \qquad \text{trace } M = 1$$

the trace being the sum of the diagonal elements of M. Computation of the average value $\bar{A}$ of an observable A in the state characterized by $M(b)$ is then accomplished by evaluation of

$$(4.3) \qquad \bar{A} = \text{trace}\,(MA) = \text{trace}\,(AM).$$

Indeed, in the state $|b\rangle$,

$$(4.4)$$
$$\bar{A} = \sum_n \sum_m (b)_m^* A_{mn}(b)_n$$
$$= \sum_n \sum_m M_{nm}(b) A_{nm} = \sum_n [M(b) A]_{nn} = \text{trace}\,(MA).$$

From the representation (4.1), it follows that this so-called density matrix representing a pure state $|b\rangle$ is idempotent,

$$(4.5) \qquad M^2 = M$$

because one has, in components, for orthonormal state vectors

$$(4.6)$$
$$(M^2)_{ik} = \sum_j M_{ij} M_{jk} = \sum_j (b)_i(b)_j^*(b)_j(b)_k^* = (b)_i(b)_k^* = M_{ik}.$$

A hermitean matrix can obviously satisfy (4.5) only if all its eigenvalues satisfy the same condition, i.e. its eigenvalues must be either 0 or $+1$. From (4.2), it follows then that only one eigenvalue is $+1$ and all

25

others are 0, because M can be brought on diagonal form by a unitary transformation, which leaves the trace invariant,

$$\text{trace}\,(U^+ M U) = \text{trace}\,(M U U^+) = \text{trace}\,M.$$

Geometrically speaking, $M(b)$ is thus a projection operator, which applied to an arbitrary vector $|a\rangle$ projects it in direction of the eigenvector $|b\rangle$ belonging to the eigenvalue $+1$ of M.

Since $|b_r\rangle$ is the eigenvector of $M(b_r)$ with eigenvalue $+1$, whereas all other vectors $|b_s\rangle$ with $s \neq r$ belong to the eigenvalue 0,

$$(4.7) \qquad M(b_r)|b_s\rangle = \left[\sum_k (b_r)_i (b_r)_k^*(b_s)_k \right] = \delta_{rs}|b_s\rangle,$$

the expansion

$$(4.8) \qquad |a\rangle = \sum_s |b_s\rangle \langle b_s|a\rangle$$

yields immediately the result

$$(4.9) \qquad M(b_r)|a\rangle = |b_r\rangle \langle b_r|a\rangle.$$

The projection operator $M(b_r)$ can thus quite literally be taken as a *measurement symbol* representing a selective measurement that accepts only those objects which possess value b_r of the attribute B and rejects all others.

The average value of the density matrix $M(b)$ in a state $|a\rangle$ is now equal to the probability for finding the value b of B if A is known to have the value a,

$$(4.10) \qquad \overline{M(b)} = \langle a|M(b)|a\rangle = \langle a|b\rangle \langle b|a\rangle = |\langle a|b\rangle|^2.$$

Since the reasoning of equation (4.4) is applicable to the hermitean operator $M(b)$ in the state $|a\rangle$, which in turn can be represented by a density matrix $M(a)$, one has the interesting result

$$(4.11) \qquad |\langle a|b\rangle|^2 = \text{trace}\,[M(a)\,M(b)].$$

The entire physical content of the quantum mechanics of pure states has thus been reduced to statements involving density matrices only.

The state

$$|a_1\rangle = \begin{pmatrix} \alpha_1 \\ \beta_1 \end{pmatrix} = \begin{pmatrix} \cos(\vartheta/2) \\ \sin(\vartheta/2)\,e^{i\varphi} \end{pmatrix}$$

describing a spin which has with certainty the value $+1$ in direction ϑ, φ will again serve as an example. The corresponding density matrix is

(4.12)

$$M = \begin{pmatrix} |\alpha_1|^2 & \alpha_1\beta_1^* \\ \beta_1\alpha_1^* & |\beta_1|^2 \end{pmatrix} = \begin{pmatrix} \cos^2(\vartheta/2) & \sin(\vartheta/2)\cos(\vartheta/2)\,e^{-i\varphi} \\ \sin(\vartheta/2)\cos(\vartheta/2)\,e^{i\varphi} & \sin^2(\vartheta/2) \end{pmatrix}$$

which obviously satisfies trace $M = 1$ and is idempotent, $M^2 = M$. A particularly instructive representation for M is obtained if one utilizes the possibility of writing any 2×2 matrix as superposition of the four linearly independent matrices

$$\sigma_1 = \begin{pmatrix} 0 & 1 \\ 1 & 0 \end{pmatrix}, \quad \sigma_2 = \begin{pmatrix} 0 & -i \\ i & 0 \end{pmatrix}, \quad \sigma_3 = \begin{pmatrix} 1 & 0 \\ 0 & -1 \end{pmatrix}, \quad I = \begin{pmatrix} 1 & 0 \\ 0 & 1 \end{pmatrix}.$$

If M is required to be hermitean and have trace unity, it must be of the form

(4.13) $$M = \tfrac{1}{2}(I + \mathbf{P}\boldsymbol{\sigma})$$

where $\mathbf{P} = (P_1, P_2, P_3)$ is a vector whose three *real* components characterize the state. The idempotence of M can be satisfied only if $\mathbf{P}$ is subject to the condition

(4.14) $$\mathbf{P}^2 = 1$$

which leaves for the description of the state two parameters, as expected. Indeed, from (4.13) and (4.14) follows, utilizing (2.46),

$$M^2 = \tfrac{1}{4}(I + \mathbf{P}\boldsymbol{\sigma})(I + \mathbf{P}\boldsymbol{\sigma})$$

(4.15) $$= \tfrac{1}{4}[I + 2\mathbf{P}\boldsymbol{\sigma} + \sum_i P_i^2\sigma_i^2 + \sum_{\substack{j,k \\ (\text{cycl.})}} P_j P_k(\sigma_j\sigma_k + \sigma_k\sigma_j)]$$

$$= \tfrac{1}{4}(I + 2\mathbf{P}\boldsymbol{\sigma} + \mathbf{P}^2 I) = \tfrac{1}{2}(I + \mathbf{P}\boldsymbol{\sigma}) = M.$$

In terms of the components α_1, β_1 the vector $\mathbf{P}$ is found by equating the elements of (4.12) and (4.13), yielding

(4.16)
$$P_1 = 2\operatorname{Re}(\beta_1\alpha_1^*) = \sin\vartheta\cos\varphi$$
$$P_2 = 2\operatorname{Im}(\beta_1\alpha_1^*) = \sin\vartheta\sin\varphi$$
$$P_3 = |\alpha_1|^2 - |\beta_1|^2 = \cos\vartheta.$$

The unit vector $\mathbf{P}$ points thus in direction ϑ, φ and may be called the polarization vector of the state. The unit length of $\mathbf{P}$, derived from the idempotence of M, reflects the total polarization of the object in a pure spin state.

So far nothing has been said that cannot be expressed in terms of state vectors and their components. However, the quantum mechanical description of states in terms of density matrices derives its particular profundity from the possibility of giving M a meaning even if the object is not in a pure state, so that the decomposition of the matrix elements of M after the fashion (4.1) is, in principle, not possible. *The density matrix description can grasp states which do not possess a state vector description.*

A *model* of such a situation is obtained if one imagines the object of measurement to consist of a statistical ensemble composed of N subsystems, which may be in different quantum states. Suppose the N subsystems are in a number of possible quantum states described each by $M_n = M(b_n)$. The average values $\overline{A}_n$ of the observable A must then be averaged once more over the ensemble before one arrives at an expression for the maximum information about A obtainable in this case, namely

$$(4.17) \qquad \overline{A} = \sum_n p_n \overline{A}_n$$

with

$$(4.18) \qquad \sum_n p_n = 1; \qquad p_n \geqslant 0$$

where p_n is the probability for finding the value $\overline{A}_n$ in the ensemble.

The formalism enabling one to compute the average value of an attribute in a pure state using the density matrix can be extended to this general case by introduction of the generalized density matrix

$$(4.19) \qquad M = \sum_n p_n M_n$$

which satisfies, because of (4.18), the condition

$$(4.20) \qquad \text{trace } M = 1$$

and yields the desired result

$$(4.21) \qquad \overline{A} = \text{trace}\,(MA) = \sum_n p_n \text{trace}\,(M_n A) = \sum_n p_n \overline{A}_n.$$

This generalized density matrix is no longer idempotent, however, unless all M_n are equal to M. *Idempotence of the density matrix is thus a criterion for the presence of a pure state.* To prove this, consider a mixture of two systems, described by density matrices M_1 and M_2 so that

$$(4.22)$$
$$M = p_1 M_1 + p_2 M_2 \qquad (0 < p_1 < 1) \quad (0 < p_2 < 1) \quad (p_1 + p_2 = 1).$$

The square of this matrix

$$(4.23) \qquad M^2 = p_1^2 M_1^2 + p_2^2 M_2^2 + p_1 p_2 (M_1 M_2 + M_2 M_1)$$

can, because of the identities

(4.24) $$M_1 M_2 + M_2 M_1 = M_1^2 + M_2^2 - (M_1 - M_2)^2$$

and

(4.25) $$p_1^2 + p_1 p_2 = p_1(p_1 + p_2) = p_1; \qquad p_2^2 + p_1 p_2 = p_2$$

be written in the form

(4.26) $$M^2 = p_1 M_1^2 + p_2 M_2^2 - p_1 p_2 (M_1 - M_2)^2.$$

Since M_1 and M_2 are idempotent, one has

(4.27) $$M - M^2 = p_1 p_2 (M_1 - M_2)^2.$$

On the right-hand side of this equation stands a positive matrix which can vanish and thus lead to idempotence of M only if $(M_1 - M_2)^2 = 0$. The square of a hermitean matrix can vanish only if all its elements vanish, and therefore $M_1 = M_2$ is the necessary condition for idempotence of M. The general case follows now by induction.

It may be worth noting here that in quantum statistical mechanics $\Sigma = -\overline{(\log M)} = -\operatorname{trace}(M \log M)$ is the entropy of the statistical ensemble represented by M. It vanishes for any pure state and only for a pure state. It is rather remarkable that one can bypass in this fashion the entire notion of "phase space," familiar from classical statistical mechanics, which is not a meaningful concept in quantum mechanics, because momenta and coordinates are, in general, incompatible attributes.

It should be stressed that the ensemble picture of a state which is not pure becomes a fictitious model when the experimental situation is such that optimum information about subsystems of the entire object is, in principle, not available. For example, consider as object a partially polarized beam of electrons, so that, in the ensemble model, one would introduce a density matrix

(4.28) $$M = \sum_n p_n M_n = \sum_n p_n \tfrac{1}{2}(I + \mathbf{P}_n \boldsymbol{\sigma}) = \tfrac{1}{2}(I + \overline{\mathbf{P}} \boldsymbol{\sigma})$$

where the average polarization

(4.29) $$\overline{\mathbf{P}} = \sum_n p_n \mathbf{P}_n$$

is no longer a unit vector, M not being idempotent. The absolute value $|\overline{\mathbf{P}}|$ represents the degree of polarization of the beam, $|\overline{\mathbf{P}}| = 0$ means the beam is unpolarized, and $|\overline{\mathbf{P}}| = 1$ designates the limiting case of full polarization. If $\overline{\mathbf{P}}$ is the only observable parameter of the object, the decomposition (4.29) becomes fictitious, because the entire experimental information available resides in

$$(4.30) \qquad M = \tfrac{1}{2}(I + \overline{\mathbf{P}}\boldsymbol{\sigma})$$

making the mental image of a set of electrons each having *definite* orientation of spin, represented by the unit vectors $\mathbf{P}_n$, fictitious and, from an operational point of view, undesirable.

Note in this connection the general result

$$\overline{\sigma}_i = \text{trace}\,(\sigma_i M)$$

$$(4.31) \qquad = \tfrac{1}{2}\,\text{trace}\,\sigma_i + \tfrac{1}{2}\,\text{trace}\,\overline{P}_i + \sigma_i\,\text{trace}\left[\sum_{\substack{k \\ (k \neq i,\, j \neq k,\, i \neq j)}} \overline{P}_k\,\sigma_j\right]$$

$$= \tfrac{1}{2}\,\text{trace}\,\overline{P}_i = \overline{P}_i.$$

NOTES

The possibility of apprehending the state of a system by the density matrix was apparently first noticed by Landau [1]. Shortly afterwards, a pair of papers was published by von Neumann [2] employing the density matrix for a very complete analysis of the probability interpretation of quantum mechanics, and of the thermodynamics of quantum mechanical ensembles. See also Dirac [3] and Pauli [4].

A comprehensive review of applications of the density matrix has been given by Fano [5].

Williams [6] uses in Chapter VIII the same formalism for a particularly lucid analysis of polarization effects in scattering processes.

REFERENCES

[1] L. Landau, *Z. Physik* **45**, 430 (1927).
[2] J. von Neumann, *Nachr. Akad. Wiss. Goettingen Math,-Phys. Kl.* pp. 245 and 273 (1927).
[3] P. A. M. Dirac, *Proc. Cambridge Phil. Soc.* **25**, 62 (1929).
[4] W. Pauli, Die Allgemeinen Prinzipien der Wellenmechanik, *in* "Encyclopedia of Physics" (S. Flügge, ed.), Vol. 5, Pt. I, p. 69. Springer, Berlin, 1958.
[5] U. Fano, *Rev. Mod. Phys.* **29**, 74 (1957).
[6] W. S. C. Williams, "An Introduction to Elementary Particles." Academic Press, New York, 1961.

The Theory of Selective Measurements

The density matrix description of physical objects can be considered as the special case of an even more general development based upon the analysis of experimental situations realized by selective measurements.

To start, introduce a symbol $M(a_i)$ representing a single selective measurement with an apparatus that accepts objects possessing the value a_i of the attribute A and rejects all others. If A is understood to represent a complete set of compatible observables, $A = A^{(1)} \times A^{(2)} \times \ldots$, then $M(a_i)$ describes a complete selective measurement, such that the object chosen possesses definite values, symbolized by a_i, for a maximum number of attributes. The selective measurements $M(a_i)$ and $M(a_k)$, performed in either order, must in the second step result in either the acceptance of all objects, namely when $i = k$, or the rejection of all objects, namely when $i \neq k$. Symbolizing with I and O the measurements that respectively accept and reject all objects, the sequence of measurements can be represented by the multiplication of the corresponding symbols, satisfying

(5.1) $$M(a_i)\, M(a_k) \;=\; M(a_k)\, M(a_i) \;=\; \delta_{ik}\, M(a_i)$$

where

(5.2) $$\delta_{ik} = \begin{cases} I & \text{for} \quad i = k \\ O & \text{for} \quad i \neq k. \end{cases}$$

If one further defines the addition of such symbols $M(a_i)$, $M(a_k)$ to mean a less specific selective measurement, admitting all objects with any of the values a_i, a_k in the summation, then the completeness of the observable A is contained in the statement

(5.3) $$\sum_i M(a_i) = I.$$

To accommodate the disturbance of the object by the measurement and the existence of incompatible sets of observables, one must consider the most general selective measurement performed with an apparatus that rejects all objects entering except those in the state b_i and permits only objects in the state a_k to leave. Symbolizing this measurement

31

process by $M(a_k, b_i)$, one should stress that the envisaged process implies *complete ignorance* about what goes on between input and output stage. Thus $M(a_k, b_i)$ is *not* meant to be identical with $M(a_k) M(b_i)$, in particular the symbol $M(a_k, a_j)$ is *not* equivalent to the symbol O, for $k \neq j$. However, the symbol $M(a_i)$ may be taken as the special case in which no change in state occurs,

$$(5.4) \qquad M(a_i) = M(a_i, a_i).$$

It is often convenient to represent a measurement symbol, such as $M(a_k, b_i)$ by a graph consisting of a "black box" symbolizing the region of ignorance between input and output stage, which latter are symbolized by a directed line entering and leaving the black box, with labels indicating the quantum numbers of input and output states, as drawn in Fig. 5.1.

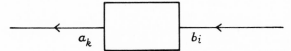

FIG. 5.1. Black box diagram representing the measurement symbol $M(a_k, b_i)$.

To be specific, let $\pm s_1$, $\pm s_2$, $\pm s_3$ denote the possible values of spin of an object in directions x, y, z, respectively, so that, for example, $-s_1$ means "spin $-\frac{1}{2}$ in direction x." The measurement $M(+s_3, -s_1)$, for example, can be realized by an experimental arrangement which is rendered graphically in Fig. 5.2, with the corresponding black box diagram drawn underneath. Everything has been drawn in one plane, omitting unimportant beam deflections, the magnets labeled according to the alignment of their magnetic fields along the respective axes. Each input and output stage corresponds to a selective Stern-Gerlach experiment, and the region of ignorance in-between represents the fate of the beam between experiments.

The compound measurement $M(a_m, b_j) M(c_k, d_i)$ admits objects in the state d_i and permits them to enter into the state a_m, and is therefore a selective measurement of the type $M(a_m, d_i)$. If C and B are incompatible, only a certain *statistically predictable* fraction of the objects leaving the stage $M(c_k, d_i)$ will be admitted into the stage $M(a_m, b_j)$ of the apparatus. Hence one has the multiplication rule

$$(5.5) \qquad M(a_m, b_j) M(c_k, d_i) = \langle b_j | c_k \rangle M(a_m, d_i)$$

where $\langle b_j | c_k \rangle$ is a number which serves to express the statistical relationship between the states b_j and c_k. It should be stressed that the notation

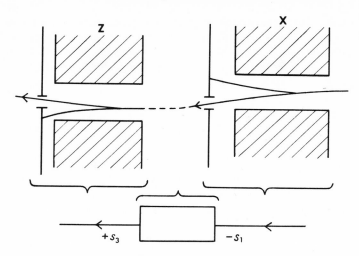

Fɪɢ. 5.2. Experimental arrangement corresponding to the black box diagram representing the measurement symbol $M(+s_3, -s_1)$.

for this number is purely conventional and does not, at this stage, imply anything about a possible interpretation in terms of geometry in an abstract space.

The value of $\langle b_j | c_k \rangle$ can be ascertained immediately for the special case of compatible observables, because then one must have

$$(5.6) \qquad M(a_m, b_j) \, M(b_k, c_i) = \delta_{jk} \, M(a_m, c_i).$$

Indeed, if $b_j \neq b_k$, then the second stage of the compound apparatus will accept none of the objects emerging from the first stage, while, if $b_j = b_k$, all such objects are admitted into the second stage, thus

$$(5.7) \qquad \langle b_j | b_k \rangle = \delta_{jk}.$$

Further special examples of (5.5) are

$$(5.8) \qquad M(a_j) \, M(b_k, c_i) = \langle a_j | b_k \rangle \, M(a_j, c_i)$$

$$(5.9) \qquad M(a_j, b_k) \, M(c_i) = \langle b_k | c_i \rangle \, M(a_j, c_i).$$

If the sequence of the stages in (5.5) is reversed, one has

$$(5.10) \qquad M(c_k, d_i) \, M(a_m, b_j) = \langle d_i | a_m \rangle \, M(c_k, b_j)$$

which is, in general, different from (5.5). *The multiplication of measurement symbols is, in general, noncommutative.* From the completeness relation (5.3) and from (5.8) and (5.9) follows

$$(5.11) \qquad \sum_j M(a_j) \, M(b_k, c_i) = M(b_k, c_i) = \sum_j \langle a_j | b_k \rangle \, M(a_j, c_i)$$

(5.12) $\sum\limits_{i} M(a_j, b_k)\, M(c_i) = M(a_j, b_k) = \sum\limits_{i} \langle b_k | c_i \rangle\, M(a_j, c_i).$

Thus measurement symbols of one type can be expressed as linear combinations of another type. By repetition of this expansion one has generally

(5.13)

$$M(c_k, d_i) = \sum\limits_{m}\sum\limits_{j} M(a_m)\, M(c_k, d_i)\, M(b_j) = \sum\limits_{m}\sum\limits_{j} \langle a_m | c_k \rangle \langle d_i | b_j \rangle\, M(a_m, b_j).$$

Because of its ability to sponsor such connections, the set of numbers $\langle a_k | b_i \rangle$ is called the "transformation function relating the A- and B-description."

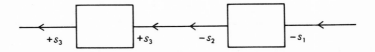

FIG. 5.3. Black box diagram representing the measurement $M(+s_3)M(-s_2, -s_1)$.

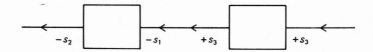

FIG. 5.4. Black box diagram representing the measurement $M(-s_2, -s_1)M(+s_3)$.

As an example involving compound selective Stern-Gerlach experiments, consider the measurement $M(+s_3)\,M(-s_2, -s_1)$, rendered graphically in Fig. 5.3. Equation (5.8) reads for this case

(5.8a) $M(+s_3)\, M(-s_2, -s_1) = \langle +s_3 | -s_2 \rangle\, M(+s_3, -s_1)$

and says that this measurement corresponds but to one channel which contributes with amplitude $\langle +s_3 | -s_2 \rangle$ to the measurement $M(+s_3, -s_1)$ which has already been rendered graphically in Fig. 5.2.

Carrying out the compound selective measurements of Fig. 5.3. in opposite order, one has a quite different situation, namely the measurement $M(-s_2, -s_1)M(+s_3)$, rendered in Fig. 5.4. Equation (5.9) is now appropriate,

(5.9a) $M(-s_2, -s_1)\, M(+s_3) = \langle -s_1 | +s_3 \rangle\, M(-s_2, +s_3)$

and gives $\langle -s_1 | +s_3 \rangle$ as the amplitude with which the particular channel contributes to the measurement $M(-s_2, +s_3)$, which is different from the one depicted in Fig. 5.2.

Transformation functions have the composition property

(5.14)
$$\sum_k \langle a_j|b_k\rangle \langle b_k|c_i\rangle = \langle a_j|c_i\rangle.$$

This follows from

(5.15)
$$\sum_k \langle a_j|b_k\rangle \langle b_k|c_i\rangle M(a_j, c_i) = \sum_k M(a_j) M(b_k) M(c_i) = M(a_j) M(c_i)$$

$$= \langle a_j|c_i\rangle M(a_j, c_i).$$

Any linear combination of measurement symbols will be called an "operator," and denoted by capital letters X, Y, ... Writing

(5.16)
$$X = \sum_i \sum_k \langle a_k|X|b_i\rangle M(a_k, b_i)$$

the expansion coefficients $\langle a_k|X|b_i\rangle$ can be arranged in an $N \times N$ matrix scheme, and will thus be called the "matrix elements" of X. Operators are accordingly elements of a linear algebra of dimensionality N^2, where N is the number of different values a_i attributable to a complete set of observables A. In this algebra the number $\langle a_i|b_k\rangle$ can be regarded as a linear numerical function of the operator $M(b_k, a_i)$, which will be called, in anticipation of an obvious matrix representation, the "trace" of $M(b_k, a_i)$,

(5.17)
$$\langle a_i|b_k\rangle = \text{trace } M(b_k, a_i).$$

This definition is consistent, because the linear relation (5.13) leads to

(5.18)
$$\text{trace } M(c_k, d_i) = \sum_m \sum_j \langle a_m|c_k\rangle \langle d_i|b_j\rangle \text{ trace } M(a_m, b_j)$$

$$= \sum_m \sum_j \langle a_m|c_k\rangle \langle d_i|b_j\rangle \langle b_j|a_m\rangle$$

$$= \sum_j \langle d_i|b_j\rangle \langle b_j|c_k\rangle = \langle d_i|c_k\rangle.$$

As special cases one obtains

(5.19)
$$\text{trace } M(a_k, a_i) = \delta_{ik}$$

(5.20)
$$\text{trace } M(a_i) = I.$$

The trace of a product of two measurement symbols is

(5.21)
$$\text{trace } [M(a_m, b_j) M(c_k, d_i)] = \langle b_j|c_k\rangle \text{ trace } M(a_m, d_i)$$

$$= \langle b_j|c_k\rangle \langle d_i|a_m\rangle$$

and one has also

(5.22) $\text{trace}\,[M(c_k, d_i)\, M(a_m, b_j)] = \langle d_i | a_m \rangle\, \text{trace}\, M(c_k, b_j)$

$$= \langle d_i | a_m \rangle \langle b_j | c_k \rangle.$$

Hence the trace of a product of two operators is independent of the order of multiplication,

(5.23) $\text{trace}\,(XY) = \text{trace}\,(YX)$

even though the operators themselves do not commute. A special case of (5.21) is

(5.24) $\text{trace}\,[M(a_i)\, M(b_k)] = \langle b_k | a_i \rangle \langle a_i | b_k \rangle.$

The general multiplication law (5.5) of measurement symbols and the definition (5.17) of the trace are obviously invariant under the transformations

(5.25) $M(a_k, b_i) \rightarrow \gamma^{-1}(a_k)\, M(a_k, b_i)\, \gamma(b_i)$

$$\langle a_k | b_i \rangle \rightarrow \gamma(a_k) \langle a_k | b_i \rangle \gamma^{-1}(b_i)$$

where the numbers $\gamma(a_k)$ and $\gamma(b_i)$ can have arbitrary nonzero values. The quantities $M(a_i)$ and $\langle a_i | a_k \rangle = \delta_{ik}$ are seen to be invariant, too.

Contact with the statistical interpretation of quantum mechanics developed in the preceding sections can now be established most naturally by considering the sequence of measurements $M(b_i)\, M(a_k)\, M(b_i)$ which differs from $M(b_i)$ because of the disturbance produced by the measurement of A in the step $M(a_k)$. Only a fraction of the objects admitted in the first stage $M(b_i)$ is transmitted through the last stage $M(b_i)$ of the apparatus. In the equation

(5.26) $M(b_i)\, M(a_k)\, M(b_i) = p(a_k, b_i)\, M(b_i)$

the number

(5.27) $p(a_k, b_i) = \langle a_k | b_i \rangle \langle b_i | a_k \rangle$

is, in contrast to $\langle a_k | b_i \rangle$, invariant under the transformation (5.25) and has the additive property

(5.28) $M(b_i)\,[M(a_k) + M(a_j)]\, M(b_i) = [p(a_k, b_i) + p(a_j, b_i)]\, M(b_i)$

so that from

(5.29) $M(b_i)\left[\sum_k M(a_k)\right] M(b_i) = M(b_i)$

follows

(5.30) $\sum_k p(a_k, b_i) = 1.$

Thus $p(a_k, b_i)$ is qualified to serve as the probability for observation of a_k if the object is known to be in the state b_i. Since a probability must be a real non-negative number, the numbers $\langle a_k | b_i \rangle$ will be subjected to the *admissible* restriction

$$(5.31) \qquad \langle b_i | a_k \rangle = \langle a_k | b_i \rangle*$$

because then

$$(5.32) \qquad p(a_k, b_i) = |\langle a_k | b_i \rangle|^2 \geqslant 0.$$

The condition (5.31) imposes on the numbers $\gamma(a_k)$, $\gamma(b_i)$ in (5.25) the restriction

$$(5.33) \qquad \gamma*(a_k) = \gamma^{-1}(a_k);$$

they must therefore be of the form

$$(5.34) \qquad \gamma(a_k) = e^{i\phi(a_k)}$$

with arbitrary real phases $\phi(a_k)$.

From the definitions (5.16) and (5.17) of operator and trace follows an expression for the matrix elements of an operator,

$$(5.35) \qquad \langle a_k | X | b_i \rangle = \text{trace}\,[X M(b_i, a_k)].$$

Indeed,

$$(5.36) \quad \text{trace}\,[X M(b_i, a_k)] = \sum_j \sum_m \langle a_m | X | b_j \rangle \, \text{trace}\,[M(a_m, b_j)\, M(b_i, a_k)]$$

$$= \sum_j \sum_m \langle a_m | X | b_j \rangle \langle b_j | b_i \rangle \, \text{trace}\, M(a_m, a_k)$$

$$= \sum_m \langle a_m | X | b_i \rangle \langle a_k | a_m \rangle = \langle a_k | X | b_i \rangle.$$

The expression for the average value of an operator X in the state b_i follows from (5.35) by specialization,

$$(5.37) \qquad \bar{X} = \langle b_i | X | b_i \rangle = \text{trace}\,[X M(b_i)].$$

The measurement symbol $M(b_i)$ is thus recognized as being identical with the density matrix $M(b_i)$ introduced in the preceding section. The justification for denoting (5.37) as an average stems from the probability definition

$$(5.38) \qquad p(a_k, b_i) = \text{trace}\,[M(a_k)\, M(b_i)]$$

which allows one to write the average value of the observable $\bar{A}$, namely

$$(5.39) \qquad \bar{A}_{b_i} = \sum_k a_k\, p(a_k, b_i)$$

in the form

$$(5.40) \qquad \overline{A}_{b_i} = \sum_k \text{trace}\,[a_k\,M(a_k)\,M(b_i)] = \text{trace}\,[A M(b_i)]$$

provided

$$(5.41) \qquad A = \sum_k a_k\,M(a_k).$$

In summary, the accompanying tabulation gives the alternative description of physically accessible quantities in the state vector and in the measurement symbol picture.

State vector picture	Measurement symbol picture
$(b_k)_r(a_i)_s^*$ or $\|b_k\rangle\langle a_i\|$	$M_{rs}(b_k, a_i)$ or $M(b_k, a_i)$
$(b_k)_r(b_k)_s^*$	$M_{rs}(b_k)$
$\langle a_k\|b_i\rangle$	trace $M(b_i, a_k)$
$\langle a_k\|a_i\rangle = \delta_{ik}$	trace $M(a_i, a_k) = \delta_{ik}$
$\|\langle a_k\|b_i\rangle\|^2 = p(a_k, b_i)$	trace $[M(b_i)\,M(a_k)] = p(a_k, b_i)$
$\langle a_k\|X\|b_i\rangle$	trace $[XM(b_i, a_k)]$
$\langle b_i\|X\|b_i\rangle = \bar{X}$	trace $[XM(b_i)] = \bar{X}$

NOTE

Sections 5 and 6 follow closely a development by Schwinger [1].

REFERENCE

[1] J. Schwinger, The algebra of microscopic measurement, *Proc. Natl. Acad. Sci. U.S.* **45**, 1542 (1959).

The Representation of Nonselective Measurements

The measurement symbols $M(a_i) = M(a_i, a_i)$ considered in the preceding section involved an apparatus which, in addition to separating the ensemble of objects into subensembles belonging to various states a_1, a_2, ..., selected one subensemble a_i and rejected all others.

It is, of course, possible to have an apparatus which performs a nonselective measurement by performing only the separation into subensembles without the selecting stage. If the separation takes place with respect to the observable B, the corresponding measurement symbol will be denoted M_b and rendered graphically as in Fig. 6.1. To obtain a

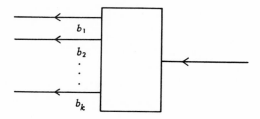

FIG. 6.1. Black box diagram representing the nonselective measurement symbol M_b.

representation of M_b in terms of selective measurement symbols $M(b_k)$, consider an object and then subject it, in succession, first to a selective measurement $M(b_k, c_i)$ and then to a selective measurement $M(a_j, b_k)$ as

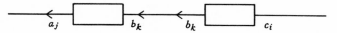

FIG. 6.2. Black box diagram representing the measurement $M(a_j, b_k) M(b_k, c_i)$.

rendered in Fig. 6.2. The probability that the object will exhibit the value b_k of B and then the value a_j of A is

$$(6.1) \qquad p(a_j, b_k, c_i) = p(a_j, b_k) \, p(b_k, c_i) = |\langle a_j | b_k \rangle \langle b_k | c_i \rangle|^2$$
$$= |\langle a_j | M(b_k) | c_i \rangle|^2$$

39

where use has been made of (5.35) and (5.9), giving

(6.2) $\langle a_j | M(b_k) | c_i \rangle = \text{trace}\,[M(c_i, a_j)\,M(b_k)] = \langle a_j | b_k \rangle \,\text{trace}\,M(c_i, b_k)$
$$= \langle a_j | b_k \rangle \langle b_k | c_i \rangle.$$

Now suppose the intermediate measurement of B were not made at all, so that the intermediate measurement symbol in the matrix element (6.2) can be replaced by the identity operation $I = \sum_k M(b_k)$, correspond-

FIG. 6.3. Black box diagram representing the measurement $M(a_j, c_i)$.

ing to the process rendered graphically in Fig. 6.3, then the corresponding probability is

(6.3) $$p(a_j, I, c_i) = |\langle a_j | c_i \rangle|^2 = |\sum_k \langle a_j | M(b_k) | c_i \rangle|^2.$$

One has in this case of "coherent" B subensembles an addition of probability amplitudes, giving rise to "interference" effects, the terms "coherent" and "interference" being coined because of the profound analogy of (6.3) with the description of interfering coherent light rays for which there is an addition of amplitudes.

On the other hand, if, in the intermediate stage, the B-separation apparatus is turned on, but without the selecting stage, as rendered

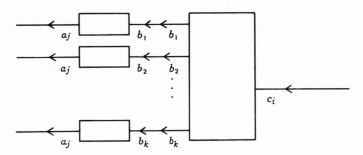

FIG. 6.4. Black box diagram representing measurement with intermediate B-separation apparatus turned on.

graphically in Fig. 6.4, then the probability for finding a_j, if c_i is the initial state, is given by

(6.4) $$p(a_j, b, c_i) = \sum_k p(a_j, b_k, c_i) = \sum_k |\langle a_j | M(b_k) | c_i \rangle|^2.$$

Thus, in this mixture of incoherent B subensembles one has an addition of probabilities themselves, the difference from (6.3) being the absence of interference terms between the different states b_k, corresponding to the addition of intensities in the optical analog.

The symbol associated with the nonselective B measurement may therefore be taken to be

$$(6.5) \qquad M_b = \sum_k e^{i\phi(b_k)} M(b_k)$$

with randomly distributed phases $\phi(b_k)$, which express the uncontrollable nature of the disturbance produced by the nonselective measurement. The probability (6.4) accordingly takes a form similar to (6.1),

$$(6.6) \qquad p(a_j, b, c_i) = |\langle a_j | M_b | c_i \rangle|^2.$$

Since the nonselective measurement does not reject objects, one must have

$$(6.7) \quad \sum_j p(a_j, b, c_i) = \sum_j \langle c_i | M_b^+ | a_j \rangle \langle a_j | M_b | c_i \rangle = \langle c_i | M_b^+ M_b | c_i \rangle = 1$$

which means that the operators M_b are unitary,

$$(6.8) \qquad M_b^+ M_b = M_b M_b^+ = I.$$

It should be noted that the selective measurement symbol $M(b_k)$ can be obtained from the nonselective symbol M_b according to (6.5) if all but one of the phases are replaced by positive infinite imaginary numbers, corresponding to an absorption of all but one of the subensembles produced by M_b.

As an example, identify the states c_i, b_k, and a_j as follows.

$$(6.9) \quad \begin{cases} |c_i\rangle = (\text{state with spin} + 1 \text{ in direction } z) = \begin{pmatrix} 1 \\ 0 \end{pmatrix} \\[2mm] |b_1\rangle = (\text{state with spin} + 1 \text{ in direction } \vartheta, \varphi) = \begin{pmatrix} \cos(\vartheta/2) \\ \sin(\vartheta/2)\, e^{i\varphi} \end{pmatrix} \\[2mm] |b_2\rangle = (\text{state with spin} - 1 \text{ in direction } \vartheta, \varphi) = \begin{pmatrix} -\sin(\vartheta/2)\, e^{-i\varphi} \\ \cos(\vartheta/2) \end{pmatrix} \\[2mm] |a_j\rangle = (\text{state with spin} + 1 \text{ in direction } x) = \frac{1}{\sqrt{2}} \begin{pmatrix} 1 \\ 1 \end{pmatrix} \end{cases}$$

so that the density matrices corresponding to the measurements of b_1 and b_2 are

$$(6.10) \quad \begin{aligned} M(b_1) &= \begin{pmatrix} \cos^2(\vartheta/2) & \sin(\vartheta/2)\cos(\vartheta/2)\, e^{-i\varphi} \\ \sin(\vartheta/2)\cos(\vartheta/2)\, e^{i\varphi} & \sin^2(\vartheta/2) \end{pmatrix} \\[2mm] M(b_2) &= \begin{pmatrix} \sin^2(\vartheta/2) & -\sin(\vartheta/2)\cos(\vartheta/2)\, e^{-i\varphi} \\ -\sin(\vartheta/2)\cos(\vartheta/2)\, e^{i\varphi} & \cos^2(\vartheta/2) \end{pmatrix} \end{aligned}$$

satisfying obviously

$$(6.11) \qquad M(b_1) + M(b_2) = \begin{pmatrix} 1 & 0 \\ 0 & 1 \end{pmatrix} = I.$$

The matrix elements relevant to the probability (6.1) are then

$$(6.12) \quad \langle a_j | M(b_1) | c_i \rangle = \frac{1}{\sqrt{2}} \overbrace{1 \quad 1} \begin{pmatrix} \cos^2(\vartheta/2) \\ \sin(\vartheta/2)\cos(\vartheta/2)\,e^{i\varphi} \end{pmatrix}$$

$$= (1/\sqrt{2})\cos(\vartheta/2)\,[\cos(\vartheta/2) + \sin(\vartheta/2)\,e^{i\varphi}]$$

$$(6.13) \quad \langle a_j | M(b_2) | c_i \rangle = \frac{1}{\sqrt{2}} \overbrace{1 \quad 1} \begin{pmatrix} \sin^2(\vartheta/2) \\ -\sin(\vartheta/2)\cos(\vartheta/2)\,e^{i\varphi} \end{pmatrix}$$

$$= (1/\sqrt{2})\sin(\vartheta/2)\,[\sin(\vartheta/2) - \cos(\vartheta/2)\,e^{i\varphi}].$$

Thus, if no measurement of spin in direction ϑ, φ is made between the selective measurements of spin $+1$ in direction z and in direction x, the probability is

$$(6.14) \qquad p(a_j, I, c_i) = |\langle a_j | c_i \rangle|^2 = |\sum_k \langle a_j | M(b_k) | c_i \rangle|^2 = \tfrac{1}{2},$$

whereas turning on the *separating* magnetic field in direction ϑ, φ *without selection* in the intermediate stage results in

$$(6.15)$$

$$p(a_j, b, c_i) = \sum_k |\langle a_j | M(b_k) | c_i \rangle|^2$$

$$= \tfrac{1}{2}\cos^2(\vartheta/2)\,(1 + \sin\vartheta\cos\varphi) + \tfrac{1}{2}\sin^2(\vartheta/2)\,(1 - \sin\vartheta\cos\varphi)$$

$$= \tfrac{1}{2}(1 + \sin\vartheta\cos\vartheta\cos\varphi).$$

NOTE

The reader's attention is drawn to a most interesting paper by Albertson [1] in which a careful analysis of the measurement process itself is given, by consistently including the measuring instrument into the quantum mechanical description *without assuming knowledge of the premeasurement state of the instrument.*

REFERENCE

[1] J. Albertson, Quantum mechanical measurement operator, *Phys. Rev.* **129**, 940 (1963).

The Fundamental Dynamical Postulate

The aim of any dynamics is to predict the outcome of experiments. The dynamical equations describing any physical object should enable one to compute the outcome of a complete measurement at time t, if the outcome of this measurement at time t_0 is known, and provided no other observations of the object are made in between the times t and t_0 which may alter the initial state established by observation at time t_0.

All physical objects which can be grasped with the language of quantum mechanics to date seem to satisfy the following fundamental postulate.

For any physical object there exists a hermitean operator $H = H^+$ so that the expectation value of any observable A satisfies the equation

$$(7.1) \quad \vec{A} = i\overline{(HA - AH)} + \overline{(\partial A/\partial t)}$$
$$= i\langle b|HA - AH|b\rangle + \langle b|\partial A/\partial t|b\rangle$$
$$= i\,\mathrm{trace}\,[(HA - AH)\,M] + \mathrm{trace}\,[(\partial A/\partial t)\,M].$$

This formulation is clearly invariant under unitary transformations. The operator H is called the Hamiltonian of the object, because in many cases it happens to be identical with the corresponding Hamiltonian in the classical description of the object, provided the canonical variables on which H depends are replaced by suitable operators. In fact, it was this correspondence with classical mechanics which first led to the discovery that postulate (7.1) constitutes a quantum mechanical description of the dynamics of some objects.

It must be borne in mind, however, that there are many physical objects for which there exists no classical analog, and the construction of a suitable Hamiltonian is often the most difficult problem encountered when description of an actual physical object is attempted. The lack of an unambiguous recipe for finding the Hamiltonian of an object is one of the major shortcomings of contemporary quantum mechanics.

The rate of change with time of the expectation value $\bar{A}$ of an observable A may be written, depending on whether $\bar{A}$ is given in the form $\bar{A} = \langle b|A|b\rangle$ or in the form $\bar{A} = \mathrm{trace}\,(AM)$, either as

$$(7.2) \quad \vec{A} = \langle b|\dot{A}|b\rangle + \langle \dot{b}|A|b\rangle + \langle b|A|\dot{b}\rangle$$

43

or as

(7.3) $\vec{A}$ = trace $(\dot{A}M)$ + trace $(A\dot{M})$.

Now there are several ways in which one may picture the evolution of a physical object described by Eq. (7.1). The following three points of view or pictures which may be attached to the representation (7.2) are particularly useful.

(I) *The State Picture.* Equation (7.1) is satisfied by assuming

(7.4) $\dot{A}_s = \partial A_s / \partial t$

so that the evolution of the state is governed by the equation

(7.5) $i|\dot{b}\rangle_s = H|b\rangle_s.$

This means in this state picture the state of the object at time t is defined by the outcome of a complete measurement on the object at that time. All observables are represented by the same operators for all times. Equation (7.5) is widely known under the name of Schroedinger's equation. The state picture is accordingly often called the Schroedinger picture. It should be noted that with respect to time this equation contains the first derivative only. Depending on the precise form of H it resembles in some cases a diffusion equation with imaginary diffusion constant.

(II) *The Operator Picture.* Equation (7.1) is satisfied by assuming

(7.6) $|\dot{b}\rangle_0 = 0$

so that the evolution of the object is governed by the operator equation

(7.7) $i\dot{A}_0 = A_0 H - H A_0 + i(\partial A_0 / \partial t).$

This means in this operator picture the state of the object is fixed as a vector defined by the outcome of all possible complete measurements on the object throughout its history. It is the representation of observables which now varies with time according to Eq. (7.7). This picture is widely known as the Heisenberg picture.

One should be careful about the symbols used to denote differentiation with respect to time. It is always understood that $\dot{\vec{A}} = \partial(\overline{A})/\partial t$; $|\dot{b}\rangle = \partial|b\rangle/\partial t$. Equation (7.7), governing the operator picture, should, strictly speaking, be written $i(DA_0/Dt) = A_0 H - H A_0 + i(\partial A_0/\partial t)$ where DA_0/Dt stands for an operator which has the expectation value $(\overline{DA_0/Dt}) = \partial(\overline{A})/\partial t$. For the Hamiltonian itself one has thus quite generally $DH/Dt = \partial H/\partial t$. Use of the symbol d/dt for differentiation with respect to time may lead to misunderstandings, because $|b\rangle$ depends in general on other variables, e.g. the position q of the object, which may also be a function of time, and in (7.5) and (7.6) the differentiation $|b\rangle$ does not include differentiation with respect to the implicit time dependence of $|b\rangle$ via the dependence on $q(t)$, say.

(III) *The Interaction Picture.* If the Hamiltonian of the object can be meaningfully split into a "free" Hamiltonian H^0, which is time independent, and an "interaction" Hamiltonian H', which may depend on time,

$$(7.8) \qquad H = H^0 + H'$$

then it is sometimes convenient to satisfy Eq. (7.1) by assuming

$$(7.9) \qquad i\dot{A}_I = A_I H^0 - H^0 A_I + i(\partial A_I/\partial t)$$

so that

$$(7.10) \qquad i|\dot{b}\rangle_I = H'|b\rangle_I.$$

This means in this interaction picture the evolution of the state is determined by the interaction Hamiltonian alone, provided the operators representing observables are made to change with time according to Eq. (7.9) which is governed entirely by the free Hamiltonian.

The choice of picture employed in the solution of any problem is dictated solely by convenience. The invariance of the fundamental dynamical postulate under changes in picture must correspond to the invariance of (7.1) under certain unitary transformations which connect the state vectors $|b\rangle_s$, $|b\rangle_0$, $|b\rangle_I$ and the operators A_s, A_0, A_I of the various pictures which describe the same physical situation.

As a first example consider the unitary transformation connecting the state picture with the operator picture,

$$(7.11) \quad |b\rangle_s = U|b\rangle_0; \qquad A_s = UA_0U^+; \qquad UU^+ = U^+U = I.$$

Since $|b\rangle_0$ is, according to Eq. (7.6), constant in time, it follows that

$$(7.12) \qquad |\dot{b}\rangle_s = \dot{U}|b\rangle_0.$$

On the other hand, Eq. (7.5) yields

$$(7.13) \qquad |\dot{b}\rangle_s = -iH|b\rangle_s = -iHU|b\rangle_0$$

and one obtains for the transformation operator U the equation

$$(7.14) \qquad i\dot{U} = HU$$

and by a similar argument applied to the inverse transformation $U^{-1} = U^+$

(7.15) $$-i\dot{U}^+ = U^+ H.$$

If H is given and does not contain the time explicitly, then Eq. (7.14) can be solved formally immediately, to yield

(7.16) $$U(t) = e^{-iHt} U(0).$$

If one adopts the initial condition $U(0) = I$, so that $|b(0)\rangle_s = |b\rangle_0$, one may write the first Eq. (7.11)

(7.17) $$|b(t)\rangle_s = e^{-iHt} |b(0)\rangle_s$$

meaning the Hamiltonian is the generator of a unitary transformation which develops, in the state picture, the state vector in time. This is the quantum mechanical analog of the well-known result of classical mechanics, where the Hamiltonian is the generator of a canonical transformation which develops the system motion in phase space. The second equation (7.11) may now be written, because of the initial condition $A_0(0) = A_s$ for the case of operators which do not contain the time t explicitly,

(7.18) $$A_0(t) = e^{iHt} A_0(0) e^{-iHt}$$

and the Hamiltonian is thus also recognized as the generator of a unitary transformation which develops, in the operator picture, the operators in time. One can thus, in the Schroedinger picture, regard the state vectors as rotating and the operators with their eigenvectors as standing still, and, in the Heisenberg picture, regard the state vectors as standing still and the operators with their eigenvectors as rotating, the sense of rotation in the abstract space being opposite in the two cases.

If $|b(0)\rangle_s$ is an eigenfunction of H with eigenvalue ω, then

$$|b(t)\rangle_s = e^{-i\omega t} |b(0)\rangle_s.$$

Knowledge of the solution of the eigenvalue problem

(7.19) $$H|b\rangle = \omega|b\rangle$$

is therefore, in principle, of importance. Unfortunately, the number of cases for which the eigenstates and eigenvalues of the Hamiltonian are known are very limited. The process of finding these solutions has often the character of a mathematical stunt.

The central mathematical problem remains the solution of the system of Eqs. (7.14). The main concern will be with

(a) those cases that can be solved easily for given H,

(b) the ramifications of those cases that cannot be solved in closed form for given H, and

(c) attempts at guessing solutions in cases for which at most some general symmetry properties of an otherwise unknown Hamiltonian are known.

As a very simple example consider the dynamics of a spin **s**, connected with the magnetic moment $\boldsymbol{\mu} = (e/m)\mathbf{s}$ of the electron, in a constant external field **B**. The Hamiltonian giving the correct quantum mechanical description of this object is simply the interaction energy

$$(7.20) \qquad H = -(\boldsymbol{\mu}\mathbf{B}) = -(e/2m)(\boldsymbol{\sigma}\mathbf{B})$$

written in terms of the spin operator $\boldsymbol{\sigma}$ introduced in Section 2, the components of **B** being treated as given parameters.

Working in a coordinate system in which **B** has a z component only, and taking as basic state vectors the eigenvectors of σ_3, the fundamental equation in the state picture (7.5) reads explicitly

$$(7.21)$$

$$i\begin{pmatrix}\dot{\alpha}\\\dot{\beta}\end{pmatrix} = -(e/2m)\,B_3\,\sigma_3\begin{pmatrix}\alpha\\\beta\end{pmatrix} = -(e/2m)\,B_3\begin{pmatrix}\alpha\\-\beta\end{pmatrix}; \qquad \begin{array}{l}\alpha = \cos(\vartheta/2)\\ \beta = \sin(\vartheta/2)\,e^{i\varphi}\end{array}$$

and has the solution

$$(7.22) \qquad \alpha(t) = \alpha(0)\,e^{+i\omega_L t}; \qquad \beta(t) = \beta(0)\,e^{-i\omega_L t}.$$

with the characteristic frequency

$$(7.23) \qquad \omega_L = (eB_3/2m)$$

so that in terms of the polar angles in accordance with (2.41)

$$(7.24) \quad (\beta/\alpha) = [\beta(0)/\alpha(0)]\,e^{-2i\omega_L t} = \tan[\vartheta(0)/2]\exp i[\varphi(0) - 2\omega_L t].$$

This means the expectation value of the spin which at time $t = 0$ points in direction $\vartheta(0)$, $\varphi(0)$ precesses around the direction of the external field with frequency $2\omega_L$.

In the operator picture the same situation is described by the fundamental equation (7.7) which reads in this example for each component of the spin, using (2.46),

$$\begin{array}{c}i\dot{\sigma}_1 = \sigma_1 H - H\sigma_1 = -\omega_L(\sigma_1\sigma_3 - \sigma_3\sigma_1) = 2i\omega_L\sigma_2\\[4pt](7.25)\qquad i\dot{\sigma}_2 = \sigma_2 H - H\sigma_2 = -\omega_L(\sigma_2\sigma_3 - \sigma_3\sigma_2) = -2i\omega_L\sigma_1\\[4pt]i\dot{\sigma}_3 = \sigma_3 H - H\sigma_3 = 0\end{array}$$

being a special case of the general equation

$$(7.26) \qquad \dot{\boldsymbol{\sigma}} = 2(\boldsymbol{\omega}_L \times \boldsymbol{\sigma})$$

which again means the expectation value of the observable $\boldsymbol{\sigma}$ precesses around the direction of the external field **B** with angular velocity $2\boldsymbol{\omega}_L = (e/m)\mathbf{B}$. This can be verified, by explicit solution, as follows.

The special case (7.25) has, according to (7.18), the solution

$$(7.27) \qquad \sigma_1(t) = e^{-i\omega_L \sigma_3 t} \sigma_1(0) e^{+i\omega_L \sigma_3 t}$$

$$(7.28) \qquad \sigma_2(t) = e^{-i\omega_L \sigma_3 t} \sigma_2(0) e^{+i\omega_L \sigma_3 t}$$

$$(7.29) \qquad \sigma_3(t) = e^{-i\omega_L \sigma_3 t} \sigma_3(0) e^{+i\omega_L \sigma_3 t} = \sigma_3(0).$$

Equation (7.29) allows one to replace in the exponents of (7.27) and (7.28) the operator σ_3 by the constant operator $\sigma_3(0)$, and one has the expansions, with the notation $[AB] = AB - BA$,

$$(7.27') \quad \sigma_1(t) = \sigma_1(0) - (i\omega_L t)/1![\sigma_3 \sigma_1]_{t=0} + (i\omega_L t)^2/2![\sigma_3[\sigma_3 \sigma_1]]_{t=0}$$
$$- (i\omega_L t)^3/3![\sigma_3[\sigma_3[\sigma_3 \sigma_1]]]_{t=0} + \ldots$$

$$(7.28') \quad \sigma_2(t) = \sigma_2(0) - (i\omega_L t)/1![\sigma_3 \sigma_2]_{t=0} + (i\omega_L t)^2/2![\sigma_3[\sigma_3 \sigma_2]]_{t=0}$$
$$- (i\omega_L t)^3/3![\sigma_3[\sigma_3[\sigma_3 \sigma_2]]]_{t=0} + \ldots.$$

The C.R.s yield

$$(7.30)$$
$$[\sigma_3 \sigma_1] = 2i\sigma_2; \qquad [\sigma_3[\sigma_3 \sigma_1]] = -(2i)^2 \sigma_1; \qquad [\sigma_3[\sigma_3[\sigma_3 \sigma_1]]] =$$
$$= -(2i)^3 \sigma_2; \ldots$$

$$(7.31)$$
$$[\sigma_3 \sigma_2] = -2i\sigma_1; \qquad [\sigma_3[\sigma_3 \sigma_2]] = -(2i)^2 \sigma_2; \qquad [\sigma_3[\sigma_3[\sigma_3 \sigma_2]]]$$
$$= (2i)^3 \sigma_1; \ldots$$

so that finally

$$(7.27'')$$
$$\sigma_1(t) = \sigma_1(0)\{1 - [(2\omega_L t)^2/2!] + \ldots\} + \sigma_2(0)\{(2\omega_L t) - [(2\omega_L t)^3/3!] + \ldots\}$$
$$= \sigma_1(0)\cos(2\omega_L t) + \sigma_2(0)\sin(2\omega_L t)$$

$$(7.28'')$$
$$\sigma_2(t) = \sigma_2(0)\{1 - [(2\omega_L t)^2/2!] + \ldots\} - \sigma_1(0)\{(2\omega_L t) - [(2\omega_L t)^3/3!] + \ldots\}$$
$$= \sigma_2(0)\cos(2\omega_L t) - \sigma_1(0)\sin(2\omega_L t).$$

The unitary operator connecting state picture and operator picture is

$$(7.32) \qquad U(t) = \exp\left[i\omega(_L\boldsymbol{\sigma}\cdot\mathbf{n})t\right]I; \qquad \mathbf{n} = \mathbf{B}/|\mathbf{B}|$$

which by expansion and utilization of the C.R.s can be written

$$(7.33) \qquad U(t) = \cos(\omega_L t) + i(\boldsymbol{\sigma}\cdot\mathbf{n})\sin(\omega_L t)$$

and represents a rotation by angle $2\omega_L t$ around $\mathbf{n}$, in accordance with the remark made at the end of Section 3. This will be taken up in a more general context in Section 12.

It is instructive to write the development of an object in time in terms of the density matrix representation (7.3). From this and the fundamental dynamical postulate (7.1) follows the identity

(7.34)

$$\text{trace}\,(A\dot{M}) + \text{trace}\,(\dot{A}M) = i\,\text{trace}\,[(HA - AH)\,M] + \text{trace}\,[(\partial A/\partial t)\,M].$$

In the state picture the condition (7.4) leads immediately to

(7.35)

$$\text{trace}\,(A_s\dot{M}_s) = i\,\text{trace}\,[(HA_s - A_sH)\,M_s] = i\,\text{trace}\,[A_s(M_sH - HM_s)]$$

which is satisfied provided

(7.36)
$$\dot{M}_s = i(M_sH - HM_s).$$

It must be stressed that M_s is a substitute for the state vector and does, therefore, depend explicitly on time, as does $|b\rangle_s$, in the state picture. The symbol denoting differentiation with respect to time of the density matrix means, as in case of the same symbol for the corresponding state vector, the partial derivative

(7.37)
$$\dot{M}_s = \partial M_s/\partial t$$

and one can thus write (7.36) in formal analogy to the concept of total differentiation

(7.38)
$$DM_s/Dt = i(HM_s - M_sH) + (\partial M_s/\partial t) = 0$$

which is the quantum mechanical analog of Liouville's theorem in classical mechanics. This equation can be verified explicitly for a pure state by using the Schroedinger equation. If the decomposition

(7.39)
$$(M)_{ik} = (b)_i(b)_k^*$$

is adopted, one obtains by partial differentiation with respect to time, using (7.5)

(7.40) $\quad (\dot{M})_{ik} = (\dot{b})_i(b)_k^* + (b)_i(\dot{b})_k^* = -i\sum_j \{H_{ij}(b)_j(b)_k^* - (b)_i(b)_j^* H_{jk}\}$

$$= -i(HM - MH)_{ik}$$

which is identical with (7.36) in components.

The transition to the operator picture is effected by satisfying (7.34) with (7.7), requiring

(7.41)
$$\dot{M}_0 = 0.$$

This can be verified directly by applying to (7.39) the unitary transformation connecting $M_s(t)$ with $M_s(0)$. From (7.17) follows

(7.42) $\quad M_s(t) = |b(t)\rangle_s \langle b(t)|_s = e^{-iHt}|b(0)\rangle_s \langle b(0)|_s e^{iHt} = e^{-iHt} M_s(0)\, e^{iHt}.$

Now, $M_s(0)$ can be identified with the constant density matrix in the operator picture, $M_s(0) = M_0$, because if an operator A_s is transformed into A_0 by

(7.43) $$A_s = U A_0 U^+$$

then the invariance of

(7.44) $$\overline{A} = \text{trace}\,(A_s\, M_s) = \text{trace}\,(U A_0\, U^+ M_s)$$

$$= \text{trace}\,(A_0\, U^+ M_s\, U) = \text{trace}\,(A_0\, M_0)$$

requires M to transform accordingly,

(7.45) $$M_0 = U^+ M_s\, U$$

which is Eq. (7.42) with $U(t) = e^{-iHt} I$.

It is interesting to note that if the eigenvectors of H,

(7.46) $$H|a_i\rangle = \omega_i |a_i\rangle \qquad \text{(say)}$$

form a complete set and can thus be used as a basis, then the matrix elements of (7.42) take the form

(7.47) $$\langle a_i | M(t) | a_k \rangle = M_{ik}(t) = M_{ik}(0)\, e^{i(\omega_k - \omega_i)t}$$

so that only energy differences, but no unobservable absolute energies, appear.

Writing for the special example of the spin magnetic moment in a fixed external field the density matrix as in (4.30)

(7.48) $$M_s = \tfrac{1}{2}[I + (\mathbf{P}\boldsymbol{\sigma})],$$

where now $\mathbf{P}(t)$ is the time dependent polarization vector characterizing the spin, then the fundamental equation (7.38) reads, with

$$H = -(e/2m)\,(\boldsymbol{\sigma}\mathbf{B}),$$

(7.49) $$-(ie/2m)\,[(\boldsymbol{\sigma}\mathbf{B})\,(\mathbf{P}\boldsymbol{\sigma}) - (\mathbf{P}\boldsymbol{\sigma})\,(\boldsymbol{\sigma}\mathbf{B})] + (\partial\mathbf{P}/\partial t)\cdot\boldsymbol{\sigma} = 0.$$

Now, by virtue of the C.R.s of the matrices σ_i one has

(7.50) $$(\boldsymbol{\sigma}\mathbf{B})\,(\mathbf{P}\boldsymbol{\sigma}) - (\mathbf{P}\boldsymbol{\sigma})\,(\boldsymbol{\sigma}\mathbf{B}) = 2i[\boldsymbol{\sigma}\cdot(\mathbf{B}\times\mathbf{P})]$$

so that (7.49) can be written

(7.51) $$[(\partial\mathbf{P}/\partial t) + (e/m)\,(\mathbf{B}\times\mathbf{P})]\cdot\boldsymbol{\sigma} = 0$$

which requires as necessary condition the *classical* equation

(7.52) $$(\partial\mathbf{P}/\partial t) = -(e/m)\,(\mathbf{B}\times\mathbf{P})$$

describing again the precession of the polarization vector around the applied field with frequency $2\boldsymbol{\omega}_L = (e/m)\,\mathbf{B}$.

The visualization of a spin by means of a precessing polarization vector is thus by all means permissible and does not lead to contradictions with the probability aspect of the Stern–Gerlach experiment.

As an example to which the interaction picture may be applied, consider a spin magnetic moment $\mu = \gamma s$ subject to the external field

(7.53) $\mathbf{B} = \mathbf{B}_0 + \mathbf{B}_1 = B_0 \mathbf{k} + B_1[\mathbf{i} \cos(\omega t) - \mathbf{j} \sin(\omega t)]$

corresponding to a fixed field B_0 in z direction and a field B_1 rotating clockwise with angular velocity

(7.54) $\boldsymbol{\omega} = -\omega \mathbf{k}$

in the (x, y) plane. The Hamiltonian can be written

(7.55) $H = -\gamma(\mathbf{sB}) = -\gamma(\mathbf{sB}_0) - \gamma(\mathbf{sB}_1) = H^0 + H'.$

In the interaction picture, the equations governing the operators are therefore identical with (7.27), (7.28), and (7.29) and one has for the spin operators the representations

$$
\left.\begin{array}{l}
\sigma_1(t) = \sigma_1(0)\cos(\omega_0 t) + \sigma_2(0)\sin(\omega_0 t) \\
\sigma_2(t) = -\sigma_1(0)\sin(\omega_0 t) + \sigma_2(0)\cos(\omega_0 t) \\
\sigma_3(t) = \sigma_3(0)
\end{array}\right\} \quad \omega_0 = 2\omega_L = \gamma B_0
$$

(7.56)

and the state vector in this picture is governed by the equation

$$
\begin{aligned}
i|\dot{b}\rangle_I = H'|b\rangle_I &= -(\gamma/2)(\boldsymbol{\sigma}\mathbf{B}_1)|b\rangle_I \\
&= -(\omega_1/2)[\sigma_1(t)\cos(\omega t) - \sigma_2(t)\sin(\omega t)]|b\rangle_I \\
&= -(\omega_1/2)\{\sigma_1(0)\cos[(\omega_0 - \omega)t] \\
&\qquad + \sigma_2(0)\sin[(\omega_0 - \omega)t]\}|b\rangle_I;
\end{aligned}
$$

(7.57)

$\omega_1 = \gamma B_1;$

which reads in components

$$
|b\rangle_I = \begin{pmatrix} \alpha \\ \beta \end{pmatrix}; \qquad \sigma_1 = \begin{pmatrix} 0 & 1 \\ 1 & 0 \end{pmatrix}; \qquad \sigma_2 = \begin{pmatrix} 0 & -i \\ i & 0 \end{pmatrix}
$$

(7.57′) $\begin{cases} \dot{\alpha} = (i\omega_1/2)\, e^{-i(\omega_0 - \omega)t}\, \beta \\ \dot{\beta} = (i\omega_1/2)\, e^{+i(\omega_0 - \omega)t}\, \alpha. \end{cases}$

By differentiation of the first of these equations, and substitution of $\dot{\beta}$ from the second, one eliminates β and obtains

(7.58) $-\ddot{\alpha} - i(\omega_0 - \omega)\dot{\alpha} - (\omega_1/2)^2 \alpha = 0$

which may be solved by

(7.59) $\alpha(t) = \alpha(0)\, e^{i(\Omega/2)t}$

provided

(7.60) $\Omega^2 + 2(\omega_0 - \omega)\Omega - \omega_1^2 = 0$

giving rise to two solutions

(7.61)
$$\Omega_1 = -(\omega_0 - \omega) + \sqrt{(\omega_0 - \omega)^2 + \omega_1^2};$$
$$\Omega_2 = -(\omega_0 - \omega) - \sqrt{(\omega_0 - \omega)^2 + \omega_1^2}$$

so that the general solution of (7.58) reads

(7.62)
$$\alpha(t) = \alpha_1(0)\, e^{i(\Omega_1/2)t} + \alpha_2(0)\, e^{i(\Omega_2/2)t}$$

from which one obtains then

(7.63)
$$\beta(t) = (2/i\omega_1)\, e^{i(\omega_0 - \omega)t}\, \dot{\alpha} = \alpha_1(0)\,(\Omega_1/\omega_1)\, e^{-i(\Omega_2/2)t} + \alpha_2(0)\,(\Omega_2/\omega_1)\, e^{-i(\Omega_2/2)t}.$$

Now suppose at time $t = 0$ the spin has with certainty the value $+1$ in z direction, i.e. $|b(0)\rangle = |b_+\rangle = \binom{1}{0}$, then

(7.64)
$$\alpha(0) = \alpha_1(0) + \alpha_2(0) = 1; \qquad \beta(0) = \alpha_1(0)\,(\Omega_1/\omega_1) + \alpha_2(0)\,(\Omega_2/\omega_1) = 0$$

giving the initial values

(7.65) $$\alpha_1(0) = \Omega_2/(\Omega_2 - \Omega_1); \qquad \alpha_2(0) = \Omega_1/(\Omega_1 - \Omega_2).$$

One can now calculate the probability for "spin flip," i.e. the probability for finding at time t the state $|b_-\rangle = \binom{0}{1}$ in which the spin has the value -1 in z direction, namely,

(7.66) $$\mathbf{P}_{\text{flip}} = |\langle b_-|b(t)\rangle|^2 = \left|\overbrace{\begin{matrix}0 & 1\end{matrix}}\binom{\alpha}{\beta}\right|^2 = |\beta(t)|^2.$$

By substitution of (7.65) into (7.63) one obtains thus, using $\Omega_1\Omega_2 = -\omega_1^2$ and $\Omega_1 - \Omega_2 = 2\sqrt{(\omega_0 - \omega)^2 + \omega_1^2}$,

(7.67)
$$\mathbf{P}_{\text{flip}} = \frac{2\Omega_1^2\Omega_2^2}{\omega_1^2(\Omega_1 - \Omega_2)^2}\left[1 - \cos\left(\frac{\Omega_1 - \Omega_2}{2}t\right)\right]$$
$$= \frac{\omega_1^2}{\omega_1^2 + (\omega_0 - \omega)^2}\sin^2\left(\frac{\sqrt{(\omega_0 - \omega)^2 + \omega_1^2}}{2}t\right).$$

The amplitude of the spin flip probability shows thus a *resonance* at $\omega = \omega_0$, which can be exploited to measure ω_0 and thus γ.

By an argument similar to that leading from (7.34) to (7.36), the density matrix in the interaction picture is found to satisfy the equation

(7.68) $$\dot{M}_I = i(M_I H' - H' M_I)$$

from which one obtains by straightforward computation after substitution of $M_I = \frac{1}{2}[I + (\mathbf{P}\boldsymbol{\sigma}_I)]$

$$\dot{P}_1 = -\omega_0 P_2 + \omega_1 P_3 \sin(\omega t)$$

(7.69)
$$\dot{P}_2 = +\omega_0 P_1 - \omega_1 P_3 \cos(\omega t)$$

$$\dot{P}_3 = -\omega_1 P_1 \sin(\omega t) + \omega_1 P_2 \cos(\omega t).$$

These equations are, of course, identical with (7.52) if **B** is given by (7.53). Their solution proceeds without difficulty. Differentiating $\dot{P}_3$ twice with respect to t and using the expressions for $\dot{P}_1$ and $\dot{P}_2$ one obtains

(7.70) $\qquad \ddot{P}_3 = -w^2 P_3 \qquad$ with $\qquad w = \sqrt{(\omega_0 - \omega)^2 + \omega_1^2}$

and thus the solution

(7.71) $\qquad\qquad P_3(t) = A\, e^{iwt} + B\, e^{-iwt} + C$

in which the integration constants A, B, C are to be determined from the initial conditions, which in the example treated above are

(7.72) $\qquad\qquad P_3(0) = 1; \qquad P_1(0) = P_2(0) = 0$

leading immediately to

(7.73) $\qquad\qquad A = B = \omega_1^2/2w^2; \qquad C = (\omega_0 - \omega)^2/w^2$

and the solution (7.71) reads in this case

(7.74) $\qquad\qquad P_3(t) = (1/w^2)\,[\omega_1^2 \cos(\omega t) + (\omega_0 - \omega)^2].$

To obtain P_{flip} by this method, one need not calculate $P_1(t)$ and $P_2(t)$. The density matrix is generally

(7.75) $\qquad M(t) = \frac{1}{2}[I + (\boldsymbol{\sigma}\mathbf{P})] = \dfrac{1}{2}\begin{pmatrix} 1+P_3 & P_1 - iP_2 \\ P_1 + iP_2 & 1 - P_3 \end{pmatrix}$

and the probability for finding the spin in any given direction ϑ, φ is $P_{\vartheta,\varphi} = \text{trace}\,[M(\vartheta,\varphi)\,M(t)]$. For spin -1 in z direction one has

$$M(\vartheta,\varphi) = M(b_-) = \begin{pmatrix} 0 & 0 \\ 0 & 1 \end{pmatrix},$$

and therefore

(7.76) $\quad P_{\text{flip}} = \frac{1}{2}\,\text{trace}\left[\begin{pmatrix} 0 & 0 \\ 0 & 1 \end{pmatrix}\begin{pmatrix} 1+P_3 & P_1 - iP_2 \\ P_1 + iP_2 & 1 - P_3 \end{pmatrix}\right] = \frac{1}{2}(1 - P_3)$

$$= (\omega_1^2/w^2)\sin^2(wt/2)$$

in agreement with (7.67).

The connection among interaction picture, state picture, and operator picture is established by unitary operators defined by

(7.77)
$$|b\rangle_s = V|b\rangle_I \qquad\qquad |b\rangle_I = W|b\rangle_0$$
$$VV^+ = V^+V = I; \qquad WW^+ = W^+W = I$$

so that

(7.78) $|b\rangle_s = VW|b\rangle_0$ i.e. $|b\rangle_s = U|b\rangle_0$ with $U = VW$.

Since one has

(7.79) $|\dot{b}\rangle_s = \dot{V}|b\rangle_I + V|\dot{b}\rangle_I = (\dot{V} - iVH')|b\rangle_I$; $|b\rangle_I = W|b\rangle_0$

and also

(7.80)
$$|\dot{b}\rangle_s = -iH|b\rangle_s = -iHV|b\rangle_I; \qquad |\dot{b}\rangle_I = -iH'|b\rangle_I = -iH' W|b\rangle_0$$

it follows that V and W must satisfy the operator equations

(7.81) $\dot{V} = i(VH' - HV);   \dot{W} = -iH' W$.

This is consistent with (7.78) because

(7.82)
$$\dot{U} = \dot{V}W + V\dot{W} = i(VH' - HV)W - iVH'W = -iHVW = -iHU$$

is identical with Eq. (7.14).

Explicit solutions can be found for V and W, for the case of spin magnetic resonance treated above, without difficulty.

NOTES

The state picture and the operator picture of quantum mechanical dynamics emerged in famous papers by Schroedinger, beginning with a short note [1] and collected in book form [2], and by Heisenberg [3], followed by papers in collaboration with Born and Jordan, whose book [4] is entirely devoted to development of the state picture. The unitary transformation connecting the two pictures was first given by Schroedinger [5].

The interaction picture, although it is implicit in many earlier works, did come into its own right through development by Tomonaga [6].

Rabi [7] first gave the exact solution of the magnetic resonance problem for the case of a spin $\frac{1}{2}$.

REFERENCES

[1] E. Schroedinger, *Ann. Physik* **79**, 361 (1926).
[2] E. Schroedinger, "Abhandlungen zur Wellenmechanik." Barth, Leipzig, 1928.
[3] W. Heisenberg, *Z. Physik* **33**, 879 (1925).
[4] M. Born and P. Jordan, "Elementare Quantenmechanik." Springer, Berlin, 1930.
[5] E. Schroedinger, *Ann. Physik* **79**, 734 (1926).
[6] S. Tomonaga, *Progr. Theoret. Phys. (Kyoto)* **1**, 27 (1946).
[7] I. I. Rabi, *Phys. Rev.* **51**, 652 (1937).

The Representation of Observables with Nondenumerably and Denumerably Infinite Ranges of Possible Values

To grasp attributes capable of a nondenumerably infinite or continuous range of possible values it is desirable to have a description which is *formally* identical with the one developed for discrete quantum numbers in Sections 1 and 2. Thus a state depending on a continuous attribute q will be denoted $|q\rangle$ and will be represented by a complex unit vector in an abstract space ("Hilbert space") of nondenumerably infinite dimensions. Formally, one may thus expect $|q\rangle$ to be an eigenvector of a suitably defined operator Q with eigenvalue q in analogy to the vector $|a_i\rangle$ in a finite dimensional space, which is an eigenvector to the operator A with eigenvalue a_i. The correspondence

$$(8.1) \qquad A|a_i\rangle = a_i|a_i\rangle \leftrightarrow Q|q\rangle = q|q\rangle$$

can be extended to all operations defined for the vector $|a_i\rangle$ in Sections 1 and 2, provided one can formally introduce in the Hilbert space the notion of orthogonality by the correspondence

$$(8.2) \qquad \langle a_i|a_k\rangle = \delta_{ik} \leftrightarrow \langle q'|q''\rangle = \delta(q'-q'')$$

and the notion of closure by the correspondence

$$(8.3) \qquad \sum_i |a_i\rangle\langle a_i| = I \leftrightarrow \int |q\rangle\,dq\langle q| = I$$

where $\delta(q)$ is the well-known delta function (see Appendix 6) and I the identity operation.

In particular, one can expand the eigenvector $|p\rangle$ of another continuous attribute P in terms of $|q\rangle$ by the correspondence

$$(8.4a) \qquad |b_k\rangle = \sum_i |a_i\rangle\langle a_i|b_k\rangle \leftrightarrow |p\rangle = \int |q''\rangle\,dq''\langle q''|p\rangle$$

and similarly

$$(8.4b) \qquad \langle c_j| = \sum_i \langle c_j|a_i\rangle\langle a_i| \leftrightarrow \langle r| = \int \langle r|q'\rangle\,dq'\langle q'|.$$

The transformation functions $\langle q'|p\rangle$ are frequently called "wave functions" or simply "ψ functions" and denoted

55

$$(8.5) \qquad\qquad \psi_p(q') = \langle q'|p\rangle.$$

Referring to the representation spanned by the eigenvectors of Q as the q representation, one can say $\psi_p(q')$ is the component of the vector $|p\rangle$ in "direction" q' in the q representation. Knowing all these components $\psi_p(q')$ is thus tantamount to knowing the vector $|p\rangle$, the components $\psi_p(q')$ *represent* $|p\rangle$ in the q representation.

The scalar product of two vectors $|p\rangle$ and $|r\rangle$ can now be evaluated in q representation in terms of their ψ functions as follows:

$$\langle r|p\rangle = \iint \langle r|q'\rangle\, dq'\langle q'|q''\rangle\, dq''\, \langle q''|p\rangle$$

$$= \iint \langle r|q'\rangle\, dq'\, \delta(q'-q'')\, dq''\langle q''|p\rangle$$

$$(8.6) \qquad = \int \langle r|q'\rangle\, dq'\langle q'|p\rangle = \int \psi_r^*(q')\,\psi_p(q')\, dq'$$

so that, in particular, the orthogonality of two states reads

$$(8.7) \qquad \langle p''|p'\rangle = \int \psi_{p''}^*(q')\,\psi_{p'}(q')\, dq' = \delta(p''-p').$$

In q representation, the operator Q can be thought of as an infinite-dimensional diagonal matrix, because its matrix elements are definable by the correspondence

$$(8.8)$$

$$\langle a_k|A|a_i\rangle = a_i\langle a_k|a_i\rangle = a_i\,\delta_{ki} \leftrightarrow \langle q''|Q|q'\rangle = q'\langle q''|q'\rangle = q'\,\delta(q'-q'').$$

The effect of any operator $f(Q)$ on an arbitrary state, $|p\rangle$ say, with components $\psi_p(q')$ in q representation is accordingly given by

$$(8.9) \qquad f(Q)|p\rangle = \int f(Q)|q''\rangle\, dq''\langle q''|p\rangle = \int |q''\rangle\, dq'' f(q'')\,\psi_p(q'')$$

so that

$$(8.10) \qquad \langle r|f(Q)|p\rangle = \int \psi_r^*(q'')f(q'')\,\psi_p(q'')\, dq''.$$

In particular, the operator $P = -i(\partial/\partial Q)$ has in q representation the matrix elements

$$(8.11) \qquad \langle q''|P|q'\rangle = -i(\partial/\partial q')\,\delta(q'-q'')$$

so that in terms of ψ-functions

$$(8.12) \qquad \langle r|P|p\rangle = \int \psi_r^*(q'')\,[-i\,\partial\psi_p(q'')/\partial q'']\, dq''.$$

It is, on first sight, rather suprising that one can pass from a representation spanned by the eigenvectors of a continuous operator, $|q\rangle$ say, to a representation spanned by the *denumerably* infinite set of eigen-vectors of a discrete operator, $|E_i\rangle$ say, where

(8.13) $\qquad H|E_i\rangle = \omega_i|E_i\rangle; \qquad i = 1, 2, \ldots,$ ad infinitum

so that

(8.14) $$|E_i\rangle = \int |q'\rangle dq' \langle q'|E_i\rangle$$

and

(8.15) $$|q'\rangle = \sum_i |E_i\rangle\langle E_i|q'\rangle$$

which requires the ψ functions to be subject to the unitarity conditions

(8.16)
$$\langle E_k|E_i\rangle = \int \langle E_k|q'\rangle dq' \langle q'|E_i\rangle = \int \psi_{E_k}^*(q')\psi_{E_i}(q') dq' = \delta_{ki}$$

(8.17)
$$\langle q''|q'\rangle = \sum_i \langle q''|E_i\rangle\langle E_i|q'\rangle = \sum_i \psi_{E_i}(q'')\psi_{E_i}^*(q') = \delta(q''-q').$$

The possibility of the transition from the E representation to the q representation and *vice versa* is thus dependent on the existence of ψ functions satisfying (8.16) and (8.17). Such existence problems can always be settled if one succeeds in actually constructing such ψ functions. This will be done for some special cases in the next section.

The measurement symbol $M(p,r)$ corresponding to the graph

can also be expressed in terms of the ψ functions according to (8.4) as

(8.18) $\qquad M(p,r) = |p\rangle\langle r| = \int\int |q''\rangle dq'' \psi_p(q'') \psi_r^*(q') dq' \langle q'|$

and its matrix elements in q representation are found to be

(8.19)
$$\langle q'''|M(p,r)|q\rangle = \int\int \delta(q'''-q'') dq'' \psi_p(q'') \psi_r^*(q') dq' \delta(q'-q'')$$
$$= \psi_p(q''') \psi_r^*(q).$$

The density matrix for a state $|p\rangle$, in particular has the form

(8.20) $\qquad \langle q'|M(p)|q\rangle = \psi_p(q')\psi_p^*(q)$

and is seen to have trace unity if the ψ functions are normalized.

NOTE

Readers plagued by mathematical scruples regarding the concept of Hilbert space will find solace and comfort in Chapter II of the monumental work by von Neumann [1].

REFERENCE

[1] J. von Neumann, "Mathematical Foundations of Quantum Mechanics." Princeton Univ. Press, Princeton, New Jersey, 1955.

Displacements of the Observer

Any actual observation of a physical object is always made in a certain coordinate system. To be precise when describing the state of the object one should thus, in principle, always include information about the position of the observer. *Any* state $|b\rangle$ must therefore contain, in addition to all the variables attributable to the object, such as spin $\mathbf{s}$, momentum $\mathbf{p}$, etc., the coordinates of the observer, x_i say, and can be considered in fact a function $|\mathbf{s}, \mathbf{p}, \ldots, x_i\rangle$.

When describing a single object it will now be assumed tentatively that one can meaningfully define an operator $\mathbf{Q}$ which represents the location of the object such that one can represent the state vector in terms of the eigenvectors of $\mathbf{Q}$, denoted $|\mathbf{q}\rangle$, satisfying $\mathbf{Q}|\mathbf{q}\rangle = \mathbf{q}|\mathbf{q}\rangle$.

Postponing questions of how to actually determine experimentally the components of the state in this "coordinate representation," occasioned by the continuous range of possible values $\mathbf{q}$ of the location $\mathbf{Q}$, making $|\mathbf{q}, \ldots\rangle$ necessarily an infinite-dimensional vector, the position variables $\mathbf{q}$ will now, by convention, be chosen to originate at the position of the observer, so that $|\mathbf{q}, \ldots\rangle$ describes the state of the object as seen by an observer (0) located at $\mathbf{q} = 0$. It must be stressed, however, that the possibility of ascribing a fixed position to the observer should be considered as a tentative hypothesis, pending investigation of whether this hypothesis is compatible with the various tasks of measurement assigned to the observer.

Disregarding then, for the time being, these profound questions regarding actual position measurements, it will now be attempted to describe the same object in terms of the state used by an observer (1) who is displaced with respect to (0). If, in particular, the origin of (1) is displaced from the origin of (0) by a vector $\mathbf{a}$ pointing from (1) to (0), as indicated in Fig. 9.1, then (1) will describe the object in terms of a state

$$(9.1) \qquad |\mathbf{q}\rangle^T = |\mathbf{q} + \mathbf{a}\rangle.$$

From the requirement that the physical attributes of the object be invariant under displacements of the observer, one can infer the existence of a unitary operator $T_\mathbf{a}$ of displacement, defined by

$$(9.2) \qquad |\mathbf{q} + \mathbf{a}\rangle = T_\mathbf{a}|\mathbf{q}\rangle.$$

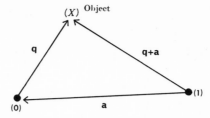

FIG. 9.1. Description of the position of an object (X) as seen by an observer (1) displaced by **a** with respect to an observer (0).

In this coordinate representation one can apply to each component the expansion

(9.3) $|\mathbf{q}+\mathbf{a}\rangle = [I + (\mathbf{a}.\,\partial/\partial\mathbf{Q}) + (\tfrac{1}{2}!)\,(\mathbf{a}.\,\partial/\partial\mathbf{Q})^2 + \ldots]|\mathbf{q}\rangle.$

Introducing an operator

(9.4) $\mathbf{P} = -i(\partial/\partial\mathbf{Q})$

one can write formally

(9.5) $T_\mathbf{a} = [I + i(\mathbf{aP}) + (i^2/2!)\,(\mathbf{aP})^2 + \ldots] = e^{i(\mathbf{aP})}.$

The inverse transformation is similarly given by

(9.6) $T_\mathbf{a}^{-1} = e^{-i(\mathbf{aP})}.$

The unitarity of $T_\mathbf{a}$, $T_\mathbf{a}^{-1} = T_\mathbf{a}^+$ requires then that $\mathbf{P}$ be a hermitean operator.

To elucidate the physical meaning of the operator $\mathbf{P}$ consider the special case in which the displacement vector **a** is a linear function of time,

(9.7) $\mathbf{a} = \mathbf{V}t.$

One envisages the observers (0) and (1) moving relative to each other with constant velocity $\mathbf{V}$. The transformation operator $T_\mathbf{a}$ is now a function of time, and represents the Galileo transformation.

Whenever in the equation

(9.8) $|b\rangle^T = T|b\rangle$

the operator T depends on time one must distinguish carefully between

   (i) $(\partial/\partial t)\,(|b\rangle^T)$ which tells how observer (1) finds *his* state description of the object varies with time, and

(ii) $(\partial|b\rangle/\partial t)^T$ which tells how observer (1) finds *observer* (0)'s state description of the same object varies with time.

Since $|b\rangle$ satisfies, in the state picture, the equation

(9.9) $$i\,\partial|b\rangle/\partial t = H|b\rangle$$

one obtains by application of T from the left

(9.10) $$i(\partial|b\rangle/\partial t)^T = H^T|b\rangle^T \quad\text{with}\quad H^T = THT^{-1}.$$

On the other hand, differentiating (9.8) with respect to time and using (9.9) one finds

(9.11)
$$i(\partial/\partial t)\,(|b\rangle^T) = iT(\partial|b\rangle/\partial t)+i\dot{T}|b\rangle = (H^T+i\dot{T}T^{-1})|b\rangle^T.$$

This means the *effective Hamiltonian* of observer (1), namely the Hamiltonian which determines the development of observer (1)'s state $|b\rangle^T$ in time, according to the fundamental dynamical postulate, is given by

(9.12) $$H_{\text{eff}} = H^T+i\dot{T}T^{-1}.$$

For the special case of the Galileo transformation (9.7) one obtains from (9.5)

(9.13) $$H_{\text{eff}} = H^T - (\mathbf{VP}).$$

Now in classical mechanics the Hamiltonian of a free particle of momentum $\mathbf{p}$ and mass m is

(9.14) $$H = (\tfrac{1}{2}m)\,\mathbf{p}^2$$

and the effective Hamiltonian after a Galileo transformation is

(9.15) $$H_{\text{eff}} = (\tfrac{1}{2}m)\,\mathbf{p}^2 - (\mathbf{Vp})$$

giving rise to the correct addition of velocities.

This correspondence suggests the identification of the operator $\mathbf{P}$ with the operator representing the momentum of the object. From this identification, it follows immediately that the position and momentum of an object are incompatible observables, because for each component P_j, Q_k one finds

(9.16) $$P_j Q_k - Q_k P_j = -iI\delta_{jk}.$$

It is instructive to work out the transformation formula for the position operator as a consequence of this commutation relation. By expansion one has, for each component,

(9.17)
$$Q^T = T_a Q T_a^{-1} = e^{iaP} Q e^{-iaP} = Q + ia[PQ] + (i^2/2!)\,a^2[P[PQ]] + \dots$$
$$= Q + a = Q + Vt$$

as expected. Differentiation of (9.17) with respect to time gives

(9.18)
$$\partial Q^T/\partial t = T_a(\partial Q/\partial t)\, T_a{}^1 + \dot{T}_a Q T_a^{-1} + T_a Q \dot{T}_a^{-1} = (\partial Q/\partial t)^T + iV(PQ^T - Q^T P)$$
$$= (\partial Q/\partial t)^T + V$$

as expected, because $(\partial Q/\partial t)^T$ tells how the observer (1) would find observer (0)'s position of the object vary with time, and Eq. (9.18) is just the "addition theorem of velocities" in accordance with the invariance under Galileo transformations.

Considering now, as an example, the Hamiltonian (9.14) of a "free" object which possesses momentum as an attribute, the state of the object satisfies, in the state picture and in coordinate representation, the equation

(9.19) $i(\partial/\partial t)|\mathbf{q}(t)\rangle = (\tfrac{1}{2}m)\,\mathbf{P}^2|\mathbf{q}(t)\rangle = -(\tfrac{1}{2}m)\,(\partial/\partial\mathbf{Q})^2|\mathbf{q}(t)\rangle.$

The state vector $|\mathbf{q}(t)\rangle$ is thus obviously not an eigenstate of H. Denoting the eigenstate of H with eigenvalue ω_i by $|E_i\rangle$, satisfying

(9.20) $H|E_i\rangle = \omega_i|E_i\rangle$

and writing in accordance with (7.17)

(9.21) $|\mathbf{q}(t)\rangle = e^{-iHt}|\mathbf{q}(0)\rangle$

one can, upon expansion of $|E_i\rangle$ in terms of $|\mathbf{q}(0)\rangle$, introducing the time-independent ψ function $\psi_{E_i}(\mathbf{q}) = \langle\mathbf{q}(0)|E_i\rangle$ by

(9.22) $|E_i\rangle = \int |\mathbf{q}'(0)\rangle d\,\mathbf{q}'\,\psi_{Ei}(\mathbf{q}'),$

cast the eigenvalue problem (9.20), upon application of (8.9), in the form

(9.23) $H|E_i\rangle = \int |\mathbf{q}'(0)\rangle\, d\mathbf{q}'(-\tfrac{1}{2}m)\,(\partial^2\psi_{E_i}/\partial\mathbf{q}'^2)$
$$= \omega_i \int |\mathbf{q}'(0)\rangle\, d\mathbf{q}'\,\psi_{E_i}(\mathbf{q}').$$

The linear independence of the components $|\mathbf{q}'(0)\rangle$ requires then that the ψ function characterizing the state $|E_i\rangle$ in q representation satisfy the eigenvalue equation

(9.24) $(-\tfrac{1}{2}m)\,\partial^2\psi_{E_i}(\mathbf{q})/\partial\mathbf{q}^2 = \omega_i\psi_{E_i}(\mathbf{q}).$

It is often convenient to expand $|\mathbf{q}(t)\rangle$ directly, with the help of a time-dependent ψ function $\varPsi$

(9.25)
$$|\mathbf{q}(t)\rangle = \sum_i |E_i\rangle\langle E_i|\mathbf{q}(t)\rangle = \sum_i |E_i\rangle\,\varPsi_{E_i}^*(\mathbf{q},t) = e^{-iHt}\sum_i |E_i\rangle\psi_{E_i}^*(\mathbf{q})$$
$$= \sum_i |E_i\rangle e^{-i\omega_i t}\psi_{E_i}^*(\mathbf{q})$$

so that the time-dependent ψ function and time-independent ψ function are related by

(9.26) $$\Psi_{E_i}(\mathbf{q}, t) = e^{i\omega_i t} \psi_{E_i}(\mathbf{q}).$$

Equation (9.24) has obviously the solution

(9.27) $$\psi_{E_i}(\mathbf{q}) = \text{const}\, e^{-i(\mathbf{k}_i\,\mathbf{q})}$$

provided $\mathbf{k}_i$ and ω_i are connected by the relation

(9.28) $$\omega_i = (\mathbf{k}_i^2/2m).$$

In this description there is thus associated with an object of energy E and momentum $\mathbf{p}$ a characteristic frequency ω and a characteristic length $|\mathbf{k}|^{-1}$

(9.29) $$\omega = E; \qquad \mathbf{k} = \mathbf{p}$$

which determine the probability amplitude for finding the object at time t at position $\mathbf{q}$, if it is known to have energy E and momentum $\mathbf{p}$ (energy E and momentum $\mathbf{p}$ being compatible, because $H\mathbf{P} - \mathbf{P}H = 0$ in this case), namely

(9.30) $$\langle \mathbf{q}(t)|E\rangle = \Psi_E(\mathbf{q}, t) = \text{const}\, e^{i[\omega t - (\mathbf{k}\mathbf{q})]}$$

where the constant has to be determined by the normalization of the probability.

In the operator picture, the development of the object described by the Hamiltonian (9.14) is given by the operator equations, using (9.16),

(9.31) $$\begin{cases} \dot{P}_k = i(HP_k - P_k H) = 0 \\ \dot{Q}_j = i(HQ_j - Q_j H) = (i/2m)(P^2 Q_j - Q_j P^2) = P_j/m \end{cases}$$

so that the expectation values of the momentum and coordinates of a free object satisfy the classical relationships

(9.32) $$\bar{\dot{P}} = 0; \qquad \bar{P} = m\bar{\dot{Q}}$$

This correspondence to classical mechanics is ultimately the justification for considering the Hamiltonian (9.14) as the "correct" Hamiltonian to be inserted in the fundamental dynamical postulate of quantum mechanics for a free object.

NOTES

The historical event which marks the advent of quantum mechanics was publication of a paper by de Broglie [1], who first noticed that by association of a frequency ω and a length $|k|^{-1}$ with energy and momentum of a particle, one can describe its propagation in a language devised originally to describe the propagation of a signal

$$f(q,t) = \int\limits_{-\infty}^{+\infty} \exp\left[i(\omega t - kq)\right]f(k)\,dk; \qquad \omega = \sqrt{m^2+k^2},$$

made up out of superposition of ψ functions.

Since the phase velocity of the ψ function, $V(\text{phase}) = \omega/k$, is obviously larger than the speed of light, as long as $m \neq 0$, one cannot identify propagation of the ψ function with the motion of some observable material, and the probability interpretation of ψ is, indeed, widely accepted today.

However, a temptation presents itself: The ψ functions making up the signal $f(q,t)$ are by analogy with interference patterns observed, for example, on the surface of liquids suggestive of some underlying medium or "ether." Such is the human urge to hang on to the familiar that a number of distinguished physicists, de Broglie among them, have plunged into elaborate, often desperate, efforts to reconstruct from the observable features of matter the hydrodynamics, as it were, of that hypothetical, not directly observable, ether which is envisaged as the stage below the so-called elementary particles in the hierarchy of nature.

From a strictly operational point of view all such attempts have thus far proven to be sterile, because they have not led to a single experiment which could be used as a crucial test. A comprehensive list of references to this approach is contained in the work of Takabayasi [2].

The proof that expectation values of momentum and coordinate of single objects satisfy the classical relationships was given by Ehrenfest [3].

REFERENCES

[1] L. de Broglie, *Nature* **112**, 540 (1923) [see also *Ann. Phys. (Paris)* (10) **3**, 22 (1925)].
[2] T. Takabayasi, *Progr. Theoret. Phys. (Kyoto), Suppl.* **4**, 1 (1957).
[3] P. Ehrenfest, *Z. Physik* **45**, 455 (1927).

Uncertainties and the Relations between Them

A concept basic for any theory of measurement is the uncertainty ΔA in the experimentally realizable knowledge of an observable A if the object is in a given state $|b\rangle$. Define ΔA as the root of the mean square deviation from the expectation value

(10.1)
$$(\Delta A)^2 = \overline{(A - \bar{A})^2} = \overline{A^2} - (\bar{A})^2.$$

Introducing the notation "norm of $|b\rangle$"

(10.2)
$$\|b\| = \sqrt{\langle b|b\rangle}$$

one may write, using the hermitean property of A,

(10.3)
$$(\Delta A)^2 = \overline{[A - \langle b|A|b\rangle]^2} = \langle b|[A - \langle b|A|b\rangle]^2|b\rangle$$
$$= \langle [A - \langle b|A|b\rangle]b|[A - \langle b|A|b\rangle]b\rangle$$
$$= \|[A - \langle b|A|b\rangle]b\|^2.$$

The following fundamental theorem holds.

If the operators P and Q representing two observables satisfy the relation

(10.4)
$$PQ - QP = -iI$$

then

(10.5)
$$\Delta P \Delta Q \geqslant \tfrac{1}{2}.$$

The inequality (10.5) is called an uncertainty relation.

To prove it introduce the abbreviations

(10.6) $\hat{P} = P - \bar{P} = P - \langle b|P|b\rangle; \qquad \hat{Q} = Q - \bar{Q} = Q - \langle b|Q|b\rangle,$

notice that again

(10.7)
$$\hat{P}\hat{Q} - \hat{Q}\hat{P} = -iI$$

and because of (10.3)

(10.8)
$$\Delta P \Delta Q = \|\hat{P}b\| \cdot \|\hat{Q}b\|.$$

65

Now apply to the right-hand side of (10.8) the well-known inequality

(10.9) $$\|u\| \cdot \|v\| \geqslant |\langle u|v\rangle|,$$

in which the equality sign holds only if $|u\rangle$ and $|v\rangle$ differ at most by a constant complex factor, and which is the generalization of the simple geometrical fact that the projection of two vectors on each other is never larger than the product of their lengths. One obtains

(10.10) $$\Delta P \Delta Q \geqslant |\langle \hat{P}b|\hat{Q}b\rangle|$$

and therefore also

(10.11) $$\Delta P \Delta Q \geqslant \mathrm{Im} \langle \hat{P}b|\hat{Q}b\rangle.$$

By employing the identity $\mathrm{Im}\,(\alpha + i\beta) = -(i/2)\,[(\alpha + i\beta) - (\alpha - i\beta)]$, using $\langle u|v\rangle^* = \langle v|u\rangle$, and supposing $|b\rangle$ to be normalized, one finds

(10.12)
$$\Delta P \Delta Q \geqslant -(i/2)\,(\langle \hat{P}b|\hat{Q}b\rangle - \langle \hat{Q}b|\hat{P}b\rangle) = -(i/2)\,(\langle \hat{Q}\hat{P}b|b\rangle - \langle \hat{P}\hat{Q}b|b\rangle)$$
$$= \tfrac{1}{2}\langle i(\hat{P}\hat{Q} - \hat{Q}\hat{P})\,b|b\rangle$$
$$= \tfrac{1}{2}\|b\|^2 = \tfrac{1}{2}$$

which completes the proof. It follows further from this derivation that the product of the two uncertainties will be a minimum if the state $|b\rangle$ satisfies the condition

(10.13) $$\hat{P}|b\rangle = i\gamma\hat{Q}|b\rangle; \qquad \gamma \quad \text{real and} \quad > 0.$$

Such states are called optimum states.

One finds, sometimes, statements to the effect that all pairs of canonical variables in classical mechanics are represented in quantum mechanics by operators satisfying (10.4) and are therefore subject to uncertainty relations (10.5). Such statements must be approached with caution, because the uncertainty relation (10.5) refers to simultaneous measurement of two observables, the object being in a single state $|b\rangle$. They may, therefore, possibly apply to the momentum and position of a single object. Now in a certain formal sense one may treat the time variable and the energy of an object in classical mechanics as canonical variables. It would, however, be quite wrong to infer from this that one cannot determine the energy of an object exactly at a given instant of time. A detailed study of the nature of energy measurement shows that the law of conservation of energy can be verified by two successive energy measurements ω_1 and ω_2 at times t_1 and t_2 only to an accuracy $|\omega_1 - \omega_2|(t_2 - t_1) > 1$, but this refers to measurements in which the object is in different states, namely $|b(t_1)\rangle$ and $|b(t_2)\rangle$, so that the proof given above is not applicable, and this "uncertainty relation for energy" has a different origin. For details the reader is referred to a particularly thorough discussion in the work by Messiah quoted at the end of this section.

Now, identifying P and Q again with the momentum and position operator of a single object, Eq. (10.13) governing the optimum states reads, in q representation,

$$(10.14) \qquad [-i(\partial/\partial Q) - \bar{P}]|b\rangle = i\gamma(Q - \bar{Q})|b\rangle$$

so that the corresponding ψ function $\psi_b(q) = \langle q|b\rangle$ satisfies the first order differential equation

$$(10.15) \qquad [-i(d/dq) - \bar{P}]\psi_b(q) = i\gamma(q - \bar{Q})\psi_b(q)$$

yielding by integration

(10.16)

$$\psi_b(q) = \exp \int^q (-\gamma q + \gamma\bar{Q} + i\bar{P})\, dq = C \exp\left[-(\gamma/2)\, q^2 + \gamma\bar{Q}q + i\bar{P}q\right]$$

or

$$(10.17) \qquad \psi_b(q) = C' \exp\left[-(\gamma/2)(q - \bar{Q})^2 + i\bar{P}q\right]$$

with some constant C', to be determined by the normalization

$$(10.18) \qquad \langle b|b\rangle = \int_{-\infty}^{+\infty} |\psi_b(q)|^2\, dq = 1$$

which gives

$$(10.19) \qquad |C'|^2 \int_{-\infty}^{+\infty} \exp\left[-\gamma(q - \bar{Q})^2\right] dq = |C'|^2 \sqrt{(\pi/\gamma)} = 1$$

so that finally

$$(10.20) \qquad \psi_b(q) = (\gamma/\pi)^{1/4} \exp\left[-(\gamma/2)(q - \bar{Q})^2 + i\bar{P}q\right].$$

The optimum state vector describes thus a Gaussian probability distribution around the expectation value $\bar{Q}$ as center. It is now easy to compute $\bar{P}$, ΔP, ΔQ for the optimum state (10.20).

An instructive exercise consists of transforming $|b\rangle$ in the q representation into $|b\rangle$ in the p representation. Quite generally these representations are defined by the eigenvalue problems

$$(10.21) \qquad P|p\rangle = p|p\rangle; \qquad Q|q\rangle = q|q\rangle,$$

respectively, and there should exist a unitary operator U so that

$$(10.22) \qquad |p\rangle = U|q\rangle; \qquad |q\rangle = U^+|p\rangle,$$

in components

(10.23) $|p'\rangle = \int \langle p'|U|q''\rangle \, dq''|q''\rangle ; \qquad |q''\rangle = \int \langle q''|U^+|p'\rangle \, dp'|p'\rangle$

which transforms any operator $A_{(q)}$ in q representation into the corresponding operator $A_{(p)}$ in p representation, according to

(10.24) $A_{(p)} = U A_{(q)} U^+; \qquad A_{(q)} = U^+ A_{(p)} U,$

in components

(10.25) $\langle p''|A|p'\rangle = \iint \langle p''|U|q''\rangle \, dq'' \langle q''|A|q'\rangle \, dq' \langle q'|U^+|p'\rangle.$

From the definitions (10.23), which may alternatively be written as expansions into ψ functions

(10.26) $|p'\rangle = \int |q''\rangle \, dq'' \, \psi_{p'}(q''); \qquad |q''\rangle = \int |p'\rangle \, dp' \, \psi_{p'}^*(q''),$

it follows that the matrix elements of U are identical with the ψ functions

(10.27)
$\langle p'|U|q''\rangle = \langle q''|p'\rangle = \psi_{p'}(q''); \qquad \langle q''|U^+|p'\rangle = \langle p'|q''\rangle = \psi_{p'}^*(q'').$

By applying the representation $P = -i(\partial/\partial Q)$ to the first equation (10.21), one finds that $\psi_{p'}^*(q'')$ satisfies the differential equation

(10.28) $-i[\partial\psi_{p'}^*(q'')/\partial q''] = p' \, \psi_{p'}^*(q'')$

having the solution

(10.29) $\psi_{p'}^*(q'') = \mathrm{const}\, e^{ip' q''} = \langle q''|U^+|p'\rangle.$

The constant of integration can be determined from the unitarity of U,

(10.30) $\langle p''|UU^+|p'\rangle = \langle p''|U|q'\rangle \, dq' \langle q'|U^+|p''\rangle$

$= |\mathrm{const}|^2 \int \exp[i(p'' - p')q'] \, dq'$

$= 2\pi |\mathrm{const}|^2 \, \delta(p' - p'') = \delta(p' - p'')$

so that finally

(10.31) $\langle p'|U|q''\rangle = (1/\sqrt{2\pi})\, e^{-ip' q''}.$

The components of $|p\rangle$ and $|q\rangle$ are thus connected by the Fourier transformation

(10.32)
$|p'\rangle = (1/\sqrt{2\pi}) \int e^{-ip' q''}|q''\rangle \, dq''; \qquad |q''\rangle = (1/\sqrt{2\pi}) \int e^{+iq'' p'}|p'\rangle \, dp'.$

In particular, one obtains the ψ function of $|b\rangle$ in p representation from (10.20) by the integration

$$(10.33) \quad \psi_b(p) = (1/\sqrt{2\pi})(\gamma/\pi)^{1/4} \int_{-\infty}^{+\infty} \exp\left[-(\gamma/2)(q-\bar{Q})^2 + i(\bar{P}-p)q\right]dq.$$

NOTES

The classic source for discussion of the uncertainty relations is the work by Heisenberg [1].

Messiah [2] should be consulted, with particular reference to Chapters IV and VIII, on the time-energy uncertainty relation.

Landau and Peierls [3] have given a provocative discussion of some implications of the time-energy uncertainty relation, and have pointed out, in particular, its consequences for the measurement of particle momentum in the relativistic case, the measurement of electromagnetic fields, and the measurement of position of particles that do not or do have mass.

REFERENCES

[1] W. Heisenberg, "Die Physikalischen Prinzipien der Quantentheorie." Hirzel, Leipzig, 1930.
[2] A. Messiah, "Quantum Mechanics," Vol. 1, North-Holland Publ. Co., Amsterdam, 1961.
[3] L. Landau and R. Peierls, *Z. Physik* **69**, 56 (1931).

A Digression on Superfluidity

The general transformation formula (9.13) is, of course, applicable to any object possessing energy and momentum and is not confined to the special case of a classical "particle" whose energy, as a function of its momentum $\mathbf{k}$, is of the form $\omega = \mathbf{k}^2/2m$. In particular, the object may be a collective excitation or "quasi particle" which is considered the carrier of excitation energy and momentum in a fluid. The energy of such a quasi particle, $\tilde{\omega}(k)$ as a function of its linear momentum may, for example, have the form indicated in Fig. 11.1. If this is the case, then the

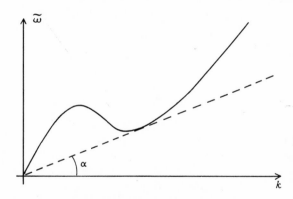

FIG. 11.1. Quasi-particle energy spectrum as proposed by Landau.

fluid in its ground state, when traveling through a capillary, cannot lose energy to the walls of the capillary and thus experience friction, unless the velocity exceeds a certain critical value $V_c = \tan\alpha$, where $\tan\alpha$ is the slope of the straight line drawn from the origin which just touches the curve $\tilde{\omega}(k)$. Historically, it was the existence of superfluid helium II exhibiting just such a critical velocity which led Landau to conjecture the existence of quasi particles in helium II with an energy spectrum of the type indicated in Fig. 11.1.

To understand the origin of this superfluid behavior in the presence of such a quasi particle energy spectrum, consider a fluid moving with

71

velocity $\mathbf{V}_0$ through a capillary, and label all quantities in this coordinate system S with no suffixes on them (Fig. 11.2). Alternatively, the same

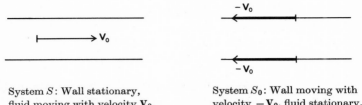

System S: Wall stationary,   System S_0: Wall moving with
fluid moving with velocity $\mathbf{V}_0$.   velocity $-\mathbf{V}_0$, fluid stationary.

FIG. 11.2. Coordinate transformation to observer moving with the fluid in a capillary.

experimental situation can be described by an observer whose coordinate system S_0 is tied to the fluid, so that in this case the walls move with velocity $-\mathbf{V}_0$. In frame S_0 all quantities will be labeled with a subscript "zero" as suffix.

Now suppose the fluid is at a temperature of absolute zero, i.e. it is in its ground state and there are no quasi particles present.* The initial kinetic energy of the fluid is

in S: in S_0:

$$(11.1) \qquad E(\text{initial}) = \tfrac{1}{2}MV_0^2 \qquad E_0(\text{initial}) = 0$$

where M is the mass of the entire fluid. When viscosity is present the fluid should start to move in S_0 and thus start to lose energy to the walls in S. In the excitation theory, fluid motion in S_0 can appear only if internal motions of the fluid are gradually excited, i.e. if quasi particles begin to appear, owing to interaction with the wall.

Suppose then that one quasi particle of momentum $\mathbf{k}_0$ and energy $\tilde{\omega}_0(k_0)$ has somehow appeared by interaction with the wall in system S_0 so that the energy of the fluid in S_0 is now

$$(11.2) \qquad\qquad E_0 = \tilde{\omega}_0(k_0) \qquad (\text{in } S_0).$$

To find the energy of that single quasi particle in system S one can now use the general transformation formula (9.13). Thus, if S moves with

---

* The absence of quasi particles in the ground state is a hypothesis which is strictly tenable in absence of interaction between the molecules of the fluid only. See the more detailed examination of this point in Section 30. The argument in the present section can, however, be modified to accommodate possible presence of a pool of quasi particles which do not disappear even at a temperature of absolute zero.

respect to S_0 with velocity $\mathbf{V}_0$, then the effective Hamiltonian H_{eff} of an object of momentum $\mathbf{P}_0$ is, in system S, given by

$$(11.3) \qquad H_{\text{eff}} = H_0^T + (\mathbf{P}_0\,\mathbf{V}_0)$$

where $H_0^T = UH_0U^{-1}$, with $U = \exp(i\mathbf{P}_0\mathbf{V}_0t)$. For "free" quasi particles, H_0 is coordinate independent and thus $H_0^T = H_0$ and the change in effective energy is entirely due to the kinetic effect of relative motion. One finds thus for the energy of the quasi particle in S

$$(11.4) \qquad \tilde{\omega}(k_0) = \tilde{\omega}_0(k_0) + (\mathbf{k}_0\,\mathbf{V}_0)$$

and therefore the total energy of the fluid in S becomes now

$$(11.5) \qquad E = \tilde{\omega}_0(k_0) + (\mathbf{k}_0\,\mathbf{V}_0) + \tfrac{1}{2}MV_0^2 \qquad \text{(in } S\text{).}$$

Now in order that viscosity be observed, the change in energy in S must be negative, which according to (11.1) and (11.5) is

$$(11.6) \qquad E - E(\text{initial}) = \tilde{\omega}_0(k_0) + (\mathbf{k}_0\,\mathbf{V}_0) < 0.$$

For a given value of k_0, this quantity has its minimum when the vectors $\mathbf{k}_0$ and $\mathbf{V}_0$ are antiparallel, corresponding to a quasi particle having

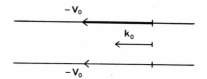

FIG. 11.3. Quasi particle of momentum $\mathbf{k}_0$ antiparallel to $\mathbf{V}_0$, described in S_0.

momentum parallel to the motion of the wall in S_0 as indicated in Fig. 11.3. One must, therefore, have certainly

$$(1.17) \qquad \tilde{\omega}_0(k_0) - k_0\,V_0 < 0 \qquad \text{or} \qquad V_0 > \tilde{\omega}_0(k_0)/k_0.$$

This inequality, which is necessary for the existence of viscosity, must be satisfied for at least some value k_0 of a possible quasi particle. Now, $\tilde{\omega}_0(k_0)/k_0$ is the slope of the straight line connecting origin and the point $\tilde{\omega}_0(k_0)$ on the excitation curve. Its minimum value corresponds just to $\tan\alpha$ as indicated in Fig. 11.1. If $\tan\alpha \neq 0$, then, for velocities $V_0 < V_c = \tan\alpha$, quasi particles cannot appear in the fluid and the fluid will be superfluid and not slow down by friction. The condition $\tan\alpha \neq 0$ is obviously equivalent to the requirement that $\tilde{\omega}(k)$ does not have vanishing slope at the origin. Therefore, in particular, whenever the lowest quasiparticles are phonons, for which $\tilde{\omega}(k) = c_0 k$ (c_0 is the velocity

of sound), superfluidity will occur. To prove that a fluid becomes superfluid at low temperatures one has thus to prove that only phonons can be
excited at sufficiently low temperatures. This will be done for some
special cases in Section 30.

As was first shown by Landau, phonons account for part of the inertia
of a fluid in which they are present. If phonons were the only excitations
available, then one can, in principle, obtain the transition temperature
separating superfluid from normal fluid behavior by computing the
temperature at which the inertia of the phonons is equal to that of the
entire fluid. In liquid helium II the contribution from the "roton" part
of the excitation spectrum, corresponding to excitations lying near the

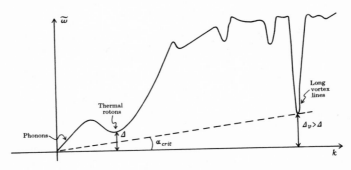

FIG. 11.4. Possible actual excitation spectrum in liquid helium II.

minimum of the curve in Fig. 11.1, dominates at temperatures above
0.6°K, and the transition to normal behavior can be accounted for by
thermal excitation of rotons with an energy $\tilde{\omega} = \varDelta \cong 9°$K. This value of $\varDelta$
leads unfortunately to a value of α, if computed after the fashion of
Fig. 11.1, which gives critical velocities far above the observed critical
velocities.

This difficulty may possibly be resolved, if the mechanism of onset of
viscosity at the critical velocity is due to an actual excitation spectrum
of a form drawn in Fig. 11.4, corresponding to possible existence of a
number of quasistable excitations of the roton type, but of higher effective mass (i.e. narrower excitation line) and higher excitation energy
$\varDelta_v > \varDelta$, so that although they may not be excited thermally in any numbers at superfluid temperatures, they may be excited *mechanically*,
owing to the lower value of α as compared to the value of α in Fig. 11.1.
Such quasi particles might be identical with the "long" rotons or vortex
lines of a type first conjectured by Onsager and Feynman in connection
with the problem posed by the critical velocity in helium II. Although

existence of what is labeled in Fig. 11.4 "thermal rotons" can be proven, and will be proven in Section 30, from first principles for special systems, a strict quantum mechanical xeistence proof for "long" rotons or vortex lines belongs to the large class of as yet unsolved problems.

The excitation of long vortex lines by friction can be looked upon as a mechanical "pumping" giving rise to occupation of energy levels Δ_v above the levels Δ which remain practically empty at temperatures below $0.6°K$, and, thus, to a situation analogous to the states of negative temperature familiar from similar situations used in the construction of masers. This raises the interesting question of whether one cannot induce transitions $\Delta_v \rightarrow \Delta$ *mechanically* by feeding into liquid helium II, moving through a capillary at velocities above the critical velocity, oscillations of a frequency in resonance with the energy level difference $\Delta_v - \Delta$, leading to the possible amplification of such oscillations by stimulated emission. This amplification should be absent as long as the liquid moves through the capillary at velocities below the critical velocity. In this connection, it might be worth noting that the excitation spectrum envisaged in Fig. 11.4 may be much too simple, and there may exist not just one but many types of vortex lines with levels Δ_v, all higher than Δ, and transitions may be inducible corresponding to transformation of one type of vortex line into another similar type, involving an energy change much smaller than the one corresponding to the entire breakup of a vortex line into thermal rotons.

NOTES

Landau [1] first proposed the existence of a quasi-particle spectrum, as in Fig. 11.1, to explain the superfluidity of liquid helium II.

Onsager [2] pointed out the possible existence of quasi-stable excitations other than thermal rotons envisaged by Landau, and Feynman [3] used the concept of quantized vortex lines of the type suggested by Onsager to give a quantitative account of the critical velocity in liquid helium II.

For a review of experimental work on vortex lines in liquid helium II see Vinen [4].

REFERENCES

[1] L. Landau, *J. Phys. USSR* **5**, 71 (1941).
[2] L. Onsager, *Nuovo Cimento* **6**, *Suppl.* **2**, 249 (1949).
[3] R. P. Feynman, *Progr. Low Temp. Phys.* **1**, Chapter II (1955).
[4] W. F. Vinen, *Progr. Low Temp. Phys.* **3**, Chapter I (1961).

Rotations of the Observer

In analogy to the development of Section 9, consider next the rotation of an observer (1) with respect to an observer (0) by angle ϕ around a given axis $\mathbf{n}$. It is claimed that the respective descriptions of an object in terms of states $|b\rangle^T$ and $|b\rangle$ are connected by the unitary transformation

(12.1)
$$|b\rangle^T = T_{\mathbf{n},\phi}|b\rangle; \qquad T_{\mathbf{n},\phi} = e^{i\mathbf{n}\cdot\phi(\mathbf{J})}$$

where $\mathbf{J}$ is a hermitean operator J_1, J_2, J_3 representing the angular momentum of the object.

To verify this consider first a single object without spin, whose angular momentum is purely orbital and shall be represented, in correspondence with the expressions known from classical mechanics, by the three operators

(12.2)
$$
\begin{aligned}
J_1 &= Q_2 P_3 - Q_3 P_2 = -i[Q_2(\partial/\partial Q_3) - Q_3(\partial/\partial Q_2)]\\
J_2 &= Q_3 P_1 - Q_1 P_3 = -i[Q_3(\partial/\partial Q_1) - Q_1(\partial/\partial Q_3)]\\
J_3 &= Q_1 P_2 - Q_2 P_1 = -i[Q_1(\partial/\partial Q_2) - Q_2(\partial/\partial Q_1)]
\end{aligned}
$$

with the understanding that the eigenstates of $\mathbf{Q}$, namely $|q\rangle$, form a complete set and may be used as a basis for the description of the object. One verifies that these operators satisfy the C.R.s

(12.3)
$$J_1 J_2 - J_2 J_1 = iJ_3 \quad \text{(cyclically)} \qquad \text{or} \qquad [\mathbf{J} \times \mathbf{J}] = i\mathbf{J}$$

which are formally identical with the C.R.s (2.47) found for the spin angular momentum operators s_i.

Now let the axis of rotation be the q_3-axis, so that the coordinates of the object as seen by the two observers are connected by the transformation

(12.4)
$$
\begin{aligned}
Q_1^T &= Q_1 \cos\phi + Q_2 \sin\phi; & Q_1 &= Q_1^T \cos\phi - Q_2^T \sin\phi; & \partial Q_1/\partial\phi &= -Q_2\\
Q_2^T &= -Q_1 \sin\phi + Q_2 \cos\phi; & Q_2 &= Q_1^T \sin\phi + Q_2^T \cos\phi; & \partial Q_2/\partial\phi &= Q_1\\
Q_3^T &= Q_3; & Q_3 &= Q_3^T; & \partial Q_3/\partial\phi &= 0.
\end{aligned}
$$

The expansion of the transformed state,

(12.5)
$$|\mathbf{q}\rangle^T = |q_1^T, q_2^T, q_3^T\rangle = |\mathbf{q}\rangle + \phi(d|\mathbf{q}\rangle/d\phi)_{\phi=0} + (\phi^2/2!)(d^2|\mathbf{q}\rangle/d\phi^2)_{\phi=0} + \cdots$$

may now be written, using

(12.6)
$$(d|\mathbf{q}\rangle/d\phi)_{\phi=0} = \sum_i (\partial|\mathbf{q}\rangle/\partial Q_i)(\partial Q_i/\partial\phi)$$
$$= [Q_1(\partial/\partial Q_2) - Q_2(\partial/\partial Q_1)]|\mathbf{q}\rangle = iJ_3|\mathbf{q}\rangle$$

in the form

(12.7)
$$|\mathbf{q}\rangle^T = \sum_{n=0}^{\infty} (\phi^n/n!)(iJ_3)^n|\mathbf{q}\rangle = e^{i\phi J_3}|\mathbf{q}\rangle$$

which verifies (12.1) for this special case.

If the object is a spin pointing in direction ϑ, φ on the unit sphere so that

(12.8)
$$|b\rangle = \begin{pmatrix} \cos(\vartheta/2) \\ \sin(\vartheta/2)\,e^{i\varphi} \end{pmatrix}$$

one expects the transformation operator for rotation around the q_3-axis to be given in terms of $s_3 = \sigma_3/2$ by

(12.9)
$$T_\phi = e^{(i/2)\phi\sigma_3}.$$

The simple form of σ_3 allows summation of the infinite series represented by T_ϕ. Using

$$\sigma_3 = \begin{pmatrix} 1 & 0 \\ 0 & -1 \end{pmatrix} \quad \text{and} \quad \sigma_3^2 = \begin{pmatrix} 1 & 0 \\ 0 & 1 \end{pmatrix} = I,$$

one finds

(12.10)
$$T_\phi = [1 + (1/2!)(i\phi/2)^2 + \ldots]I + [(i\phi/2) + (1/3!)(i\phi/2)^3 + \ldots]\sigma_3$$
$$= \cos(\phi/2)I + i\sin(\phi/2)\sigma_3 = \begin{pmatrix} e^{i\phi/2} & 0 \\ 0 & e^{-i\phi/2} \end{pmatrix}$$

so that

(12.11)
$$|b\rangle^T = e^{i\phi/2}\begin{pmatrix} \cos(\vartheta/2) \\ \sin(\vartheta/2)\,e^{i(\varphi-\phi)} \end{pmatrix}$$

which differs, apart from a physically unobservable common phase factor, from $|b\rangle$ by the change in azimuth angle $\varphi^T = \varphi - \phi$. This verifies again the suitability of the unitary transformation (12.1) for this special case.

Of particular interest is rotation with uniform angular velocity $\boldsymbol{\omega}$ so that

(12.12)
$$\mathbf{n}\phi = \boldsymbol{\omega}t$$

and the transformation operator becomes a function of time

(12.13)
$$T_{\boldsymbol{\omega}} = e^{i(\boldsymbol{\omega}\cdot\mathbf{J})t}.$$

The effective Hamiltonian for the rotating observer will then be given, in analogy to the form encountered in case of uniform linear motion, by

(12.14)
$$H_{\text{eff}} = H^T + i\dot{T}T^{-1} = H^T - (\boldsymbol{\omega}\cdot\mathbf{J}).$$

It is understood that $\boldsymbol{\omega}$ is the vector of angular velocity of observer (1) as described by observer (0). The angular velocity of observer (0) with respect to observer (1) is then obviously $-\boldsymbol{\omega}$.

In the following, the transformation formulae for the components of angular momentum are needed. They are, supposing the rotation is characterized by angular velocity $\boldsymbol{\omega} = -\omega\mathbf{k}$,

(12.15)
$$\begin{aligned}
J_1^T &= e^{-i\omega J_3 t} J_1 e^{i\omega J_3 t} = J_1 \cos(\omega t) + J_2 \sin(\omega t) \\
J_2^T &= e^{-i\omega J_3 t} J_2 e^{i\omega J_3 t} = -J_1 \sin(\omega t) + J_2 \cos(\omega t) \\
J_3^T &= e^{-i\omega J_3 t} J_3 e^{i\omega J_3 t} = J_3
\end{aligned}$$

and follow by straightforward computation from the C.R.s (12.3). For example, with the usual abbreviation $[J_n J_m] = J_n J_m - J_m J_n$,

(12.16)

$$\begin{aligned}
e^{-i\omega J_3 t} J_1 e^{i\omega J_3 t} &= J_1 + (i\omega t)[J_1 J_3] + (i\omega t)^2/2![[J_1 J_3]J_3] + \dots \\
&= J_1[1 - (\omega t)^2/2! + \dots)] + J_2[\omega t - (\omega t)^3/3! + \dots)] \\
&= J_1 \cos(\omega t) + J_2 \sin(\omega t).
\end{aligned}$$

For any three operators J_1, J_2, J_3 satisfying the C.R.s (12.3) one can prove the following theorem:

There exist simultaneous eigenstates of one of the three operators, J_3 say, and of $J^2 = J_1^2 + J_2^2 + J_3^2$, which are characterized by two quantum numbers j and m so that

(12.17)

$$J^2|j,m\rangle = j(j+1)|j,m\rangle \qquad \text{and} \qquad J_3|j,m\rangle = m|j,m\rangle$$

where j has the possible values $j = 0, \frac{1}{2}, 1, \frac{3}{2}, \dots$, and where for given j the quantum number m may assume the $(2j+1)$ possible values $-j$, $-j+1, \dots, j-1, j$. The proof of this theorem is given in Appendix 1.

As an application of the concepts developed for rotating observers, consider again the case of a spin magnetic moment in an external field

(12.18)
$$\mathbf{B} = B_0\mathbf{k} + B_1[\mathbf{i}\cos(\omega t) - \mathbf{j}\sin(\omega t)]; \qquad \boldsymbol{\omega} = -\omega\mathbf{k}$$

giving rise to spin magnetic resonance as explained in Section 7. The dynamical equation governing the magnetic moment $\boldsymbol{\mu} = \gamma\mathbf{s}$ in the laboratory system reads in the state picture

(12.19)
$$i|\dot{b}\rangle = -\gamma(\mathbf{sB})|b\rangle.$$

As a first step towards solving (12.19), which has a time dependent Hamiltonian $H = -\gamma(\mathbf{sB})$, transform to a system (1) rotating around the laboratory system with angular velocity $\boldsymbol{\omega}$. The transformed state

(12.20)
$$|b\rangle' = e^{i(\boldsymbol{\omega}\mathbf{s})t}|b\rangle$$

satisfies then the equation

(12.21)
$$i\partial|b\rangle'/\partial t = H'_{\text{eff}}|b\rangle'$$

with

(12.22) $$H'_{\text{eff}} = H' - (\boldsymbol{\omega}\mathbf{s}) = -\gamma\, e^{i(\boldsymbol{\omega}\mathbf{s})t}(\mathbf{sB})\, e^{-i(\boldsymbol{\omega}\mathbf{s})t} - (\boldsymbol{\omega}\mathbf{s}).$$

This can be evaluated using formulae (12.15) as follows,

(12.23)
$$e^{i(\boldsymbol{\omega}\mathbf{s})t}(\mathbf{sB})\, e^{-i(\boldsymbol{\omega}\mathbf{s})t} = e^{-i\omega s_3 t}[s_3 B_0 + B_1(s_1\cos\omega t - s_2\sin\omega t)]\, e^{i\omega s_3}$$
$$= s_3 B_0 + s_1 B_1$$

so that the effective Hamiltonian may be written

(12.24)
$$H'_{\text{eff}} = -\gamma(\mathbf{sB}'_{\text{eff}})$$

where the effective field is now

(12.25)
$$\mathbf{B}'_{\text{eff}} = [B_0 - (\omega/\gamma)]\mathbf{k} + B_1\mathbf{i}.$$

Equation (12.21) contains thus a Hamiltonian which no longer depends on the time t. The solution of this equation can be expedited by performing a second transformation to a system (2) rotating around the direction of $\mathbf{B}'_{\text{eff}}$ with angular velocity $\mathbf{w}$

(12.26)
$$\mathbf{w} = -\gamma\mathbf{B}'_{\text{eff}}$$

having according to (12.25) the numerical value

(12.27)
$$w = \sqrt{\omega_1^2 + (\omega_0 - \omega)^2}$$

where the characteristic frequencies ω_0 and ω_1 are defined, as in Section 7, by

(12.28) $$\omega_0 = \gamma B_0; \qquad \omega_1 = \gamma B_1.$$

With this choice, the twice transformed state

(12.29) $$|b\rangle'' = e^{i(\mathbf{ws})t}|b\rangle'$$

satisfies the equation

(12.30) $$i\partial|b\rangle''/\partial t = 0$$

because the effective Hamiltonian governing the spin in system (2) is

(12.31) $$H''_{\text{eff}} = H'_{\text{eff}} - (\mathbf{ws}) = -\gamma(\mathbf{sB}'_{\text{eff}}) + \gamma(\mathbf{sB}'_{\text{eff}}) = 0$$

the effective field $\mathbf{B}'_{\text{eff}}$ being the same in both systems (1) and (2), so that H'_{eff} is invariant under the transformation (12.29). The solution of equation (12.19) is thus obtained immediately from the observation that

(12.32) $$|b(t)\rangle'' = |b(0)\rangle'' = |b(0)\rangle' = |b(0)\rangle$$

and

(12.33) $$|b(t)\rangle = e^{-i(\boldsymbol{\omega}\mathbf{s})t}e^{-i(\mathbf{ws})t}|b(t)\rangle''$$

so that

(12.34) $$|b(t)\rangle = e^{-i(\boldsymbol{\omega}\mathbf{s})t}e^{-i(\mathbf{ws})t}|b(0)\rangle.$$

For the explicit evaluation of the two transformations it is useful to introduce the angle Θ between the effective field $\mathbf{B}'_{\text{eff}}$ and $\mathbf{B}_0$,

(12.35) $$\cos\Theta = [B_0 - (\omega/\gamma)]/B'_{\text{eff}} = (\omega_0 - \omega)/w; \qquad \sin\Theta = \omega_1/w.$$

One has then

(12.36) $$(\boldsymbol{\omega}\mathbf{s}) = -\omega s_3 \qquad \text{and} \qquad (\mathbf{ws}) = -w(s_1\sin\Theta + s_3\cos\Theta).$$

To establish contact with observation consider a spin $\frac{1}{2}$ and suppose at time $t = 0$ the spin is parallel to $\mathbf{B}_0$, i.e. $|b(0)\rangle = |b_+\rangle = \binom{1}{0}$. The probability for spin flip is then [see (7.66)] with $|b_-\rangle = \binom{0}{1}$

(12.37) $$P_{\text{flip}} = |\langle b_-|b(t)\rangle|^2 = |\langle b_-|e^{-i(\boldsymbol{\omega}\mathbf{s})t}e^{-i(\mathbf{ws})t}b_+\rangle|^2.$$

To evaluate this expression observe that both $|b_+\rangle$ and $|b_-\rangle$ are eigenstates of $-i(\boldsymbol{\omega}\mathbf{s})t = (\omega/2)\sigma_3 t$ with eigenvalues $+i\omega t/2$ and $-i\omega t/2$, respectively. One may thus write

(12.38)

$$\langle b_-|e^{-i(\boldsymbol{\omega}\mathbf{s})t}e^{-i(\mathbf{ws})t}b_+\rangle = \langle e^{i(\boldsymbol{\omega}\mathbf{s})t}b_-|e^{-i(\mathbf{ws})t}b_+\rangle = e^{-i\omega t/2}\langle b_-|e^{-i(\mathbf{ws})t}b_+\rangle$$

and therefore

(12.39) $\quad P_{\text{flip}} = |\langle b_- | \exp[i(w/2)\, t(\sigma_1 \sin\Theta + \sigma_3 \cos\Theta)]\, b_+\rangle|^2.$

Expansion of the exponential operator, using $\sigma_i^{2n} = I$ and $\sigma_i^{2n+1} = \sigma_i$, gives

(12.40)
$$\exp[i(w/2)\, t(\sigma_1 \sin\Theta + \sigma_3 \cos\Theta)] = I \cos(wt/2) + i(\sigma_1 \sin\Theta + \sigma_3 \cos\Theta) \\ \times \sin(wt/2).$$

P_{flip} contains thus only matrix elements of the form

(12.41) $\quad \langle b_- | \sigma_1 b_+ \rangle = \begin{pmatrix} 0 \\ 1 \end{pmatrix} \begin{pmatrix} 0 & 1 \\ 1 & 0 \end{pmatrix} \begin{pmatrix} 1 \\ 0 \end{pmatrix} = 1; \qquad \langle b_- | \sigma_3 b_+ \rangle = 0;$

$$\langle b_- | I b_+ \rangle = 0,$$

so that finally

(12.42)
$$P_{\text{flip}} = \sin^2\Theta \, \sin^2(wt/2) = \frac{\omega_1^2}{\omega_1^2 + (\omega_0 - \omega)^2} \sin^2\left(\frac{\sqrt{\omega_1^2 + (\omega_0 - \omega)^2}}{2}\, t\right)$$

which is identical with the resonance formula (7.67) obtained earlier by a completely different method using the interaction picture.

NOTE

Rabi *et al.* [1] promulgated the use of rotating coordinates in the treatment of magnetic resonance problems.

REFERENCE

[1] I. I. Rabi, N. F. Ramsay, and J. Schwinger, *Rev. Mod. Phys.* **26**, 167 (1954).

The Connection between Invariance Properties of the Hamiltonian and Conservation Laws

Whenever the state describing a physical object is subjected to the unitary transformation

(13.1) $$T = e^{i\tau S}; \qquad T^{-1} = T^{+} = e^{-i\tau S}:$$

where S is a hermitean operator, and τ a parameter characterizing the transformation, then the Hamiltonian of the system is transformed according to

(13.2) $$H^{T} = e^{i\tau S} H e^{-i\tau S}.$$

The actual calculation of H^{T} involves in general an expansion

(13.3)
$$H^{T} = (I + i\tau S + \ldots) H (I - i\tau S + \ldots) = H + i\tau[SH] + (i\tau)^{2}/2![S[SH]] + \ldots$$

One observes that if the Hamiltonian H commutes with the generator S of the unitary transformation T, then the Hamiltonian is invariant under the transformation.

The converse cannot be inferred, i.e. one cannot conclude $H^{T} = H$ means $[SH] = 0$, except in case of an infinitesimal unitary transformation

(13.4) $$T = I + i\tau S; \qquad T^{-1} = I - i\tau S$$

where τ is now an infinitesimal parameter. The condition

(13.5) $$H^{T} = H + i\tau[SH] = H$$

is then indeed equivalent to

(13.6) $$[SH] = 0.$$

Now, it had been established earlier that the rate of change of the expectation value of any observable is determined by the commutator of the corresponding operator with the Hamiltonian. If the commutator vanishes, the expectation value of the observable will be constant in time.

One concludes: The operators representing constants of the motion of any physical system generate those infinitesimal unitary transformations which leave the Hamiltonian invariant. This corresponds to the well-known fact of classical mechanics in which the constants of the motion generate those infinitesimal canonical transformations which leave the Hamiltonian invariant.

In particular, the results of Sections 9 and 12 may be applied to yield the conclusion that the law of conservation of momentum is a consequence of the invariance of the Hamiltonian under infinitesimal translations, and the law of conservation of angular momentum is a consequence of the invariance of the Hamiltonian under infinitesimal rotations.

The Hamiltonian itself may be looked upon as the generator of an infinitesimal unitary transformation representing displacement in time. Equation (7.5) may be written for an infinitesimal time interval

$$(13.7) \qquad \frac{|b(t+\tau)\rangle_s - |b(t)\rangle_s}{\tau} = -iH|b(t)\rangle_s$$

which is identical with

$$(13.8) \qquad |b\rangle_s^T = |b(t+\tau)\rangle_s = (I + i\tau S)|b(t)\rangle_s$$

provided

$$(13.9) \qquad S = -H$$

[see also the statements following Eq. (7.16)].

The law of conservation of energy is thus a consequence of the invariance of the Hamiltonian under infinitesimal displacements in time, if H is used to represent the energy of the system. This invariance will hold whenever H does not contain the time t explicitly.

These connections between conservation laws and invariance properties of the Hamiltonian are of considerable help in selecting Hamiltonians suitable for the description of physical systems which are subjected to conservation laws. The invariance requirements act as a severe restriction on possible choices of a Hamiltonian.

As an example, consider an object having momentum $\mathbf{P}$, spin $\mathbf{S}$, and orbital angular momentum $\mathbf{L}$, and try to find the most general Hamiltonian having, apart from the familiar kinetic energy $\mathbf{P}^2/2m$, terms not higher than linear in each of $\mathbf{P}$, $\mathbf{S}$, and $\mathbf{L}$. The requirement of invariance under rotations restricts the choice to the expressions

$(13.10); (13.11); (13.12) \quad (\mathbf{P}\cdot\mathbf{L}); \qquad (\mathbf{P}\cdot\mathbf{S}); \qquad (\mathbf{S}\cdot\mathbf{L})$

$(13.13); (13.14); (13.15) \quad (\mathbf{P}\cdot[\mathbf{S}\times\mathbf{L}]); \qquad (\mathbf{S}\cdot[\mathbf{P}\times\mathbf{L}]); \qquad (\mathbf{L}\cdot[\mathbf{P}\times\mathbf{S}])$

because these terms are scalars which commute with the generator $\mathbf{L+S}$ of the rotations, as can be easily verified. If one adds the further requirement that the Hamiltonian should be invariant under reversal of all motions, then the only remaining terms are $(\mathbf{P\cdot L})$, $(\mathbf{P\cdot S})$, $(\mathbf{S\cdot L})$, because reversal of motion changes the sign of each $\mathbf{P}$, $\mathbf{S}$, $\mathbf{L}$, so that the terms (13.13)–(13.15) also change sign under this transformation and thus do not satisfy the invariance requirement. By adding the final requirement that the Hamiltonian be invariant under inversion of the coordinate system one finds that only the term $(\mathbf{S\cdot L})$ satisfies this invariance, because, under inversion of coordinates, $\mathbf{P}$ changes sign, but $\mathbf{S}$ and $\mathbf{L}$ do not, so that (13.10) and (13.11) change sign, whereas (13.12) is the only true scalar in this case. The Hamiltonian

$$(13.16) \qquad H = \beta(\mathbf{S\cdot L}); \qquad \beta \quad \text{a number,}$$

represents the energy of the so-called spin-orbit coupling which plays an important part in the establishment of energy levels in atoms and nuclei.

The transformations of reversal of motion and inversion of coordinates, which have just been introduced in a rather casual fashion, warrant a more extensive treatment, to be given in Sections 14 and 15. They differ from the displacements and rotations in that they cannot be thought of as evolving continuously from the identity transformation, and therefore the corresponding infinitesimal transformations do not exist. However, invariance of the Hamiltonian under the unitary inversion operation leads in quantum mechanics to a peculiar conservation law which has no classical analog. The consequences of invariance under reversal of motion turn out to be somewhat more subtle, because reversal of motion must be represented by an antiunitary operator, and requires a special treatment beyond the one given in the present section. It will be found that existence of anti-unitary symmetry operators may lead to the existence of so-called superselection rules which in effect again guarantee the conservation of certain quantum numbers.

It has been known for a long time that most processes which lend themselves to description in terms of classical physics are invariant under inversion of coordinates and/or under reversal of motion. The consequences of such invariances are often inconspicuous in classical physics and therefore had received little attention until some processes were discovered which, at first sight, appeared to violate inversion symmetry and the corresponding quantum mechanical conservation law, leading to a general re-examination of the symmetry properties of both quantum mechanical and classical descriptions of physical objects and processes.

NOTES

Born *et al.* [1] first recognized the vanishing commutator between an observable and the Hamiltonian as the condition for the conservation of that observable.

Weyl [2] stressed the connection with symmetry properties of the Hamiltonian.

Goldstein [3, especially Chapter 8] can be consulted for a review of the connection between invariance of the Hamiltonian under canonical transformations and conservation laws in classical mechanics.

REFERENCES

[1] M. Born, W. Heisenberg, and P. Jordan, *Z. Physik* **35**, 557 (1926).
[2] H. Weyl, *Z. Physik* **46**, 1 (1927).
[3] H. Goldstein, "Classical Mechanics." Addison-Wesley, Reading, Massachusetts, 1951.

The Invariance under Inversion of Coordinates and the Law of Conservation of Parity

Any physical process or property involving an actual object is said to have parity if the mirror image of this process or property is again a possible physical process or property involving the same actual object.

As an example consider a heavy top, spinning around an axis **a**, subject to a torque around an axis **b** perpendicular to **a**, which precesses around an axis **c** perpendicular to both **a** and **b**.

The experiment may be done using a bicycle wheel mounted on one end of a short axle, with a flexible chain supporting the other end of the axle, as indicated in Fig. 14.1. The torque is realized by the weight **W**

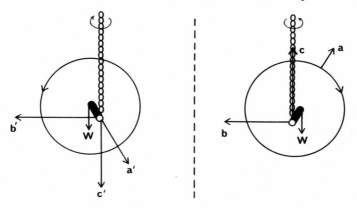

Fig. 14.1. The precession of a heavy top and the mirror image of this process.

of the wheel. Actual performance of the experiment shows that **a**, **b**, **c**, in that order, form a right-hand system if the usual conventions about right and left are adopted.

If this experiment is viewed through a mirror parallel to the plane formed by **a** and **c**, one sees the spin of the wheel reversed, $\mathbf{a}' = -\mathbf{a}$, the direction of precession reversed, $\mathbf{c}' = -\mathbf{c}$, but the direction of the weight **W** and thus the direction of the torque is unchanged, $\mathbf{b}' = \mathbf{b}$.

87

The process has parity, because the process seen in the mirror can be performed in actual fact by reversing the spin of the actual wheel, causing the actual precession produced by the same torque to also reverse its direction. The vectors **a**′, **b**′, **c**′, in this order, still form a right-hand system.

As a second example consider the magnetic field produced by an electric current in a straight conductor. If a magnetic needle is placed above the current carrying conductor, it is found to be deflected so that the N pole of the magnet points in a right-hand sense around the current **I**, if the usual conventions about right and left are adopted.

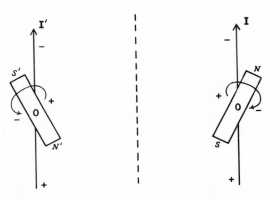

Fig. 14.2. Deflection of a magnetic dipole by a current and mirror image of this process (assuming charge to be a scalar).

On first sight one might think this effect does not have parity, because seen through a mirror parallel to the plane formed by conductor and undeflected needle the magnet is deflected in the opposite direction while the current direction remains unchanged. If one keeps in mind, however, that any magnet can be replaced by a suitable ring current, it is seen, as indicated in Fig. 14.2, that the mirror image of a magnet is a magnet with the signs of its poles reversed, provided the mirror image of a charge is again a charge of the same sign.

One can conclude the effect has parity, because the deflection seen in the mirror can be obtained with an actual current and an actual magnet.

It is important to note that purely electromagnetic phenomena have parity even if electric charge is pseudoscalar, i.e. if the mirror image of a positive charge is a negative charge and *vice versa*. As a result, the straight current **I**′ would be reversed, as indicated in Fig. 14.3, but now the mirror image of the magnet would be a magnet with the signs of the poles unchanged. The mirror image of the deflection would thus again correspond to an effect obtainable with actual current and an actual magnet.

Experiments involving electromagnetic effects only cannot be used to decide whether an electric charge is scalar or pseudoscalar. The mirror image of any purely electromagnetic process can either be realized within the actual world using charges of the same sign, or it can be realized within an "antiworld" in which the signs of all charges are reversed. In this sense the electromagnetic field has a high degree of symmetry.

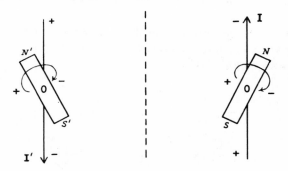

Fig. 14.3. Deflection of a magnetic dipole by a current and mirror image of this process (assuming charge to be a pseudoscalar).

Processes involving the weak interactions which cause, among other things, the β decay of nuclei, have less symmetry than purely electromagnetic processes. The mirror image of a β decay cannot be realized within the actual world containing positively charged nuclei only, but it can presumably be realized in an antiworld in which all particles involved in the decay are replaced by their antiparticles and in which the signs of all electric charges are reversed. Thus β decay does not have parity, *unless one admits experiments in the antiworld among the physically possible experiments.*

As a specific example, consider the β decay of a cobalt-60 nucleus which, under the emission of an electron and an antineutrino, goes into a nickel-60 nucleus. It has been found that the electron emitted goes off preferentially in a direction opposite to the magnetically aligned spin of the Co^{60} nucleus. Viewed through a mirror perpendicular to the direction of nuclear spin one sees the electron emitted parallel to the direction of spin, as indicated in Fig. 14.4, a process which does not happen in the actual world. The mirror image of the decay does, however, represent correctly the corresponding decay in the antiworld in which

FIG. 14.4. The β decay of Co^{60} and its mirror image (charge scalar).

the signs of all charges are reversed, as indicated in Fig. 14.5, and in which anti-Co^{60} goes, under emission of an antielectron (or positron) and a neutrino, into an anti-Ni^{60}.

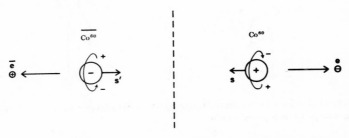

FIG. 14.5. The β decay of Co^{60} and its mirror image (charge pseudoscalar).

If one wishes to maintain reflection symmetry as a universal principle, one is thus forced to assign an electric charge the transformation charac-ter of a pseudoscalar. *From now on this will be done.* Thus, if one wishes to describe the mirror image of objects which *do* have electric charge as an intrinsic property, one must augment the operation of coordinate inversion (labeled Π) by an operation of "charge conjugation" (labeled Γ), resulting in what is called in the literature the operation of "combined inversion" (labeled $\Sigma = \Pi\Gamma$). As will be shown in Section 19, one cannot, in general, represent Γ (and thus Σ) by a unitary operator, and the sym-metry under combined inversion does not lead to a simple "law of conservation of combined parity".

It should be stressed once more, however, that no inconsistency results if charge is treated as a scalar as long as purely electromagnetic processes are considered. Thus, coordinate inversion without particle conjugation is a valid symmetry operation as long as weak interactions are excluded from consideration.

As a first quantum mechanical example consider a single object having *no* internal properties such as spin, charge, etc., and which may therefore be described completely by a state vector $|b\rangle$ depending only on the coordinates of the object, labeled $\mathbf{q}$, so that there exists a one-component ψ function $\psi_b(\mathbf{q})$ defined by

(14.1)
$$|b\rangle = \int |\mathbf{q}\rangle \, d\mathbf{q} \langle \mathbf{q}|b\rangle = \int |\mathbf{q}\rangle \, \psi_b(\mathbf{q}) \, d\mathbf{q}.$$

To investigate how one should describe the same object if it is viewed through a mirror, it is sufficient to consider only the inversion $\mathbf{q} \to -\mathbf{q}$ of the coordinates, because any reflection can be decomposed into an inversion and a proper rotation. Inversion of coordinates requires description of the same object in terms of a transformed state

(14.2)
$$|b\rangle^T = \int |\mathbf{q}\rangle \, \psi_b(-\mathbf{q}) \, d\mathbf{q}.$$

An attempt will now be made to connect $|b\rangle^T$ with the untransformed state by a linear unitary operator Π defined as

(14.3)
$$|b\rangle^T = \Pi|b\rangle$$

so that Π represents the operation $\mathbf{q} \to -\mathbf{q}$ carried out on the ψ function in (14.1). To ensure that the parity of an object is embodied in the quantum mechanical description, it suffices to require the Hamiltonian of the object to be invariant under the transformation,

(14.4)
$$H = \Pi H \Pi^{-1},$$

because then the transformed state satisfies the same dynamical equation as the untransformed state, i.e. in the state picture from

(14.5)

$$H|b\rangle = i|\dot{b}\rangle \quad \text{follows} \quad \Pi H \Pi^{-1} \Pi|b\rangle = i\Pi|\dot{b}\rangle \quad \text{or} \quad H|b\rangle^T = i|\dot{b}\rangle^T$$

so that if $|b\rangle$ is a possible state, then $|b\rangle^T$ is also a possible state of the same object.

Equation (14.4) contains the "law of conservation of parity," because it may be written

(14.6)
$$\Pi H - H \Pi = 0$$

which means the expectation value of Π is constant in time. This law is borne out by all experiments which do not involve weak interactions. On the other hand, since the Hamiltonian describing weak interaction processes such as the Co^{60} decay is apparently not invariant under pure coordinate inversion, one may refer to the peculiar correlations of the decay products as indicated in Fig. 14.4 as being caused by "non conservation of parity" in that case.

Of particular interest are the eigenstates of the operator Π, defined by

(14.7) $\Pi|u\rangle = P|u\rangle;$ P a number.

The eigenvalues can be determined immediately if $|u\rangle$ is required to be a unique function of the coordinates, because in that case application of Π twice on $|u\rangle$ must be the identity operation (up to an arbitrary phase factor which by convention is chosen to be $+1$)

(14.8) $\Pi^2|u\rangle = P^2|u\rangle = |u\rangle.$

Hence there are two eigenvalues of Π, namely the roots of

(14.9) $P^2 = 1$ i.e. $P_1 = +1$ and $P_2 = -1.$

Eigenstates of Π with eigenvalue $+1$ are called states of even parity denoted $|u_+\rangle$, and eigenstates with eigenvalue -1 are called states of odd parity denoted $|u_-\rangle$,

(14.10) $\Pi|u_+\rangle = +|u_+\rangle$ and $\Pi|u_-\rangle = -|u_-\rangle.$

It is seen that the parity of a single object, describable by a unique state vector, is a dichotomic variable, and, in accordance with the concepts developed in Section 1, one may therefore use the representation

(14.11) $\Pi = \begin{pmatrix} 1 & 0 \\ 0 & -1 \end{pmatrix};$ $|u_+\rangle = \begin{pmatrix} 1 \\ 0 \end{pmatrix};$ $|u_-\rangle = \begin{pmatrix} 0 \\ 1 \end{pmatrix}.$

A state vector $|b\rangle$ need not be an eigenstate of Π, but it can always be decomposed

(14.12) $|b\rangle = |u_+\rangle\langle u_+|b\rangle + |u_-\rangle\langle u_-|b\rangle = \psi_b(+)\begin{pmatrix} 1 \\ 0 \end{pmatrix} + \psi_b(-)\begin{pmatrix} 0 \\ 1 \end{pmatrix}$

so that

(14.13) $|b\rangle^T = \Pi|b\rangle = \psi_b(+)\begin{pmatrix} 1 \\ 0 \end{pmatrix} - \psi_b(-)\begin{pmatrix} 0 \\ 1 \end{pmatrix}.$

It is understood that $\psi_b(+)$ and $\psi_b(-)$ are ψ functions only with respect to the two-dimensional parity space; with respect to all other attributes they are still state vectors, with $n/2$ components if n is the number of components of $|b\rangle$. One should, therefore, if one wishes to adhere to quite impeccable notation, write the direct product $|\psi_b(+)\rangle \times \binom{1}{0}$, etc.

A state with both $\psi_b(+) \neq 0$ and $\psi_b(-) \neq 0$ is called a state of mixed parity. One can always choose two linear combinations which span the same space as the vectors $|b\rangle$ and $|b\rangle^T$, and which have definite parity, namely

(14.14)

$$\psi_b(+)\begin{pmatrix} 1 \\ 0 \end{pmatrix} = \tfrac{1}{2}(|b\rangle + |b\rangle^T)$$

$$\psi_b(-)\begin{pmatrix} 0 \\ 1 \end{pmatrix} = \tfrac{1}{2}(|b\rangle - |b\rangle^T).$$

Eigenstates of the Hamiltonian H belonging to nondegenerate eigenvalues ω are always eigenstates of Π, i.e. have definite parity. Indeed, if

$$(14.15) \qquad H|u\rangle = \omega|u\rangle$$

then also

$$(14.16) \quad \Pi H \Pi^{-1} \Pi|u\rangle = \omega \Pi|u\rangle \qquad \text{or} \qquad H\Pi|u\rangle = \omega\Pi|u\rangle.$$

For nondegenerate ω this can be true only if

$$(14.17) \qquad \Pi|u\rangle = P|u\rangle; \qquad P \quad \text{a number,}$$

which by definition (14.7) shows that $|u\rangle$ is an eigenstate of Π.

The eigenstates of orbital angular momentum $|l, m\rangle$ of an object also have definite parity. In polar coordinates these states read (see Appendix 1)

$$(14.18) \qquad |l, m\rangle = \int |\vartheta, \varphi\rangle \, d\Omega \, Y_{lm}(\vartheta, \varphi)$$

where the ψ function

$$(14.19) \qquad Y_{lm}(\vartheta, \varphi) \propto \sin^m \vartheta \, (\cos^{l-m} \vartheta + a \cos^{l-m-2} \vartheta + \ldots) \, e^{im\varphi}.$$

Now the inversion $\mathbf{q} \to -\mathbf{q}$ corresponds in polar coordinates to the transformation $r, \vartheta, \varphi \to r, \pi - \vartheta, \pi + \varphi$ so that

$$(14.20)$$

$$\Pi Y_{lm}(\vartheta, \varphi) \propto \sin^m(\pi - \vartheta) \left[\cos^{l-m}(\pi - \vartheta) + a \cos^{l-m-2}(\pi - \vartheta) + \ldots \right] e^{im(\pi + \varphi)}.$$

Since for

$$(14.21) \qquad
\begin{aligned}
l \text{ even}: \quad & \cos^{l-m}(\pi - \vartheta) = \begin{cases} + \cos^{l-m} \vartheta & \text{if } m \text{ even} \\ - \cos^{l-m} \vartheta & \text{if } m \text{ odd} \end{cases} \\
l \text{ odd}: \quad & \cos^{l-m}(\pi - \vartheta) = \begin{cases} - \cos^{l-m} \vartheta & \text{if } m \text{ even} \\ + \cos^{l-m} \vartheta & \text{if } m \text{ odd} \end{cases}
\end{aligned}$$

and since further

$$(14.22) \qquad \sin^m(\pi - \vartheta) \, e^{im(\pi + \varphi)} = \begin{cases} + \sin^m \vartheta \, e^{im\varphi} & \text{if } m \text{ even} \\ - \sin^m \vartheta \, e^{im\varphi} & \text{if } m \text{ odd} \end{cases}$$

one has

$$(14.23) \qquad \Pi Y_{lm}(\vartheta, \varphi) = (-1)^l Y_{lm}(\vartheta, \varphi).$$

This means a state of odd l has odd parity and a state of even l has even parity, regardless of what the value of m may be.

One can classify generally all observables as even or odd depending on whether their operators do not or do change sign under inversion of coordinates. Any even operator $A(+)$ is thus defined as

(14.24) $$\Pi A(+)\Pi^{-1} = A(+)$$

and any odd operator $B(-)$ is defined as

(14.25) $$\Pi B(-)\Pi^{-1} = -B(-)$$

independent of the transformation properties of $A(+)$ or $B(-)$ under rotations or translations of coordinates. It is easily verified that $A(+)$ and $B(-)$ may be represented in the representation (14.11) by

(14.26) $$A(+) = A\begin{pmatrix}1 & 0 \\ 0 & 1\end{pmatrix}$$

(14.27) $$B(-) = B\begin{pmatrix}0 & 1 \\ 1 & 0\end{pmatrix}$$

where now A and B contain the operations on all other variables of the object, and the 2×2 matrices operate on the parity variable alone. In any general state (14.12), one finds, thus, for the expectation values of $A(+)$ and $B(-)$ the expressions

(14.28) $\overline{A(+)} = \langle b|A(+)|b\rangle = \langle\psi_b(+)|A|\psi_b(+)\rangle + \langle\psi_b(-)|A|\psi_b(-)\rangle$

(14.29) $\overline{B(-)} = \langle b|B(-)|b\rangle = \langle\psi_b(+)|B|\psi_b(-)\rangle + \langle\psi_b(-)|B|\psi_b(+)\rangle.$

From (14.29) one reads immediately the *fundamental theorem*:

The expectation value of any odd observable vanishes in any state of definite parity, i.e. in a state for which either $\psi_b(+) = 0$ or $\psi_b(-) = 0$.

Example of an odd observable is the linear momentum of an object, examples of even observables are angular momentum and energy of an object.

From the foregoing argument, one concludes that an object in a state of definite parity *may* possess even observables, such as angular momentum and/or energy, but that it *cannot* possess odd observables, such as linear momentum. In a state of mixed parity, however, an object *may* possess odd observables.

Prior to the discovery of the correlation between the magnetic moment of the Co^{60} nucleus and the direction of electron emission in its β decay (see Fig. 14.4), electric charge had conventionally been assumed to transform under inversion as a scalar (see Fig. 14.2). With this assumption, Maxwell's equations lead one to consider as odd observables, magnetic pole, electric dipole, magnetic quadrupole, etc., and to consider as even observables, magnetic dipole, electric quadrupole, etc. Accordingly, the absence of any elementary magnetic poles and of any electric dipole moment in the ground state of any object (usually considered a nondegenerate eigenstate of H and thus having definite parity) was taken as sufficiently explained by inversion invariance, which does not permit existence of odd observables, such as magnetic pole and electric dipole, in states of definite parity.

However, since the advent of the Co^{60} experiment, there is now reason to assign to electric charge the transformation property of a pseudoscalar, in accord with the explanations given in Figs. 14.3 and 14.5, and the classification of multipole moments given above is reversed. This raises the following questions:

(i) If indeed electric charge, magnetic dipole, etc., are, respectively, pseudoscalar, vector, etc., i.e. odd observables, how is it possible that many so-called elementary particles and other simple objects possess these observables in their ground state?

(ii) If magnetic pole, electric dipole, etc., are even observables, why are they never observed in the ground state of any "elementary" object? This question has its root in the generally valid observation that nature usually realizes any possibility open to her on general principle, an observation which is sometimes stated aphoristically as, "Anything that is not forbidden is compulsory."

A consistent, though not necessarily correct, set of possible answers to these two questions is the following:

(i) The very fact that there are ground states of objects with non-vanishing electric charge means such ground states cannot be states of definite parity. This is possible provided the ground state is *degenerate* with respect to the energy of the object. Such a situation may arise quite generally if to every ground state of energy ω_0 describing an object with given charge there exists another such state of the same energy ω_0 describing the same kind of object, but with opposite charge. If this point of view is accepted, then the only "elementary" particles that *may* possess definite parity are the neutral photon, the neutral pion, and the neutral kaons. Should any of these particles indeed have definite parity, then it cannot possess odd observables such as magnetic dipole moment, etc. The conditions under which the concept of the intrinsic parity of elementary particles is meaningful will be examined in more detail in Section 26.

(ii) The nonexistence of magnetic dipole, electric dipole, etc., in "elementary" systems *may* be due to some other invariance requirements, such as invariance under reversal of motion, which will be taken up in the following Section 15, and/or invariance under particle conjugation, to be taken up in Section 28.

The division of states into states of even and odd parity, corresponding to eigenvalues $+1$ and -1 of the operator Π, was based on the requirement that the state vector of an object having no intrinsic spatial properties, except position, be a unique function of the coordinates. This means Π^2 can differ from the unit operation only by a phase factor (a constant of modulus 1) which had been chosen such that $\Pi^2 = I$. If one tries to assign parity to objects having spin $j = \frac{1}{2}$, one encounters an additional ambiguity stemming from the double-valuedness of spin states as a function of the coordinates. Depending on whether one considers application of Π twice as amounting to a rotation of angle $n2\pi$ with n even, or amounting to that rotation with n odd, one obtains

(14.30)
$$\Pi'^2|u'\rangle = P'^2|u'\rangle = |u'\rangle \qquad \text{or} \qquad \Pi''^2|u''\rangle = P''^2|u''\rangle = -|u''\rangle$$

each case corresponding to a possible representation in which

(14.31) $P_1' = +1;\ \ P_2' = -1 \qquad \text{or} \qquad P_1'' = +i;\ \ P_2'' = -i.$

Accordingly one may represent the operator Π by either

(14.32) $\Pi' = \begin{pmatrix} 1 & 0 \\ 0 & -1 \end{pmatrix} \qquad \text{or} \qquad \Pi'' = \begin{pmatrix} i & 0 \\ 0 & -i \end{pmatrix}$

if the parity states are represented as usual by $|u_+'\rangle = |u_+''\rangle = \binom{1}{0}$ and $|u_-'\rangle = |u_-''\rangle = \binom{0}{1}$. Since a phase factor is always at one's disposal in any unitary operator, however, no physical restriction is obtained if the representation Π' is used, but it should be kept in mind that this is a convention.

As will be shown in Section 26, the only experimental information available about the parity of two interacting objects is whether they have the same parity or whether they have opposite parity. It is apparently impossible to devise an experiment which would allow determination of the absolute parity of an object. One must therefore make the further convention of assigning to some object a certain parity and then determining the parity of all other objects relative to that reference object. The standard convention is to assign even parity to the vacuum state.

NOTES

Wigner [1] gave the first clear formulation of the quantum mechanical law of conservation of parity as derived from inversion symmetry, pointing out that it has no analog in classical mechanics and that it is the origin of selection rules discovered in atomic spectra by Laporte [2].

(Incidentally, Wigner did not use the term "parity" in his original paper, and Pauli in his famous encyclopedia article calls this quantity the "signature." The present author has been unable to ascertain who invented the term "parity" for the eigenvalue of the inversion operator Π.)

Weyl [3] gives an account of the somewhat inconspicuous role played by inversion symmetry in classical physics.

Wu *et al.* [4] discovered the nonconservation of parity in β-decay of Co^{60} by performing an experiment suggested by Lee and Yang [5]. The possibility of re-establishing reflection symmetry in case of parity non-conservation through admission of combined inversion as a symmetry operation had already been suggested by Yang [6] in his report to the International Conference in Theoretical Physics, Seattle, 1956. See also Landau [7] and footnote 9 in the paper by Wick *et al.* [8].

REFERENCES

[1] E. Wigner, Über die Erhaltungssätze in der Quantenmechanik, *Nachr. Akad. Wiss. Goettingen Math.-Phys. Kl.* p. 375 (1927).

[2] O. Laporte, *Z. Physik* **23**, 135 (1924).

[3] H. Weyl, "Symmetry." Princeton Univ. Press, Princeton, New Jersey, 1952.

[4] C. S. Wu, E. Ambler, R. W. Hayward, D. D. Hoppes, and P. R. Hudson, *Phys. Rev.* **105**, 1413 (1957).

[5] T. D. Lee and C. N. Yang, *Phys. Rev.* **104**, 254 (1956).

[6] C. N. Yang, *Rev. Mod. Phys.* **29**, 231 (1957).

[7] L. D. Landau, *Nucl. Phys.* **3**, 127 (1957).

[8] G. C. Wick, A. S. Wightman, and E. P. Wigner, *Phys. Rev.* **88**, 101 (1952).

Invariance under Reversal of Motion

Another symmetry property with which all objects and processes occurring in nature seem to be endowed is "reversality." Any physical process or property involving an actual object is said to have reversality if after reversal of all motions one obtains again a *possible* physical process or property involving the same object.

If reversality were not a universal property of physical reality, there should exist stationary situations which are invariant with respect to all symmetry operations except reversal of motion. For example, if one were to observe in a β-decay process a state in which the spin of the nucleus s, the velocity of the emitted electron v_e, and the velocity of the emitted γ-ray v_γ are correlated as indicated in Fig. 15.1, so that v_γ, v_e,

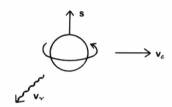

FIG. 15.1. Example of a state which does not have reversality.

and s, in this order, form a right-hand system, one would have to infer that this state does not have reversality, unless there exists an, as yet undiscovered, other form of matter, metamatter (say), distinct from antimatter, which is in every respect the motion-reversed image of ordinary matter. Indeed, if one reverses in Fig. 15.1 all motions, one obtains a state indicated in Fig. 15.2, which is essentially different from Fig. 15.1 because the reversed motions $\tilde{v}_\gamma$, $\tilde{v}_e$, and $\tilde{s}$, in this order, form a left-hand system contrary to the hypothetical experimental situation of Fig. 15.1. It should be noted that the state pictured in Fig. 15.1 has parity, because its mirror image, as indicated in Fig. 15.3, is again a state in which the reflected motions v'_γ, v'_e, s', in this order, form a right-hand system.

To date all experiments designed to detect states of the kind indicated in Fig. 15.1 have had negative results. In particular, the β decay of Co^{60} as described in Fig. 14.4 has reversality, because the reversed state, in which the direction of both nuclear spin and electron velocity are

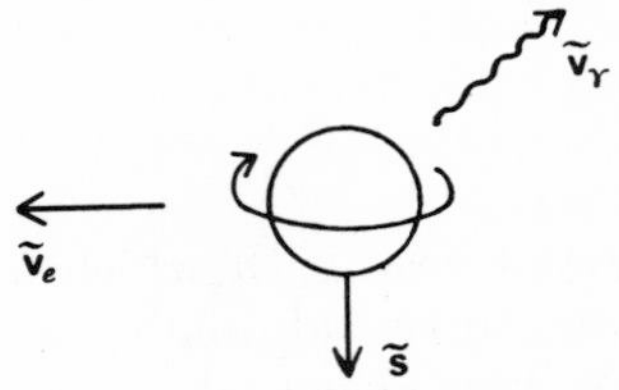

FIG. 15.2. The state obtained by reversal of motion in the state described by Fig. 15.1.

reversed, is again a possible state with the electron moving in the direction opposite to the direction of nuclear spin.

The operation of reversal of motion for a nonstationary process can be visualized by imagining a moving picture taken of the object under consideration, and then having the film reeled off backwards. Since this operation is formally equivalent to changing, in all expressions depending

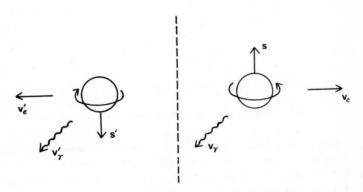

FIG. 15.3. The state described by Fig. 15.1 and its mirror image.

on time, the sign of the time variable, the operation may be referred to as "time reversal." Thus reversal of motion means not only the reversing of all motions in a given state, but also the reversing of the sequence of states.

In collision processes the reversality of both initial and final state may lead to a very general relation between the process and its "inverse" process, in which *only* the sequence of final and initial state is inter-changed. This relation is known as the "principle of detailed balance" and will be taken up in Section 28.

One must not confuse reversality of a process with the thermodynamic notion of reversibility of a process. A process will have reversality as long as the process with all motions reversed is in principle a possible process, however improbable it may be. The mechanical mixing of a deck of cards has reversality although thermodynamically it would have to be classified as an irreversible process. It is an interesting and open question whether the breakdown of invariance under reversal of motion, and thus of the principle of detailed balance, if ever observed, would have any thermodynamical consequences at all, and if so, how far reaching these would be.

Observables can quite generally be classified as time-even or time-odd depending on whether they do not or do change sign under time reversal. By definition of time reversal all even derivatives of the coordinate Q of an object, Q, $\ddot{Q}$, etc., are time-even, and all odd derivatives, $\dot{Q}$, $\dddot{Q}$, etc., are time-odd. If the usual assignment of time-evenness to the mass of an object is taken for granted, one can infer from this that linear momentum $\mathbf{P}$ and angular momentum $\mathbf{J}$ are time-odd, whereas the energy H is a time-even observable.

Regarding the transformation properties of electromagnetic observables under time reversal, Maxwell's equations contain only the information that electric charge and magnetic pole must transform oppositely under time reversal, so that if one of them is time-even the other one must be time-odd. In absence of any experimental information of other than electromagnetic origin which could decide whether electric charge is time-even or time-odd, charge can be considered as time-odd, in accordance with a suggestion by Feynman who has proposed looking upon positrons as electrons "running backward in time," but it should be stressed that at this stage this is purely conventional. (In fact, evidence will be presented later indicating that electronic charge should be considered as time-even, because the lepton number characterizing electrons and positrons, as well as neutrinos and antineutrinos, is time-even.) Accordingly, electric field $\mathbf{E}$ and electric dipole moment $\mathbf{p}$ may tentatively be considered as time-odd observables, and magnetic field $\mathbf{B}$ and magnetic dipole moment $\mathbf{\mu}$ as time-even observables.

Although the definition of time-even and time-odd hermitean operators, representing time-even and time-odd observables in quantum mechanics, makes sense, the division of states into states of even and odd "reversality," in analogy to the division of states into states of even and odd parity, is obscure for the following reasons.

It is clear that the transformed state under time reversal Θ,

(15.1) $$|b\rangle^T = \Theta|b\rangle$$

cannot be connected with $|b\rangle$ by a linear unitary operator, because if one demands that time-even energy H, and time-odd linear momentum **P** and angular momentum **J**, transforms according to

(15.2) $$\Theta H \Theta^{-1} = H; \qquad \Theta \mathbf{P} \Theta^{-1} = -\mathbf{P}; \qquad \Theta \mathbf{J} \Theta^{-1} = -\mathbf{J},$$

then the fundamental dynamical equation

(15.3) $$H|b\rangle = i\, \partial|b\rangle/\partial t$$

and the canonical C.R.s

(15.4) $$[PQ] = i; \qquad [\mathbf{J} \times \mathbf{J}] = i\mathbf{J}$$

are not invariant under the corresponding unitary transformation, since only one side of each equation changes sign.

One can, however, satisfy the invariance requirements demanded by the reversality of physical reality with an antiunitary operator

(15.5) $$\Theta = TK$$

where T is a unitary operator representing the transformation $t \to -t$, and K is the antilinear operator of complex conjugation, applied to all numbers in the state vectors, so that in terms of the ψ function characterizing the state $|b\rangle$

(15.6)
$$\Theta|b\rangle = \Theta \int |q\rangle\, dq\, \psi_b(q,t) = \int |q\rangle\, dq\, T\psi_b^*(q,t) = \int |q\rangle\, dq\, \psi_b^*(q,-t).$$

The notion of eigenstates of the reversal operator Θ, in analogy to the eigenstates of the linear inversion operator Π, is thus obviously not a sensible concept in general. The fundamental dynamical equation satisfied by $|b\rangle^T$ follows now from (15.3) by application of (15.5),

(15.7) $$\Theta H \Theta^{-1} \Theta|b\rangle = \Theta i\, \partial|b\rangle/\partial t = -i\Theta\, \partial|b\rangle/\partial t = i(\partial/\partial t)\,(\Theta|b\rangle)$$

which for $\Theta H \Theta^{-1} = H$ can be written

(15.8) $$H|b\rangle^T = i\, \partial|b\rangle^T/\partial t$$

and is seen to be identical with Eq. (15.3) satisfied by $|b\rangle$. The canonical C.R.s (15.4) are also invariant under (15.5), because the operator K changes the sign of the imaginary unit which appears on the right-hand side of each Eq. (15.4), so that application of Θ in accordance with the transformation properties (15.2) of **P** and **J** leads to no change in these equations.

Although a quantum number associated with Θ, analogous to parity, does not make sense in general, one can characterize a state vector $|b\rangle$ by the eigenvalue of the operator Θ^2, which is a linear unitary operator. The physical requirement that the operation of time reversal Θ carried out twice should result in the same state,

$$(15.9) \qquad \Theta^2|b\rangle = \epsilon|b\rangle; \qquad |\epsilon| = 1,$$

means that *all states must be eigenstates of* Θ^2. Since the operation of complex conjugation carried out twice is equivalent to the identity operation, $K^2 = I$, one can write

$$(15.10) \qquad \Theta^2 = TKTK = TT^*K^2 = TT^* = \epsilon I.$$

From the unitarity of T follows further that

$$(15.11) \qquad \epsilon^2 = 1$$

because for a unitary operator one has $T^{-1} = T^+ = \tilde{T}^*$, so that the last equation (15.10) can be written

$$(15.12) \qquad T^* = \epsilon\tilde{T}^*$$

and by transposing this equation once again one obtains

$$(15.13) \qquad \tilde{T}^* = \epsilon T^* = \epsilon^2\tilde{T}^*$$

which can be true only if (15.11) holds. The possible eigenvalues of Θ^2 are therefore $\epsilon = +1$ and $\epsilon = -1$. Multiplication of Θ with a phase $e^{i\alpha}$ does not affect Θ^2, of course, because $e^{i\alpha}\Theta e^{i\alpha}\Theta = e^{i\alpha}e^{-i\alpha}\Theta^2 = \Theta^2$.

It can be shown quite generally that states describing an object having integer total angular momentum j and no other internal attributes belong to the eigenvalue $\epsilon = +1$ of Θ^2, and that states describing an object having half-odd integer total angular momentum j and no other internal attributes belong to the eigenvalue $\epsilon = -1$ of Θ^2. (For the purpose of the present work it actually suffices to verify this statement only for the values $j = 0, \frac{1}{2}, 1$.) To do so, use will be made of a construction of the operator Θ in the representation in which the component operators of angular momentum J_1, J_3 have real matrix elements, and the component operator J_2 has pure imaginary matrix elements, as explained in Appendix 1.

The transformation equation (15.2) for **J**, in components,

$$(15.14) \quad \Theta J_1 + J_1\Theta = 0; \qquad \Theta J_2 + J_2\Theta = 0; \qquad \Theta J_3 + J_3\Theta = 0$$

requires the unitary operator T defined by (15.5) to satisfy the relations

$$(15.15) \quad TJ_1 + J_1T = 0; \qquad TJ_2 - J_2T = 0; \qquad TJ_3 + J_3T = 0$$

which are solved by a rotation of amount π around the 2-axis, as is intuitively obvious,

(15.16) $T = e^{i\pi J_2}; \qquad T^+ = T^{-1} = e^{-i\pi J_2}.$

Indeed, by expansion and use of the C.R.s for the J_i after the fashion of (12.16) one finds

$$TJ_1 T^{-1} = e^{i\pi J_2} J_1 e^{-i\pi J_2} = J_1 \cos\pi + J_3 \sin\pi = -J_1$$

(15.17) $$TJ_2 T^{-1} = e^{i\pi J_2} J_2 e^{-i\pi J_2} = J_2$$

$$TJ_3 T^{-1} = e^{i\pi J_2} J_3 e^{-i\pi J_2} = J_3 \cos\pi - J_1 \sin\pi = -J_3.$$

In this representation the operator Θ^2 takes now the form

(15.18)
$$\Theta^2 = TT^* = e^{i\pi J_2} e^{-i\pi J_2^*} = e^{2i\pi J_2} = I + 2i\pi J_2 + [(2i\pi)^2/2!]J_2^2 + \dots$$

Now consider an object in an eigenstate of the total angular momentum J^2 characterized by the quantum number j so that $J^2 = j(j+1)I$. It is always possible to perform a unitary operation which leaves Θ^2 invariant, but makes J_2 a diagonal matrix, so that it can be replaced by its eigenvalues which differ from the value j of the total angular momentum only by integers. Thus, since $\exp[2\pi i(\text{integer})] = 1$, Θ^2 is identical with $\exp[2\pi ij]$ which is equal to $+1$ for integer j and -1 for half-odd integer j.*

It is perhaps instructive to verify this general result for the special cases $j = 0, j = \frac{1}{2}, j = 1$. For $j = 0$ one has trivially

(15.19) $\Theta^2(j = 0) = I.$

For $j = \frac{1}{2}$ one has $J_2 = \frac{1}{2}\begin{pmatrix} 0 & -i \\ i & 0 \end{pmatrix}$ and thus $J_2^2 = \frac{1}{2}^2 I$, so that

(15.20) $\Theta^2(j = \frac{1}{2}) = I\cos\pi + 2iJ_2\sin\pi = -I.$

For $j = 1$ one has

$$J_2 = \frac{1}{\sqrt{2}}\begin{pmatrix} 0 & -i & 0 \\ i & 0 & -i \\ 0 & i & 0 \end{pmatrix}; \qquad J_2^2 = \frac{1}{2}\begin{pmatrix} 1 & 0 & -1 \\ 0 & 2 & 0 \\ -1 & 0 & 1 \end{pmatrix}; \qquad J_2^3 = J_2,$$

so that

(15.21) $\Theta^2(j = 1) = I + iJ_2\sin(2\pi) - J_2^2[1 - \cos(2\pi)] = I$

and so on.

---

* The author is indebted to Mr. David Pink and to Mr. Patrick Whelan for this simple line of reasoning.

The cases $\epsilon = +1$ and $\epsilon = -1$ require now separate treatment.

(*i*) $\epsilon = +1$. The object can be described in terms of an orthogonal set $|b_+\rangle$ of eigenstates of a complete observable B, for example in terms of the eigenstates $|l,m\rangle$ of J^2 and J_3, which are simultaneously eigenstates of the operator Θ^2,

$$\text{(15.22)} \qquad \Theta^2|b_+\rangle = |b_+\rangle.$$

It follows that $|b_+\rangle^T = \Theta|b_+\rangle$ is also an eigenstate of Θ^2 with eigenvalue $+1$. So is

$$\text{(15.23)} \quad |u_{\text{even}}\rangle = c(|b_+\rangle + \Theta|b_+\rangle); \qquad c \text{ real, if} \quad \Theta|b_+\rangle \neq -|b_+\rangle;$$

unless it vanishes, and one has thus

$$\text{(15.24)} \qquad \Theta|u_{\text{even}}\rangle = |u_{\text{even}}\rangle$$

which explains the label "even". If it should happen that for a specific state, $|b'_+\rangle$ say, $\Theta|b'_+\rangle = -|b'_+\rangle$, one chooses

$$\text{(15.23')} \qquad |u_{\text{even}}\rangle = i|b'_+\rangle; \qquad \text{if} \quad \Theta|b'_+\rangle = -|b'_+\rangle;$$

and Eq. (15.24) is again valid. Similarly, one can construct from the set $|b_+\rangle$ a set $|u_{\text{odd}}\rangle$ by

$$\text{(15.25)} \quad |u_{\text{odd}}\rangle = c(|b_+\rangle - \Theta|b_+\rangle); \qquad c \text{ real, if} \quad \Theta|b_+\rangle \neq |b_+\rangle$$

$$\text{(15.25')} \qquad |u_{\text{odd}}\rangle = i|b'_+\rangle; \qquad \text{if} \quad \Theta|b'_+\rangle = |b'_+\rangle$$

which satisfies

$$\text{(15.26)} \qquad \Theta|u_{\text{odd}}\rangle = -|u_{\text{odd}}\rangle.$$

This possibility of finding even and odd states is, contrary to the situation under the inversion operation Π, *not equivalent to the possibility of labeling the states by a physically meaningful quantum number characteristic of Θ,* such as parity in case of Π, *because one can transform an even state into an odd state under Θ and* vice versa *simply by multiplication of $|u\rangle$ with a physically unobservable phase factor i.* The quantum number ϵ characteristic of Θ^2, however, is not affected by such a change in phase and should therefore be a physically meaningful label of a state.

Although the states $|u\rangle$ are not eigenstates of Θ in the usual sense, they may be referred to as "invariant states" of Θ, and one has the *general theorem: In a state invariant under Θ the expectation value of any time-odd operator vanishes.*

Indeed, if $\Theta A \Theta^{-1} = -A$, then in an invariant state $|u\rangle$, so that $\Theta|u\rangle = e^{i\alpha}|u\rangle$ with arbitrary α,

$$\text{(15.27)} \quad \bar{A} = \langle u|A|u\rangle = -\langle u|\Theta A\Theta^{-1}|u\rangle = -\langle u|e^{-i\alpha}A\,e^{i\alpha}|u\rangle = -\bar{A}$$

requiring $\bar{A} = 0$, independent of the phase α. This statement is thus invariant under unitary transformations, even though any particular state $|u\rangle$, say $|u_{\text{even}}\rangle$, allows only real orthogonal transformations $|u\rangle \to R|u\rangle$ with $\tilde{R} = R^{-1}$ if $\Theta = I$ is to remain valid. Under a unitary transformation $U^+ = U^{-1}$ the operator Θ transforms into

$$\Theta' = U\Theta U^{-1} = UT(U^{-1})^* K = U(U^{-1})^* T' K$$

with

$$T' = \tilde{U}^+ T \tilde{U} \qquad (\text{or} \quad \Theta' = U\tilde{U}T' K).$$

If Feynman's assignment of time-oddness to electric charge and electric dipole moment were correct, one could, for example, conclude from this theorem that an elementary particle which has integer spin, and no other intrinsic attributes which might invalidate the labeling with the quantum number $\epsilon = +1$, must be electrically neutral and cannot possess any electric dipole moment either.

If it should happen for a specific state $|b'_+\rangle$ that $\Theta|b'_+\rangle = |b''_+\rangle$, then also $\Theta|b''_+\rangle = |b'_+\rangle$, and the two invariant states

$$(15.28) \quad \begin{aligned} |u'_{\text{even}}\rangle &= c(|b'_+\rangle + \Theta|b'_+\rangle) = c(|b'_+\rangle + |b''_+\rangle) \\ |u''_{\text{even}}\rangle &= c(|b''_+\rangle + \Theta|b''_+\rangle) = c(|b''_+\rangle + |b'_+\rangle) \end{aligned}$$

are accidentally identical. This apparent incompleteness can always be removed by choosing $i|b'_+\rangle$ and $|b''_+\rangle$ as basis, because then the invariant states

$$(15.28') \quad \begin{aligned} |v'_{\text{even}}\rangle &= c(i|b'_+\rangle + \Theta i|b'_+\rangle) = ic(|b'_+\rangle - |b''_+\rangle) \\ |v''_{\text{even}}\rangle &= c(|b''_+\rangle + \Theta|b''_+\rangle) = c(|b''_+\rangle + |b'_+\rangle) \end{aligned}$$

are properly orthogonal.

It is perhaps instructive to have the foregoing statements illustrated for the case $j = 1$ in which one has explicitly

$$(15.29) \qquad T(j = 1) = e^{i\pi J_2} = I + iJ_2 \sin \pi - J_2^2(1 - \cos \pi)$$

$$= I - 2J_2^2 = \begin{pmatrix} 0 & 0 & 1 \\ 0 & -1 & 0 \\ 1 & 0 & 0 \end{pmatrix}.$$

Taking as basis the eigenstates of J_3, namely

$$(15.30) \quad |1, +1\rangle = \begin{pmatrix} 1 \\ 0 \\ 0 \end{pmatrix}; \qquad |1, 0\rangle = \begin{pmatrix} 0 \\ 1 \\ 0 \end{pmatrix}; \qquad |1, -1\rangle = \begin{pmatrix} 0 \\ 0 \\ 1 \end{pmatrix}$$

one finds

(15.31)

$$\Theta|1, +1\rangle = |1, -1\rangle; \qquad \Theta|1, 0\rangle = -|1, 0\rangle; \qquad \Theta|1, -1\rangle = |1, +1\rangle$$

and the accidental cases (15.23′) and (15.28) are realized. One chooses therefore as basis instead, the set with the phases

$$(15.30') \quad |1, +1\rangle = \begin{pmatrix} i \\ 0 \\ 0 \end{pmatrix}; \quad |1, 0\rangle = \begin{pmatrix} 0 \\ i \\ 0 \end{pmatrix}; \quad |1, -1\rangle = \begin{pmatrix} 0 \\ 0 \\ 1 \end{pmatrix}$$

which yield the three orthogonal invariant states

$$|u(+1)\rangle = (1/\sqrt{2})(|1, +1\rangle + \Theta|1, +1\rangle)$$

$$= (1/\sqrt{2})(|1, +1\rangle - i|1, -1\rangle) = \frac{1}{\sqrt{2}}\begin{pmatrix} i \\ 0 \\ -i \end{pmatrix}$$

$$(15.32) \quad |u(0)\rangle = |1, 0\rangle = \begin{pmatrix} 0 \\ i \\ 0 \end{pmatrix}$$

$$|u(-1)\rangle = (1/\sqrt{2})(|1, -1\rangle + \Theta|1, -1\rangle)$$

$$= (1/\sqrt{2})(|1, -1\rangle - i|1, +1\rangle) = \frac{1}{\sqrt{2}}\begin{pmatrix} 1 \\ 0 \\ 1 \end{pmatrix}$$

The fact that the matrix T is symmetrical for $j = 1$, as given in Eq. (15.29), is a special case of the general fact that T is symmetrical whenever $\epsilon = +1$. This follows from $\Theta^2 = TT^* = I$ by multiplication with the transposed operator $\tilde{T} = T^{+*}$ yielding

$$(15.33) \quad TT^*T^{+*} = \tilde{T} \quad \text{and} \quad T = \tilde{T} \quad \text{for} \quad \epsilon = +1$$

by the unitary of T. Similarly, for the case $\epsilon = -1$, which will be taken up now, one obtains

$$(15.34) \quad T = -\tilde{T} \quad \text{for} \quad \epsilon = -1$$

which means in this case T must be a skew-symmetrical matrix.

   (*ii*) $\epsilon = -1$. The object can be described in terms of an orthogonal set $|b_-\rangle$ of eigenstates of a complete observable B, for example in terms of the eigenstates $|j, m\rangle$ of J^2 and J_3, which are simultaneous eigenstates of Θ^2,

$$(15.35) \qquad\qquad \Theta^2|b_-\rangle = -|b_-\rangle.$$

It follows again, as in the case $\epsilon = +1$, that $|b_-\rangle^T = \Theta|b_-\rangle$ is also eigenstate of Θ^2 with eigenvalue -1. However, $|b_-\rangle^T$ *is always orthogonal to* $|b_-\rangle$, because

$$(15.36)$$

$$\langle b_-|\Theta b_-\rangle = \langle \Theta^2 b_-|\Theta b_-\rangle = -\langle b_-|\Theta b_-\rangle \qquad \text{(and therefore} = 0)$$

where the first equality sign follows from $\langle a|b\rangle = \langle \Theta b|\Theta a\rangle$ and the second from (15.35).

Denoting the two orthogonal complements of any state $|b_-\rangle$ by $|b_-'\rangle$ and $|b_-''\rangle$ and defining them with a convention of phase so that

$$(15.37) \quad |b_-''\rangle = i\Theta|b_-'\rangle \quad \text{and} \quad |b_-'\rangle = i^*\Theta|b_-''\rangle = -i\Theta|b_-''\rangle$$

one can thus say the eigenvectors of Θ^2 are twofold degenerate, and the transition from the set $|b_-'\rangle$ to the set $|b_-''\rangle$ cannot be effected by a unitary transformation. *To accommodate the reversality of an object one requires therefore, in the case $\epsilon = -1$, for its complete description an additional two dimensions in the abstract state vector space.* One can satisfy Eqs. (15.37) by the $2n$-dimensional representation

$$(15.37') \quad |b_-'\rangle = |b\rangle \times \begin{pmatrix} 1 \\ 0 \end{pmatrix}; \qquad |b_-''\rangle = |b\rangle^T \times \begin{pmatrix} 0 \\ i \end{pmatrix}; \qquad |b\rangle^T = TK|b\rangle$$

where n is the dimension of the space spanned by the eigenvectors $|b\rangle$ of B, provided in reversality space, spanned by the vectors $\begin{pmatrix} 1 \\ 0 \end{pmatrix}$ and $\begin{pmatrix} 0 \\ i \end{pmatrix}$, the operator Θ acquires an additional 2×2 matrix and is thus altogether of the form

$$(15.38) \quad \Theta = T_- K; \qquad T_- = T \times \begin{pmatrix} 0 & -1 \\ 1 & 0 \end{pmatrix}$$

where T represents the operation $t \to -t$ applied to the components of $|b\rangle$ only. Indeed,

$$i\Theta|b_-'\rangle = i|b\rangle^T \begin{pmatrix} 0 & -1 \\ 1 & 0 \end{pmatrix}\begin{pmatrix} 1 \\ 0 \end{pmatrix} = i|b\rangle^T \begin{pmatrix} 0 \\ 1 \end{pmatrix} = |b_-''\rangle$$

(15.37)

$$-i\Theta|b_-''\rangle = -i|b\rangle\begin{pmatrix} 0 & -1 \\ 1 & 0 \end{pmatrix}\begin{pmatrix} 0 \\ -i \end{pmatrix} = -i|b\rangle\begin{pmatrix} i \\ 0 \end{pmatrix} = |b_-'\rangle.$$

Operators that are even or odd under reversal of motion also acquire in reversality space an additional 2×2 matrix, which for hermitean operators must be taken as the unit matrix,

$$(15.39) \quad A(\text{even}) = A \times \begin{pmatrix} 1 & 0 \\ 0 & 1 \end{pmatrix}; \qquad B(\text{odd}) = B \times \begin{pmatrix} 1 & 0 \\ 0 & 1 \end{pmatrix}$$

because then

(15.40)

$$\Theta A(\text{even}) - A(\text{even})\Theta = \begin{pmatrix} 0 & -(\Theta A - A\Theta) \\ \Theta A - A\Theta & 0 \end{pmatrix}$$

$$= 0 \qquad \text{provided} \quad \Theta A - A\Theta = 0$$

and

$$(15.41) \quad \Theta B(\text{odd}) + B(\text{odd}) \, \Theta = \begin{pmatrix} 0 & \Theta B + B\Theta \\ -(B\Theta + \Theta B) & 0 \end{pmatrix}$$

$$= 0 \qquad \text{provided} \quad \Theta B + B\Theta = 0.$$

For the purpose of illustration consider the case $j = \tfrac{1}{2}$ in which one has
(15.42)

$$T(j = \tfrac{1}{2}) = e^{i\pi J_2} = I \cos(\pi/2) + 2iJ_2 \sin(\pi/2) = 2iJ_2 = \begin{pmatrix} 0 & 1 \\ -1 & 0 \end{pmatrix}.$$

Taking as basis the eigenstates of J_3, namely

$$(15.43) \quad |b_1\rangle = |\tfrac{1}{2}, +\tfrac{1}{2}\rangle = \begin{pmatrix} 1 \\ 0 \end{pmatrix}; \qquad |b_2\rangle = |\tfrac{1}{2}, -\tfrac{1}{2}\rangle = \begin{pmatrix} 0 \\ 1 \end{pmatrix}$$

one has the representations

$$|b'_{1-}\rangle = \begin{pmatrix} 1 \\ 0 \end{pmatrix} \times \begin{pmatrix} 1 \\ 0 \end{pmatrix} = \begin{pmatrix} 1 \\ 0 \\ 0 \\ 0 \end{pmatrix}; \qquad |b'_{2-}\rangle = \begin{pmatrix} 0 \\ 1 \end{pmatrix} \times \begin{pmatrix} 1 \\ 0 \end{pmatrix} = \begin{pmatrix} 0 \\ 0 \\ 1 \\ 0 \end{pmatrix}$$

(15.44)

$$|b''_{1-}\rangle = \begin{pmatrix} 1 \\ 0 \end{pmatrix} \times \begin{pmatrix} 0 \\ i \end{pmatrix} = \begin{pmatrix} 0 \\ i \\ 0 \\ 0 \end{pmatrix}; \qquad |b''_{2-}\rangle = \begin{pmatrix} 0 \\ 1 \end{pmatrix} \times \begin{pmatrix} 0 \\ i \end{pmatrix} = \begin{pmatrix} 0 \\ 0 \\ 0 \\ i \end{pmatrix}$$

$$T_- = \begin{pmatrix} 0 & 1 \\ -1 & 0 \end{pmatrix} \times \begin{pmatrix} 0 & -1 \\ 1 & 0 \end{pmatrix} = \begin{pmatrix} 0 & 0 & 0 & -1 \\ 0 & 0 & 1 & 0 \\ 0 & 1 & 0 & 0 \\ -1 & 0 & 0 & 0 \end{pmatrix}.$$

Transition to a different set of basis vectors $|a_-\rangle$, say, can be carried out by writing the vectors $|a'_{j-}\rangle$ as linear combinations

$$(15.45) \qquad |a'_{j-}\rangle = \sum_k V_{jk} |b'_{k-}\rangle + \sum_k W_{jk} |b''_{k-}\rangle$$

so that the set $|a''_{j-}\rangle$, if it is to satisfy (15.37), is accordingly expanded in the fashion

$$(15.46) \qquad |a''_{j-}\rangle = i\Theta |a'_{j-}\rangle = -\sum_k W^*_{jk} |b'_{k-}\rangle + \sum_k V^*_{jk} |b''_{k-}\rangle.$$

Hence, in reversality space the transformation matrix S connecting the set

$$\begin{pmatrix} |a\rangle \\ |a\rangle^T \end{pmatrix}$$

with the set

$$\begin{pmatrix} |b\rangle \\ |b\rangle^T \end{pmatrix}, \qquad \text{so that} \quad \begin{pmatrix} |a\rangle \\ |a\rangle^T \end{pmatrix} = S \begin{pmatrix} |b\rangle \\ |b\rangle^T \end{pmatrix},$$

has obviously the form

(15.47) $S = \begin{pmatrix} V & W \\ -W* & V* \end{pmatrix}$; V and W n-dimensional matrices.

In order that $|a\rangle$, $|a\rangle^T$ be orthonormal if $|b\rangle$, $|b\rangle^T$ are orthonormal, S must be unitary,

(15.48) $SS^+ = I$ or

$$\begin{pmatrix} V & W \\ -W* & V* \end{pmatrix}\begin{pmatrix} V^+ & -W^{*+} \\ W^+ & V^{*+} \end{pmatrix} = \begin{pmatrix} VV^+ + WW^+ & -V\widetilde{W} + W\widetilde{V} \\ -W*V^+ + V*W^+ & W*W^{*+} + V*V^{*+} \end{pmatrix} = \begin{pmatrix} 1 & 0 \\ 0 & 1 \end{pmatrix}$$

requiring

(15.49) $VV^+ + WW^+ = I$ and $V\widetilde{W} = W\widetilde{V}$.

It is a well-known fact of algebra that if (15.49) is satisfied, S becomes a "symplectic" matrix, which means it leaves $T = \begin{pmatrix} 0 & -1 \\ 1 & 0 \end{pmatrix}$ invariant, in the sense $ST\widetilde{S} = T$. In other words, any other set $|a\rangle$, $|a\rangle^T$ can be obtained from the basis $|b\rangle$, $|b\rangle^T$ by a unitary symplectic transformation.

Finally, since any even hermitean operator $A = A^+$, $\Theta A\Theta^{-1} = A$ satisfies generally, because of $\Theta^2 = -I = -\Theta^{-1}\Theta$ and $\langle a|b\rangle = \langle \Theta b|\Theta a\rangle$,

$$\begin{aligned}(15.50) \qquad \langle b'_-|Ab'_-\rangle &= \langle \Theta Ab'_-|\Theta b'_-\rangle = \langle \Theta A\Theta^{-1}\Theta b'_-|\Theta b'_-\rangle \\ &= \langle A\Theta b'_-|\Theta b'_-\rangle = \langle Ab''_-|b''_-\rangle = \langle b''_-|Ab''_-\rangle\end{aligned}$$

and an odd hermitean operator $B = B^+$, $\Theta B\Theta^{-1} = -B$ satisfies, by the same argument,

(15.51) $\langle b'_-|Bb'_-\rangle = -\langle b''_-|Bb''_-\rangle$

it follows that in a state

(15.52) $|b_{\text{even}-}\rangle = c(|b'_-\rangle + |b''_-\rangle) = c(|b'_-\rangle + i\Theta|b'_-\rangle)$

the expectation value of any odd operator vanishes, and in a state

(15.53) $|b_{\text{odd}-}\rangle = c(|b'_-\rangle - |b''_-\rangle)$

the expectation value of any even operator vanishes.

The fact that, in the case where $\Theta^2 = -I$, a state and its time-reversed analog are always orthogonal—and that all hermitean operators, be they even or odd, are necessarily represented by diagonal matrices in reversality space according to (15.39)—can be summarized by saying *a "superselection rule" holds*: *There are no observables which have matrix elements connecting the orthogonal states* $|b'_-\rangle$ *and* $|b''_-\rangle$.

Despite the impossibility of introducing the concept of "reversality" of a state in analogy to the concept of parity, it is, in principle, possible to introduce the concept of "relative reversality" of a state and its time-reversed analog, provided one can have composite physical objects made up out of an object and its time-reversed analog, so that the state of the composite object is a direct product $|c\rangle = |b\rangle \times |b\rangle^T$. Such a state can be an eigenstate of Θ in the sense that if $\Theta|c\rangle = \tau|c\rangle$, where τ is a number, then the definition of τ is not affected by multiplication of $|b\rangle$ with a phase factor $e^{i\alpha}$ because, in that case, $|b\rangle^T$ has to be multiplied with $e^{-i\alpha}$. Unless one can take serious certain speculations, touched upon in Section 17, which make the muon-neutrino the time-reversed analog of the ordinary neutrino, no such states $|c\rangle$ seem to be realized in nature. However, a similar situation arises with respect to the antiunitary operation of particle conjugation, making it possible, for example, to consider the positronium state as a direct product of a particle state and its particle conjugate state, giving rise to the concept of the "relative conjugality" of electron and positron, exhibited as "conjugality" of positronium, as will be explained in Section 28.

It is interesting to speculate whether there is any physical attribute which removes the basic degeneracy of the states $|b_-\rangle$ with respect to all other attributes B. Such an attribute might be the fermion number F, which has the value $+1$ for any fermion and the value -1 for any anti-fermion, is additive, and is, by experimental evidence, strictly conserved in all known interactions. Thus if $|b'_-\rangle$ represents a fermion state satisfying

(15.54)
$$F|b'_-\rangle = +|b'_-\rangle$$

then $|b''_-\rangle$ would represent an antifermion state, satisfying

(15.55)
$$F|b''_-\rangle = -|b''_-\rangle.$$

The operator F would act in the space spanned by the vectors $|b\rangle$ as identity and thus have in reversality space the representation

(15.56)
$$F = I \times \begin{pmatrix} 1 & 0 \\ 0 & -1 \end{pmatrix}$$

which guarantees that Eqs. (15.54) and (15.55) hold, and makes F an operator odd under time reversal,

(15.57)
$$FT_- + T_- F = I \times \left[\begin{pmatrix} 1 & 0 \\ 0 & -1 \end{pmatrix} \begin{pmatrix} 0 & -1 \\ 1 & 0 \end{pmatrix} + \begin{pmatrix} 0 & -1 \\ 1 & 0 \end{pmatrix} \begin{pmatrix} 1 & 0 \\ 0 & -1 \end{pmatrix} \right]$$

$$= I \times \left[\begin{pmatrix} 0 & -1 \\ -1 & 0 \end{pmatrix} + \begin{pmatrix} 0 & 1 \\ 1 & 0 \end{pmatrix} \right] = 0.$$

Thus, if $|b'_-\rangle$ represents an electron characterized by the quantum numbers $(\mathbf{k}, j = \frac{1}{2}, m = +\frac{1}{2})$, the time reversed state $|b''_-\rangle$ describes a positron having the quantum numbers $(-\mathbf{k}, j = \frac{1}{2}, m = -\frac{1}{2})$. This is precisely the suggestion made by Feynman, who proposed to describe positrons as electrons "running backwards in time." If this point of view is adopted, then the strict conservation of F becomes a consequence of the super-selection rule which does not permit the transformation of a fermion state into an antifermion state and *vice versa*.

Unfortunately, this point of view, though very attractive, is not more than a tentative hypothesis based on rather tenuous ground, because it is not at all certain that the degeneracy of states belonging to the quantum number $\Theta^2 = -1$ is always the origin of the superselection rule leading to conservation of fermion number. In fact, there is yet another symmetry operation of apparent universal validity, namely the combined inversion

$$\Sigma = \Pi \Gamma \tag{15.58}$$

made up out of the unitary operation of coordinate inversion Π and the operator Γ which represents the conversion of particles into antiparticles and *vice versa*. In the case of fermions, Σ is antiunitary, as will be shown in Section 19, and therefore gives rise to yet another dichotomic quantum number, namely the eigenvalue of Σ^2 which may be $+1$ or -1, as in the case of the operation Θ^2. Thus, if a fermion belongs to the quantum number $\Sigma^2 = -1$, yet another superselection rule would separate the halfspaces $|b\rangle$ and $\Sigma|b\rangle$, and could be made responsible for the strict conservation of yet another attribute which removes the degeneracy between the state and its combined inverted state. Altogether it should be possible to classify elementary particles into "types" which are labeled according to the values of $\epsilon_\Theta = \Theta^2$ and $\epsilon_\Sigma = \Sigma^2$ as $|b_{\epsilon_\Sigma, \epsilon_\Theta}\rangle$, and of which there must then exist four types, namely $|b_{++}\rangle$, $|b_{+-}\rangle$, $|b_{-+}\rangle$, $|b_{--}\rangle$.

Experimentally, there is strong evidence that fermion number is separately conserved for baryons and for two kinds of leptons. One may speak of a law of conservation of baryon number B, of a law of conservation of lepton number L, and of a law of conservation of muon number L_μ.

There are some reasons to believe that the conservation of the lepton number L, associated with the ordinary neutrinos ($L = +1$) and anti-neutrinos ($L = -1$) emitted in various β decays of nuclei, is a consequence of the superselection rule arising from the quantum number $\epsilon_\Sigma = -1$ attributed to neutrino and antineutrino (as well as to electron and positron), because the antineutrino state $|\bar\nu\rangle$ can be obtained from the neutrino state $|\nu\rangle$ by combined inversion, $|\bar\nu\rangle = \Sigma|\nu\rangle$, and similarly the positron state $|e^+\rangle$ should be considered as obtained from the electron state $|e^-\rangle$ by combined inversion, as will be shown in Sections 17 and 19. Muon number L_μ and baryon number B, on the other hand, can be consistently associated with the superselection rule flowing from time-reversal symmetry, as will be shown in Sections 17 and 29.

In order that an attempt at classification of elementary particles into "types" can be made in proper detail, it is found convenient to develop first a formalism which allows the description of states with a variable number of particles present.

NOTES

Kramers [1] first noticed a peculiar twofold degeneracy of states describing an odd number of electrons in absence of external magnetic fields, which was recognized by Wigner [2], as a consequence of time-reversal symmetry in case of systems of half-odd integer angular momentum. This paper also marks the advent of anti-unitary symmetry operations in quantum mechanics.

Zocher and Török [3] have given a review of the on first sight inconspicuous consequences of time-reversal symmetry in classical physics.

Wigner [4] has developed the formalism of anti-unitary operators in full generality, and has also adduced [5] experimental criteria which permit to decide, in principle, whether a symmetry operation must be represented by a unitary or an anti-unitary operator in quantum mechanics.

Feynman [6] has promulgated the view to look upon positrons as electrons "running backward in time."

Wick *et al.* [7] have coined the term "superselection rule" for situations in which there are neither spontaneous transitions between states belonging to two subspaces, nor measurable quantities with finite matrix elements between these states.

Wigner [8] has pointed out that one might be able to salvage, should the need arise, symmetry with respect to reversal of motion by introduction of the concept of metamatter, distinct from antimatter.

REFERENCES

[1] H. A. Kramers, *Proc. Acad. Sci. Amsterdam* **A33**, 959 (1930).
[2] E. Wigner, *Nachr. Akad. Wiss. Goettingen Math.-Phys. Kl.* p. 546 (1932).
[3] H. Zocher and C. Török, *Proc. Natl. Acad. Sci. U.S.* **39**, 681 (1953).
[4] E. Wigner, *J. Math. Phys.* **1**, 409 (1960).
[5] E. Wigner, *J. Math. Phys.* **1**, 414 (1960).
[6] R. P. Feynman, "Theory of Fundamental Processes," p. 26 (with further references at the end of the book). Benjamin, New York, 1961.
[7] G. C. Wick, A. S. Wightman, and E. Wigner, *Phys. Rev.* **88**, 101 (1952).
[8] E. P. Wigner, *Rev. Mod. Phys.* **29**, 255 (1957).

The Particle Concept in Quantum Mechanics

The formalism of quantum mechanics lends itself naturally to an adaptation toward the description of "particles," which are thought of as carriers of various observable attributes, to be elaborated in the following Sections 17 and 18.

It should be understood that the particle concept in quantum mechanics is an abstraction rather far removed from the naive particle concept of classical physics, in which a particle may be visualized, with impunity, as a kind of small ball, capable of travelling as a sort of coherent body along a specifiable path in space and time. The very incompatibility of observable position and momentum has led in quantum mechanics to an erosion of such naive pictures, and the word "particle" stands for a quantum mechanical state characterized by a set of quantum numbers which are associated, in principle, with an identifiable event such as the momentum transfer in a "collision," or with a sequence of events such as the vapor trail in a cloud chamber.

This gain in abstraction, purchased with the loss of a naively satisfying picture, has brought under the domain of the quantum mechanical particle concept phenomena which earlier were thought of as belonging to the classical field concept. Thus, large sections of the field dynamics of gases, fluids, and solids have been grasped quantum mechanically through introduction of the concept of the exciton, the phonon, the roton, etc., and it has become possible to describe many properties of macroscopic bodies with the same formalism that was introduced originally for the purpose of describing the so-called elementary particles, the fermions and the bosons.

A remnant of the classical distinction between "actual" particles such as electrons, and "actual" fields such as the velocity field in a fluid, is the quantum mechanical distinction between particles and quasi particles. There is, at present, still a conceptual division between the "vacuum" from which the various elementary particles are thought to arise, and the "quasi vacuum" from which the various quasi particles or excitons of solid, liquid, or gas state may arise.

By generally accepted usage, the elementary particle vacuum is imagined as a completely empty state. The quasi vacuum, in contrast, is always thought of as some kind of full state, either as the ground state of a lattice made up out of actually present atoms, or as the electronic Fermi sea without holes or gaps, etc. However, *the formalism*, to be developed later on, *does not reflect this distinction between vacuum and quasi vacuum, particles and quasi particles*. Both vacuum and quasi vacuum can always be represented by a state vector denoted $|0\rangle$, indicating a state completely empty of particles and quasi particles, respectively.

In this connection it seems worth recalling the curious historical fact that, at a certain stage in the development of the theory of elementary particles, Dirac found it convenient to introduce as the vacuum state for the electron-positron particle system a far from empty state, namely the state in which all negative energy levels were filled without holes or gaps, and which thus resembled what today would be called a quasi-vacuum state. That Dirac could do so without getting involved in serious inconsistencies is rather remarkable, and the formal reasons for Dirac's alternative will be taken up in some detail in Section 17.

In any case, the distinction between particles and quasi particles in recent years has shown a tendency to become blurred, and as a result there has been a vigorous cross-fertilization of elementary particle theory and theories of the solid, liquid, and gas states of matter.

It is, of course, not intended here to deny by these remarks the usefulness of the description of macroscopic bodies in terms of atoms making up, say, a solid lattice. After all, lattice atoms are rather manifest in experiments involving relatively high energies, such as in X-ray diffraction patterns produced by solids, so that in this experimental sense one may say lattice atoms do, in fact, exist.

Nevertheless, it seems by no means established that one cannot, in principle, account for *all* observations made on macroscopic bodies in terms of quasi particles. Should such a program turn out to be realizable, one can logically conceive a situation in which it might become feasible and even profitable to abandon the concept of atoms entirely and describe matter entirely in terms of what are now called quasi particles.

In this vein, one can speculate about the possibility of developing a theory of elementary particles in terms of an underlying substratum or *Urmaterie*, which might play for the elementary particle vacuum the same role as that played by the atoms of macroscopic bodies for the quasi vacuum, which is a substratum from which the quasi particles arise. Dirac's original notion regarding the electron-positron vacuum may conceivably have been abandoned too rashly.

Returning now to the established knowledge about particles and quasi particles, there is a categorical statement one can make about the occupation number of fermion or boson states which cannot be made, in the same categorical fashion, about quasi particles, as follows.

Suppose one knows what constitutes a complete set of quantum numbers τ for a physical object, so that specific values of τ characterize the object completely. For example, in case of an elementary particle describable in terms of momentum **k**, spin **s**, and lepton number L, τ may stand for the six quantum numbers $(\mathbf{k}, j, m, L)$. Since the notion of particle implies the existence of objects which can, in principle, be counted, one may infer the existence of occupation states $|0_\tau\rangle$, $|1_\tau\rangle$, $|2_\tau\rangle$, ... which are eigenstates of a suitably chosen operator representing the observable number of particles having property τ, so that these states describe situations in which, respectively, 0, 1, 2, ... particles of property τ are present.

For reasons only partly understood to date, all particles with half-odd integer spin, called fermions, satisfy the

Exclusion Principle: The number of fermions N_τ in a given, *complete*, quantum state τ is restricted to either 0 or 1.

The number n_τ of particles with integer spin, called bosons, in a given quantum state τ is apparently unrestricted, except that n_τ must be $\geqslant 0$.

No categorical statement of this kind can be made for quasi particles. Restrictions on the occupation number of quasi-particle states, if any, will have to be stated separately for each case.

NOTES

Pauli [1] discovered the exclusion principle. Readers interested in the history of the profound contributions to quantum mechanics made by Pauli will find abundant food for thought in Fierz and Weisskopf [2].

The connection between validity of the exclusion principle and spin, which is difficult to understand, has been the subject of work by Pauli [3] and Lüders and Zumino [4].

Dirac [5] first conceived of the vacuum as a state which need not be empty.

A more recent attempt to fashion a description of elementary particles after a quasi-particle model is contained in the work by Nambu and Jona-Lasinio [6]. An earlier, more naive, attempt by Kaempffer [7] turned out to be abortive.

REFERENCES

[1] W. Pauli, *Z. Physik* **31**, 765 (1925).
[2] M. Fierz and V. F. Weisskopf, (ed.), "Theoretical Physics in the Twentieth Century." Wiley (Interscience), New York, 1960.
[3] W. Pauli, *Phys. Rev.* **58**, 716 (1940).
[4] G. Lüders and B. Zumino, *Phys. Rev.* **110**, 1450 (1958).
[5] P. A. M. Dirac, *Proc. Roy. Soc.* **A126**, 360 (1930).
[6] Y. Nambu and G. Jona-Lasinio, *Phys. Rev.* **122**, 345 (1961); **124**, 246 (1961).
[7] F. A. Kaempffer, *Can. J. Phys.* **31**, 165 (1953); **32**, 259 (1954).

Fermion States

The exclusion principle implies that the number of fermions N_τ in a given complete quantum state τ is a dichotomic variable. The only existing occupation states are $|0_\tau\rangle$ and $|1_\tau\rangle$, which may be represented in a two-dimensional abstract space as

$$(17.1) \qquad |0_\tau\rangle = \begin{pmatrix} 1 \\ 0 \end{pmatrix}_\tau; \qquad |1_\tau\rangle = \begin{pmatrix} 0 \\ 1 \end{pmatrix}_\tau.$$

These state vectors form an orthonormal set. A suitable operator representing the observable fermion number $N(\tau)$ is obviously, in this space,

$$(17.2) \qquad N(\tau) = \begin{pmatrix} 0 & 0 \\ 0 & 1 \end{pmatrix}_\tau,$$

since it has the required property

$$(17.3) \qquad N(\tau)|0_\tau\rangle = 0_\tau|0_\tau\rangle; \qquad N(\tau)|1_\tau\rangle = 1_\tau|1_\tau\rangle.$$

A particularly useful concept is that of the annihilation operator $a(\tau)$ and the creation operator $a^+(\tau)$ which connect the states $|0_\tau\rangle$ and $|1_\tau\rangle$ according to

$$(17.4) \qquad a(\tau)|1_\tau\rangle = c_\tau|0_\tau\rangle; \qquad a^+(\tau)|0_\tau\rangle = c_\tau^*|1_\tau\rangle$$

where c_τ is some complex number. Consistency with the exclusion principle requires further the relations

$$(17.5) \qquad a(\tau)|0_\tau\rangle = 0; \qquad a^+(\tau)|1_\tau\rangle = 0.$$

A suitable set of such operators is

$$(17.6) \qquad a(\tau) = c_\tau \begin{pmatrix} 0 & 1 \\ 0 & 0 \end{pmatrix}_\tau; \qquad a^+(\tau) = c_\tau^* \begin{pmatrix} 0 & 0 \\ 1 & 0 \end{pmatrix}_\tau$$

which justifies the notation adopted in (17.4) making $a^+(\tau)$ the hermitean conjugate of $a(\tau)$. The requirement that successive application of one creation operator and one annihilation operator in either order should restore the original state subjects the factor c_τ to the condition

(17.7) $$|c_\tau|^2 = 1$$

so that indeed

(17.8)

$$a^+(\tau)\,a(\tau)|1_\tau\rangle = c_\tau a^+(\tau)|0_\tau\rangle = |c_\tau|^2|1_\tau\rangle = |1_\tau\rangle;   \text{etc.}$$

With this choice of c_τ one finds the operators $a(\tau)$ and $a^+(\tau)$ connected with $N(\tau)$ by the equations

$$a^+(\tau)\,a(\tau) = |c_\tau|^2\begin{pmatrix} 0 & 0 \\ 0 & 1 \end{pmatrix}_\tau = N(\tau);$$

(17.9)

$$a(\tau)\,a^+(\tau) = |c_\tau|^2\begin{pmatrix} 1 & 0 \\ 0 & 1 \end{pmatrix}_\tau = I(\tau) - N(\tau)$$

where $I(\tau)$ stands for the identity operator

(17.10) $$I(\tau) = \begin{pmatrix} 1 & 0 \\ 0 & 1 \end{pmatrix}_\tau.$$

The two equations (17.9) added yield the anticommutation relation

(17.11) $$a(\tau)\,a^+(\tau) + a^+(\tau)\,a(\tau) = I(\tau).$$

The description can now be extended to envelop all quantum states $\tau_1, \tau_2, \ldots, \tau_m, \ldots$ accessible to the fermions under consideration. The occupation state of a fermion system is defined in the product space spanned by the occupation states for the various quantum states as

(17.12)

$$|N_{\tau_1}, N_{\tau_2}, \ldots, N_{\tau_m}, \ldots\rangle = |N_{\tau_1}\rangle \times |N_{\tau_2}\rangle \times \ldots \times |N_{\tau_m}\rangle \times \ldots$$

For convenience it will now be agreed to write the occupation numbers in a certain order, corresponding to a certain sequence of labeling the quantum states τ_m. Thus τ_m will always be written to the left of τ_n if $m < n$. Any physically observable effect must, of course, be unaffected by any change in convention.

An important special occupation state is the vacuum state defined by

(17.13) $$|0\rangle = |0_{\tau_1}, 0_{\tau_2}, \ldots, 0_{\tau_m}, \ldots\rangle$$

which describes, thus, a completely empty state. To this state there exists, at least for fermions, an interesting counterpart, namely the full state defined by

(17.14) $$|1\rangle = |1_{\tau_1}, 1_{\tau_2}, \ldots, 1_{\tau_m}, \ldots\rangle.$$

It should now be possible to construct creation operators $a^+(\tau_m)$ which raise fermions from vacuum into the various quantum states τ_m,

(17.15) $$a^+(\tau_m)|\ldots 0_{\tau_m}\ldots\rangle = c^*_{\tau_m}|\ldots 1_{\tau_m}\ldots\rangle$$

and annihilation operators $a(\tau_m)$ which remove fermions from the full state of the various quantum numbers τ_m,

(17.16) $$a(\tau_m)|\ldots 1_{\tau_m}\ldots\rangle = c_{\tau_m}|\ldots 0_{\tau_m}\ldots\rangle.$$

In this fashion any occupation state (17.12) can either be raised from vacuum by successive application of the various creation operators or be obtained from the full state by successive application of the appropriate annihilation operators.

By suitable choice of the factors c_{τ_m} one can represent, in the product space (17.12), the operators $a^+(\tau_m)$ and $a(\tau_m)$ by matrices satisfying the anti-C.R.s

(17.17)

$$\{a(\tau_m)\,a(\tau_{m'})\} = \{a^+(\tau_m)\,a^+(\tau_{m'})\} = 0; \qquad \{a(\tau_m)\,a^+(\tau_{m'})\} = I(\tau_m)\,\delta_{mm'}$$

where $\{ab\} = ab + ba$, and which contain relation (17.11) as a special case. This is accomplished by putting

(17.18) $$c_{\tau_m} = \prod_{n=1}^{m-1}(1 - 2N_{\tau_n})$$

so that condition (17.7) is again satisfied, c_{τ_m} being $+1$ or -1 depending on whether the number of occupied states $N_{\tau_m} = 1$ with $n < m$ is even or odd, because the occupied states each contribute the factor -1 to (17.18) whereas the empty states each contribute the factor $+1$. By noting the identity

(17.19) $$I(\tau_n) - 2N(\tau_n) = \begin{pmatrix} 1 & 0 \\ 0 & -1 \end{pmatrix}_{\tau_n} = I_3(\tau_n) \qquad \text{(say)}$$

one can write down immediately the matrix representation for $a(\tau_m)$ and $a^+(\tau_m)$,

(17.20)

$$a(\tau_m) = \begin{pmatrix} 1 & 0 \\ 0 & -1 \end{pmatrix}_{\tau_1} \times \ldots \times \begin{pmatrix} 1 & 0 \\ 0 & -1 \end{pmatrix}_{\tau_{m-1}} \times \begin{pmatrix} 0 & 1 \\ 0 & 0 \end{pmatrix}_{\tau_m} \times \begin{pmatrix} 1 & 0 \\ 0 & 1 \end{pmatrix}_{\tau_{m+1}} \times \ldots$$

(17.21) $$a^+(\tau_m) = \prod_{n=1}^{m-1} I_3(\tau_n) \times \begin{pmatrix} 0 & 0 \\ 1 & 0 \end{pmatrix}_{\tau_m} \times \prod_{n=m+1}^{\infty} I(\tau_n)$$

the understanding being that a matrix labeled τ_n operates solely on the subspace belonging to the quantum state τ_n in the general product space (17.12).

The representation used here has the interesting property of being invariant under a unitary transformation

(17.22)

$$U = U^+ = \begin{pmatrix} 0 & 1 \\ 1 & 0 \end{pmatrix}_{\tau_1} \times \dots \times \begin{pmatrix} 0 & 1 \\ 1 & 0 \end{pmatrix}_{\tau_n} \times \dots = \prod_{n=1}^{\infty} I_1(\tau_n) \qquad \text{(say)}$$

which, in effect, interchanges the roles played by occupied quantum states and empty quantum states or "holes." Indeed, for any n,

(17.23)

$$N'(\tau) = UN(\tau) U^+ = \begin{pmatrix} 0 & 1 \\ 1 & 0 \end{pmatrix}\begin{pmatrix} 0 & 0 \\ 0 & 1 \end{pmatrix}\begin{pmatrix} 0 & 1 \\ 1 & 0 \end{pmatrix} = \begin{pmatrix} 1 & 0 \\ 0 & 0 \end{pmatrix} = I(\tau) - N(\tau)$$

so that

(17.24) $$U|0\rangle = |1\rangle ; \qquad U|1\rangle = |0\rangle$$

and

(17.25)

$$a'(\tau_m) = Ua(\tau_m) U^+ = (-1)^{m-1} a^+(\tau_m); \qquad a^{+'}(\tau_m) = (-1)^{m-1} a(\tau_m).$$

Equations (17.25) can be written explicitly

(17.26)

$$a'(\tau_m) = c'_{\tau_m} \begin{pmatrix} 0 & 0 \\ 1 & 0 \end{pmatrix}_{\tau_m} ; \qquad a^{+'}(\tau_m) = c'^*_{\tau_m} \begin{pmatrix} 0 & 1 \\ 0 & 0 \end{pmatrix}_{\tau_m}$$

with

(17.27)

$$c'_{\tau_m} = \prod_{n=1}^{m-1} (1 - 2N'_{\tau_n}) = \prod_{n=1}^{m-1} (2N_{\tau_n} - 1) = (-1)^{m-1} c_{\tau_m}.$$

It is this peculiar symmetry between occupied states raised from vacuum and corresponding holes in the full state which enabled Dirac to consider positrons as holes in a filled "sea" of negative energy electrons.

For application to the theory of quasi particles it is often convenient to perform a unitary transformation $U(\tau_m)$ in which, for some of the quantum states only, namely those belonging to τ_n with $n < m$, the roles of occupied and empty states are interchanged. The corresponding operator is then

(17.28)

$$U(\tau_m) = U^+(\tau_m) = \begin{pmatrix} 0 & 1 \\ 1 & 0 \end{pmatrix}_{\tau_1} \times \dots \times \begin{pmatrix} 0 & 1 \\ 1 & 0 \end{pmatrix}_{\tau_{m-1}} \times \begin{pmatrix} 1 & 0 \\ 0 & 1 \end{pmatrix}_{\tau_m} \times \dots$$

As a first example, consider a fermion in a simultaneous eigenstate of momentum $\mathbf{k}$ and spin quantum number $s = +1$ or $s = -1$, denoted $|\mathbf{k}, s\rangle$, which is obtained from vacuum by application of the creation operator $a^+(\mathbf{k}, s)$, so that, including the two-dimensional occupation space, the total state vector of such a single fermion state reads explicitly

(17.29) $\qquad |1_{\mathbf{k},s}\rangle = a^+(\mathbf{k},s)|0\rangle = |\mathbf{k},s\rangle \times \begin{pmatrix} 0 \\ 1 \end{pmatrix}.$

The simplest particles that can be described completely by their momentum $\mathbf{k}$ and spin s are the neutrino and the antineutrino, which possess no mass so that energy Ω and momentum $\mathbf{k}$ satisfy

(17.30) $\qquad \Omega^2 - \mathbf{k}^2 = 0$

or, *if one requires energy to be a positive number,*

(17.31) $\qquad \Omega = |\mathbf{k}|.$

The neutrinos and antineutrinos arising in the β decay of nuclei differ experimentally in that the spin of the neutrino is always aligned parallel with a direction opposite to its momentum (spin and momentum are "antiparallel"): the neutrino is left-handed,

(17.32) $\qquad (\mathbf{s}\cdot\mathbf{k}) = -k = -|\mathbf{k}|,$ for neutrino,

whereas the antineutrino is right-handed, in the sense

(17.33) $\qquad (\mathbf{s}\cdot\mathbf{k}) = +|\mathbf{k}|,$ for antineutrino.

Equation (17.31) can therefore be written

(17.34) $\qquad \Omega = \begin{cases} -(\mathbf{s}\cdot\mathbf{k}) & \text{for neutrino} \\ +(\mathbf{s}\cdot\mathbf{k}) & \text{for antineutrino} \end{cases}$

Accordingly, suitable Hamiltonians for the description of ν and $\bar{\nu}$ should be

(17.35) $\qquad H_\nu = -(\boldsymbol{\sigma}\cdot\mathbf{P}); \qquad H_{\bar{\nu}} = +(\boldsymbol{\sigma}\cdot\mathbf{P}),$

with the understanding that, upon resolution into components in spin space,

(17.36)

$$H|\mathbf{k},s\rangle = \mp(\boldsymbol{\sigma}\cdot\mathbf{P})|\mathbf{k},s\rangle = |\mathbf{k}|\,|\mathbf{k},s\rangle \quad \text{means}$$

$$\mp \begin{pmatrix} P_3 & P_1 - iP_2 \\ P_1 + iP_2 & -P_3 \end{pmatrix} \begin{pmatrix} |\mathbf{k},s\rangle_1 \\ |\mathbf{k},s\rangle_2 \end{pmatrix} = |\mathbf{k}| \begin{pmatrix} |\mathbf{k},s\rangle_1 \\ |\mathbf{k},s\rangle_2 \end{pmatrix}$$

In coordinate representation, the eigenstate $|\mathbf{k},s\rangle$ can be characterized by two-component ψ functions [see Eq. (9.30)]

(17.37) $\qquad \langle \mathbf{q}(t)|\mathbf{k},s\rangle = \Psi_{\mathbf{k},s}(\mathbf{q},t) = e^{+i\Omega t}\psi_{\mathbf{k},s}(\mathbf{q})$

so that

$$(17.38) \qquad\qquad |\mathbf{k}, s\rangle = \int |\mathbf{q}(t)\rangle \, \Psi_{\mathbf{k}, s}(\mathbf{q}, t) \, d\mathbf{q}.$$

By the same reasoning as the one employed in the derivation of (9.30) from (9.20), these ψ functions must satisfy, if one uses $\mathbf{P} = -i(\partial/\partial\mathbf{Q})$ and $H = +i(\partial/\partial t)$ and denotes the ψ functions for neutrino and antineutrino by ϕ and π, respectively,

(17.39)

$$H_\nu \phi(\mathbf{q}) = +i[\boldsymbol{\sigma}(\partial/\partial\mathbf{q})]\,\phi(\mathbf{q})$$

$$= +i \begin{pmatrix} (\partial/\partial q_3) & (\partial/\partial q_1) - i(\partial/\partial q_2) \\ (\partial/\partial q_1) + i(\partial/\partial q_2) & -(\partial/\partial q_3) \end{pmatrix} \begin{pmatrix} \phi_1 \\ \phi_2 \end{pmatrix} = -|\mathbf{k}| \begin{pmatrix} \phi_1 \\ \phi_2 \end{pmatrix}$$

(17.40)

$$H_{\bar\nu} \pi(\mathbf{q}) = -i[\boldsymbol{\sigma}(\partial/\partial\mathbf{q})]\,\pi(\mathbf{q})$$

$$= -i \begin{pmatrix} (\partial/\partial q_3) & (\partial/\partial q_1) - i(\partial/\partial q_2) \\ (\partial/\partial q_1) + i(\partial/\partial q_2) & -(\partial/\partial q_3) \end{pmatrix} \begin{pmatrix} \pi_1 \\ \pi_2 \end{pmatrix} = -|\mathbf{k}| \begin{pmatrix} \pi_1 \\ \pi_2 \end{pmatrix}.$$

The solutions of (17.39) can be obtained by writing

$$(17.41) \qquad\qquad \phi_n(\mathbf{q}) = A_n \, e^{-i\mathbf{k}\mathbf{q}}$$

with amplitudes to be determined from the linear equations

$$(17.42) \qquad \begin{aligned} +k_3 A_1 + (k_1 - ik_2) A_2 &= -|k| A_1 \\ +(k_1 + ik_2) A_1 - k_3 A_2 &= -|k| A_2 \end{aligned}$$

and the normalization condition

$$(17.43) \qquad \int_V \phi^*(\mathbf{q})\,\phi(\mathbf{q})\,d\mathbf{q} = (|A_1|^2 + |A_2|^2)\,V = 1.$$

The necessary and sufficient condition for existence of a nontrivial solution of (17.42) is the vanishing of the coefficient determinant

$$(17.44) \qquad \begin{vmatrix} +k_3 + |k| & +(k_1 - ik_2) \\ +(k_1 + ik_2) & -k_3 + |k| \end{vmatrix} = \mathbf{k}^2 - k_1^2 - k_2^2 - k_3^2 = 0$$

which is identically true and affirms that the solutions $\phi(\mathbf{q}_n)$ are indeed of the form (17.41). One obtains thus, with suitable phase convention,

(17.45)

$$A_1 = -\frac{(k_1 - ik_2)}{\sqrt{V}\sqrt{2|k|(|k| + k_3)}} \; ; \qquad A_2 = +\frac{(|k| + k_3)}{\sqrt{V}\sqrt{2|k|(|k| + k_3)}}$$

These amplitudes show dramatically the correlation between directions of spin and momentum of the neutrino if they are written in terms of the polar angles ϑ, φ describing the direction of $\mathbf{k}$, $k_1 = |k| \sin \vartheta \cos \varphi$, $k_2 = |k| \sin \vartheta \sin \varphi$, $k_3 = |k| \cos \vartheta$, so that

(17.46)

$$A_1 = -\frac{\sin \vartheta\, e^{-i\varphi}}{\sqrt{V}\sqrt{2(1 + \cos \vartheta)}} = -\frac{1}{\sqrt{V}} \sin (\vartheta/2)\, e^{-i\varphi};$$

$$A_2 = \frac{1}{\sqrt{V}} \cos (\vartheta/2)$$

because now the function

(17.47) $$\phi(\mathbf{q}) = \frac{1}{\sqrt{V}} \begin{pmatrix} -\sin (\vartheta/2)\, e^{-i\varphi} \\ \cos (\vartheta/2) \end{pmatrix} e^{-i\mathbf{kq}}$$

contains the spin state (2.42) describing a spin -1 in direction ϑ, φ of the momentum.

The solutions of (17.40) are obtained by writing in the same fashion

(17.48) $$\pi_n(\mathbf{q}) = B_n\, e^{-i\mathbf{kq}}$$

yielding, with suitable convention of phase,

(17.49)

$$B_1 = \frac{(|k| + k_3)}{\sqrt{V}\sqrt{2|k|(|k| + k_3)}}; \qquad B_2 = \frac{(k_1 + ik_2)}{\sqrt{V}\sqrt{2|k|(|k| + k_3)}}$$

so that in polar coordinates

(17.50) $$\pi(\mathbf{q}) = \frac{1}{\sqrt{V}} \begin{pmatrix} \cos (\vartheta/2) \\ \sin (\vartheta/2)\, e^{i\varphi} \end{pmatrix} e^{-i\mathbf{kq}}$$

the antineutrino function π contains the spin state (2.41) describing the spin $+1$ in direction ϑ, φ of the momentum, as expected.

Neutrinos and antineutrinos have an attribute, called the lepton number L, which is strictly conserved in all transitions involving creation or annihilation of these and other particles classified as leptons. By convention, one attributes the value $L = +1$ to neutrino ν, electron e^-, and muon μ^-, and the value $L = -1$ to antineutrino $\bar{\nu}$, antielectron (or positron) e^+, and antimuon μ^+. Since no exception to this apparent conservation law is known, one is tempted to interpret it as consequence of some superselection rule associated with the existence of some anti-unitary symmetry operator whose square is $-I$.

One such symmetry operator is the operator of time reversal Θ which according to the development of Section 15 has indeed the property

$\Theta^2 = -I$ for leptons of spin $\frac{1}{2}$ provided they possess no intrinsic attributes other than spin which might invalidate this property. However, this operator is not a possible candidate for generation of the desired superselection rule preventing any neutrino state going over into an antineutrino state and *vice versa*, because the operation of time reversal, changing the sign of *both* spin and momentum, does not change the handedness. The very existence of neutrinos and antineutrinos with definite but opposite handedness forces one, then, to draw a far-reaching conclusion regarding Feynman's suggestion to consider antiparticles as particles "running backwards in time": This suggestion *cannot* be adopted to distinguish formally the neutrinos and antineutrinos emitted in various β decays.

Fortunately, there is an operator which changes the handedness of a particle, namely the operator of coordinate inversion Π, changing the sign of momentum only and allowing it therefore to be represented in spin space by the identity, so that altogether

$$(17.51) \qquad \Pi = I \times \Pi_D$$

where Π_D operates on the dynamical variables such as $\mathbf{P}$, H and represents the operation $\mathbf{q} \to -\mathbf{q}$ in any ψ function. In accord with a remark made at the end of Section 14, Π_D is by convention assumed to be real, so that $\Pi_D^2 = I$.

As it stands, Π does not represent a symmetry operation on either neutrino or antineutrino states, because of the definite handedness of these particles, so that applied to a neutrino state, for example, Π produces a state not realized in nature. One can, however, repair this defect by introducing an operator of "particle conjugation" Γ which, applied to any neutrino state, converts it into the corresponding antineutrino state without affecting the dynamical attributes, so that the operator of "combined inversion" (in the older literature often labeled PC)

$$(17.52) \qquad \Sigma = \Pi\Gamma$$

is now a symmetry operator of the neutrino-antineutrino system. Moreover, *if Γ is assumed to be antiunitary with*

$$(17.53) \qquad \Gamma^2 = -I$$

then one can treat both neutrino and antineutrino so that, for these particles, conservation of lepton number L becomes the consequence of a superselection rule separating the two halves of the inversion space spanned by the states $|\nu\rangle$ and $|\bar{\nu}\rangle = \Sigma|\nu\rangle$.

A representation of Γ may be obtained by considering the merger of the two two-dimensional spaces spanned by the eigenfunctions of

$-(\mathbf{\sigma P})$ and $(\mathbf{\sigma P})$, so that a neutrino-antineutrino system can be described by a four-component ψ function

$$(17.54) \qquad \psi(\mathbf{q}) = \begin{pmatrix} \phi(\mathbf{q}) \\ \pi(\mathbf{q}) \end{pmatrix}$$

In this space, Γ can be represented by

$$(17.55) \qquad \Gamma = \begin{pmatrix} 0 & 1 \\ -1 & 0 \end{pmatrix} K$$

and Π by the identity, together giving for the operator of inversion the representation

$$(17.56) \qquad \Sigma = \begin{pmatrix} 0 & 1 \\ -1 & 0 \end{pmatrix} \Pi_D K.$$

One has then indeed

$$(17.57) \qquad \Sigma^2 = -I.$$

The lepton number L takes in this space the form

$$(17.58) \qquad L = \begin{pmatrix} 1 & 0 \\ 0 & -1 \end{pmatrix}$$

and one can write the expressions (17.35) as a single Hamiltonian

$$(17.59) \qquad H = -L(\mathbf{\sigma P}) = \begin{pmatrix} -(\mathbf{\sigma P}) & 0 \\ 0 & (\mathbf{\sigma P}) \end{pmatrix}$$

satisfying

$$(17.60) \qquad H\psi(\mathbf{q}) = -|\mathbf{k}|\,\psi(\mathbf{q}).$$

These considerations do not prove that the operation of particle conjugation Γ *must* be antiunitary for neutrinos. Such a proof will be given later in Section 19 for electrons and positrons. All that has been shown thus far is the consistency of this assumption with the point of view from which the antineutrino is revealed as a neutrino "seen through a mirror." Furthermore, these considerations do not prove that a particle corresponding to a time-reversed left-handed neutrino cannot exist in nature. All one can say is that such a particle, if it exists, cannot be identified with the right-handed antineutrino emitted in the β decay of nuclei.

There is mounting experimental evidence for the existence of another kind of neutrino, the so-called muon-neutrino ν_μ and its antiparticle $\bar{\nu}_\mu$, having the same handedness as ν and $\bar{\nu}$ respectively, and being emitted in reactions involving muons, for example,

$$\pi^+ \rightarrow \mu^+ + \nu_\mu; \qquad \pi^- \rightarrow \mu^- + \bar{\nu}_\mu;$$
$$\mu^+ \rightarrow e^+ + \bar{\nu}_\mu + \nu; \qquad \mu^- \rightarrow e^- + \nu_\mu + \bar{\nu}.$$

Apparently, there is a second lepton number L_μ (the "muon number"), which is additive and conserved independently of L, accounting for the absence of processes such as

$$\mu^- \nrightarrow e^- + \gamma; \qquad \mu^- + p \nrightarrow n + e^-.$$

If the assignments of lepton numbers laid down in Table 17.1 are adopted, then the transition $\mu^- \rightarrow e^- + \nu_\mu + \bar{\nu}$, for example, will be allowed because the lepton numbers on both sides of the reaction are $L = +1 = +1+1-1$ and $L_\mu = -1 = +1-1-1$.

TABLE 17.1

ASSIGNMENT OF LEPTON NUMBERS TO THE VARIOUS EXISTING LEPTONS

	μ^+	μ^-	e^+	e^-	$\bar{\nu}$	ν	$\bar{\nu}_\mu$	ν_μ
L	-1	$+1$	-1	$+1$	-1	$+1$	-1	$+1$
L_μ	$+1$	-1	-1	$+1$	-1	$+1$	$+1$	-1

It is this muon-neutrino ν_μ which can be accommodated by Feynman's suggestion, if one adjoins the reversality space, so that, for example, after the fashion of (15.37) the muon-neutrino state is obtained by time reversal from the neutrino state, and the antimuon-neutrino state by time reversal from the antineutrino state,

$$(17.61) \qquad |\nu_\mu\rangle = i\Theta|\nu\rangle; \qquad |\nu\rangle = -i\Theta|\nu_\mu\rangle$$

making the muon-neutrino a left-handed particle in accordance with observations carried out on the decay $\pi^+ \rightarrow \mu^+ + \nu_\mu$.

The state $|\bar{\nu}_\mu\rangle$ will again result from application of the operation of combined inversion Σ to the state $|\nu_\mu\rangle$, and conservation of L among the muon-neutrinos will be guaranteed again because of $\Sigma^2 = -I$. The conservation of L_μ, on the other hand, will follow from the superselection rule owing to the property $\Theta^2 = -I$ of the antiunitary time reversal operation Θ.

This discussion raises a number of seemingly perplexing questions regarding the transformation properties of electron, muon, and nucleon states under time reversal Θ and under combined inversion Σ. If the assignment of lepton numbers L and L_μ proposed above is adopted, consistency would require one to consider positrons as *spatial* inverses of electrons, contrary to Feynman's original intention, and only muons could possibly be identified with time-reversed electrons. The already baffling mass difference between muon and electron would, in this case, acquire an additional degree of mystery through the necessary transformation $m_\mu = \Theta m_e \Theta^{-1}$. The existence of only one baryon number assignable to nucleons and antinucleons, on the other hand, would suggest that for nucleons only one antiunitary symmetry operation has the property that its square is equal to $-I$. This turns out to be possible because of intrinsic attributes other than spin shared by baryons. In Section 29 reasons will be given for the suggestion that the conservation of baryon number B is indeed a superselection rule following from invariance under time reversal.

NOTES

Jordan and Wigner [1] invented the representation of creation and annihilation operators for fermions.

Dirac [2] has pointed out the possibility of interchanging the concepts of "occupied" and "unoccupied" fermion states.

Weyl [3] invented the two-component equation which was employed to describe the neutrino by Lee and Yang [4]. This equation had been rejected, prior to the discovery of nonconservation of parity in weak interactions, by Pauli [5], on the grounds that it violated the reflection symmetry of nature. The neutrino had first been proposed as a hypothetical particle by Pauli [6]. The name "neutrino" is apparently due to Fermi.

Konopinski and Mahmoud [7] first suggested the existence of a law of conservation of leptons. (See also Lee and Yang [4].)

For assignment of the second lepton number L_μ see, for example, Horn [8].

REFERENCES

[1] P. Jordan and E. Wigner, *Z. Physik* **47**, 631 (1928).
[2] P. A. M. Dirac, "Quantum Mechanics," 4th ed., §65. Oxford Univ. Press, London and New York, 1958.
[3] H. Weyl, *Z. Physik* **56**, 330 (1929).
[4] T. D. Lee and C. N. Yang, *Phys. Rev.* **105**, 1671 (1957).
[5] W. Pauli, *in* "Handbuch der Physik" (S. Flügge, ed.), Vol. 24, p. 226. Springer, Berlin, 1933.
[6] W. Pauli, Discussion at the American Physical Society meeting, Pasadena, 1931.
[7] E. Konopinski and H. M. Mahmoud, *Phys. Rev.* **92**, 1045 (1953).
[8] D. Horn, *Phys. Letters* **2**, 303 (1962).

Boson States

Since the occupation number of a boson state characterized by the set of complete quantum numbers τ is unlimited, the eigenstates of the boson number $n(\tau)$ will span a denumerably infinite-dimensional space and may be taken to be represented by the orthonormal set

$$(18.1) \qquad |0_\tau\rangle = \begin{pmatrix} 1 \\ 0 \\ 0 \\ \cdot \\ \cdot \end{pmatrix}; \qquad |1_\tau\rangle = \begin{pmatrix} 0 \\ 1 \\ 0 \\ \cdot \\ \cdot \end{pmatrix}; \qquad \ldots;$$

$$|n_\tau\rangle = \begin{pmatrix} 0 \\ \cdot \\ 1 \\ 0 \\ \cdot \end{pmatrix} \left.\begin{matrix} \\ \\ \end{matrix}\right\} n_\tau - 1 \text{ components } 0$$

In this space the operator representing the observable boson number $n(\tau)$ is then diagonal and reads

$$(18.2) \qquad n(\tau) = \begin{pmatrix} 0 & 0 & 0 & \cdots \\ 0 & 1 & 0 & \\ 0 & 0 & 2 & \\ \cdot & & & \cdot \\ \cdot & & & \cdot \\ \cdot & & & \cdot \end{pmatrix}_\tau$$

A suitable set of operators representing, respectively, annihilation and creation of a boson in quantum state τ is

$$(18.3) \qquad b(\tau) = \begin{pmatrix} 0 & \sqrt{1} & 0 & 0 & \\ 0 & 0 & \sqrt{2} & 0 & \\ 0 & 0 & 0 & \sqrt{3} & \\ 0 & 0 & 0 & 0 & \\ & & & & \cdot \\ & & & & \cdot \\ & & & & \cdot \end{pmatrix}_\tau$$

$$b^+(\tau) = \begin{pmatrix} 0 & 0 & 0 & 0 \\ \sqrt{1} & 0 & 0 & 0 \\ 0 & \sqrt{2} & 0 & 0 \\ 0 & 0 & \sqrt{3} & 0 \\ & & & & \ddots \end{pmatrix}_\tau$$

having the property

$$(18.4) \quad b(\tau)\,|n_\tau\rangle = \sqrt{n_\tau}\,|n_\tau-1\rangle; \qquad b^+(\tau)\,|n_\tau\rangle = \sqrt{n_\tau+1}\,|n_\tau+1\rangle.$$

The normalization implied by (18.4) insures that any state with $n_\tau < 0$ is identically zero, and that the operator $n(\tau)$ is connected with $b(\tau)$ and $b^+(\tau)$ by relations analogous to (17.9), namely

$$(18.5) \qquad b^+(\tau)\,b(\tau) = n(\tau); \qquad b(\tau)\,b^+(\tau) = I(\tau)+n(\tau)$$

which may be subtracted to yield the C.R.

$$(18.6) \qquad b(\tau)\,b^+(\tau) - b^+(\tau)\,b(\tau) = I(\tau)$$

The treatment can now be extended to envelop all quantum states $\tau_1, \tau_2, \ldots, \tau_m, \ldots$ accessible to the bosons under consideration. The occupation state of a boson system is defined in the product space spanned by the occupation states for the various quantum states as

$$(18.7) \quad |n_{\tau_1}, n_{\tau_2}, \ldots, n_{\tau_m}, \ldots\rangle = |n_{\tau_1}\rangle \times |n_{\tau_2}\rangle \times \ldots \times |n_{\tau_m}\rangle \times \ldots$$

It is easy to construct creation operators $b^+(\tau_m)$ which raise, from any state, another boson into the quantum state τ_m, thus

$$(18.8) \qquad b^+(\tau_m)\,|\ldots n_{\tau_m}\ldots\rangle = \sqrt{n_{\tau_m}+1}\,|\ldots n_{\tau_m}+1\ldots\rangle$$

and similarly annihilation operators $b(\tau_m)$ having the property

$$(18.9) \qquad b(\tau_m)\,|\ldots n_{\tau_m}\ldots\rangle = \sqrt{n_{\tau_m}}\,|\ldots n_{\tau_m}-1\ldots\rangle.$$

They are simply represented by

$$(18.10) \quad b(\tau_m) = \prod_{n=1}^{m-1} I(\tau_n) \times \begin{pmatrix} 0 & \sqrt{1} & & \\ & & \sqrt{2} & \\ & & & \ddots \\ & & & & \ddots \end{pmatrix}_{\tau_m} \times \prod_{n=m+1}^{\infty} I(\tau_n)$$

$$(18.11) \quad b^+(\tau_m) = \prod_{n=1}^{m-1} I(\tau_n) \times \begin{pmatrix} 0 & & & \\ & \sqrt{1} & & \\ & & \sqrt{2} & \\ & & & \ddots \end{pmatrix}_{\tau_m} \times \prod_{n=m+1}^{\infty} I(\tau_n)$$

which satisfy the C.R.s, as is easily verified,

(18.12)

$$[b(\tau_m) b(\tau_{m'})] = [b^+(\tau_m) b^+(\tau_{m'})] = 0; \qquad [b(\tau_m) b^+(\tau_{m'})] = I(\tau_m) \delta_{mm'}$$

where $[ab] = ab - ba$. Equations (18.12) contain (18.6) as a special case.

A very simple kind of boson is the photon, which, like the neutrino and antineutrino, has no mass and no internal dynamical attributes except momentum $\varkappa$ and an *abstract dichotomic* polarization variable S_3 which will be labeled $S_3 = +1$ for right-handed and $S_3 = -1$ for left-handed circular polarization in the direction of $\varkappa$. The photon differs fundamentally from those fermions, however, because it belongs to the eigenstates $|j, m\rangle$ of angular momentum with integer j. There is no meaningful distinction between a photon and its antiparticle; photons are observed in all linear combinations of right-handed and left-handed polarization.

Experimentally, momentum $\varkappa$ and polarization S_3 of a photon are compatible observables, and there should thus exist a representation in which a right-hand circular polarized photon is raised from vacuum by application of an operator $b^+(\varkappa, R)$

$$(18.13) \qquad |1_{\varkappa, R}\rangle = b^+(\varkappa, R) |0\rangle = |\varkappa, R\rangle \times \begin{pmatrix} 0 \\ 1 \\ 0 \\ \vdots \end{pmatrix}$$

and in which a left-hand circular polarized photon is raised from vacuum by application of $b^+(\varkappa, L)$,

$$(18.14) \qquad |1_{\varkappa, L}\rangle = b^+(\varkappa, L) |0\rangle = |\varkappa, L\rangle \times \begin{pmatrix} 0 \\ 1 \\ 0 \\ \vdots \end{pmatrix}$$

where $|\varkappa, R\rangle$ and $|\varkappa, L\rangle$ are the simultaneous eigenstates of momentum $\mathbf{P}$ and polarization S_3, so that they may be decomposed into

$$(18.15) \quad |\varkappa, R\rangle = |\varkappa, +1\rangle \times |R\rangle; \qquad |\varkappa, L\rangle = |\varkappa, -1\rangle \times |L\rangle$$

where $|R\rangle$ and $|L\rangle$ are the two-dimensional eigenstates of the dichotomic polarization S_3, which, according to the general development of Sections 1 and 2, may be written

$$(18.16) \qquad |R\rangle = \begin{pmatrix} 1 \\ 0 \end{pmatrix}; \qquad |L\rangle = \begin{pmatrix} 0 \\ 1 \end{pmatrix}; \qquad S_3 = \begin{pmatrix} 1 & 0 \\ 0 & -1 \end{pmatrix},$$

and where $|\varkappa, \pm 1\rangle$ are the eigenstates of momentum $\mathbf{P}$ satisfying

$$(18.17) \qquad\qquad \mathbf{P}|\varkappa, \pm 1\rangle = \varkappa|\varkappa, \pm 1\rangle.$$

Since the photon has no mass, the energy ω is, as in the case of the massless fermions,

$$(18.18) \qquad\qquad \omega = |\varkappa|$$

and since its spin $j = 1$ is aligned parallel to $\varkappa$ for $S_3 = +1$ and antiparallel to $\varkappa$ for $S_3 = -1$, the vectors $|\varkappa, \pm 1\rangle$ must be eigenstates of the Hamiltonians

(18.19)

$$H_+ = +(\mathbf{JP}) \qquad \text{for} \quad S_3 = +1 \qquad \text{i.e.} \qquad (\mathbf{JP})|\varkappa, +1\rangle = |\varkappa|\,|\varkappa, +1\rangle$$

(18.20)

$$H_- = -(\mathbf{JP}) \qquad \text{for} \quad S_3 = -1 \qquad \text{i.e.} \quad -(\mathbf{JP})|\varkappa, -1\rangle = |\varkappa|\,|\varkappa, -1\rangle,$$

respectively, where $\mathbf{J}$ are the component operators of angular momentum for $j = 1$, namely (see Appendix 1)

$$(18.21) \quad J_1 = \frac{1}{\sqrt{2}}\begin{pmatrix} 0 & 1 & 0 \\ 1 & 0 & 1 \\ 0 & 1 & 0 \end{pmatrix}; \qquad J_2 = \frac{1}{\sqrt{2}}\begin{pmatrix} 0 & -i & 0 \\ i & 0 & -i \\ 0 & i & 0 \end{pmatrix};$$

$$J_3 = \begin{pmatrix} 1 & 0 & 0 \\ 0 & 0 & 0 \\ 0 & 0 & -1 \end{pmatrix}.$$

In the space spanned by the eigenvectors of S_3 (18.16) one can thus write the Hamiltonian

$$(18.22) \qquad\qquad H = \begin{pmatrix} (\mathbf{JP}) & 0 \\ 0 & -(\mathbf{JP}) \end{pmatrix} = S_3 \times (\mathbf{JP}).$$

Applied to a general photon state of momentum $\varkappa$, which is a linear combination

(18.23)

$$|\varkappa, S\rangle = \alpha(S)|\varkappa, R\rangle + \beta(S)|\varkappa, L\rangle; \qquad |\alpha|^2 + |\beta|^2 = 1,$$

the Hamiltonian then always has the eigenvalue (18.18),

(18.24) $$H|\varkappa, S\rangle = \omega|\varkappa, S\rangle.$$

It is firmly understood that the energy ω is never anything but a positive number.

An eigenstate of momentum **P** can, according to an insight gained in Section 14, never be a state of definite parity. It is intuitively clear that the operation of inversion of coordinates Π will make a photon of momentum $\varkappa$ and polarization S_3 appear to look like a photon of momentum $-\varkappa$ and polarization $-S_3$, the mirror image of a right-handed screw being a left-handed screw, and *vice versa*. In the space (18.16) the operator of inversion is accordingly of the form

(18.25) $$\Pi_S = \begin{pmatrix} 0 & 1 \\ 1 & 0 \end{pmatrix}$$

so that $|R\rangle$ and $|L\rangle$ are converted into each other and S_3 is odd under inversions,

(18.26)
$$\Pi_S|R\rangle = |L\rangle; \qquad \Pi_S|L\rangle = |R\rangle; \qquad \Pi_S S_3 + S_3 \Pi_S = 0.$$

In the total space (18.15) one has then the representation

(18.27) $$\Pi = \Pi_D \times \Pi_S = \begin{pmatrix} 0 & \Pi_D \\ \Pi_D & 0 \end{pmatrix}$$

where Π_D operates on the dynamical variables according to

(18.28) $$\Pi_D \mathbf{J} \Pi_D^{-1} = \mathbf{J}; \qquad \Pi_D \mathbf{P} \Pi_D^{-1} = -\mathbf{P},$$

leaving thus the Hamiltonian (18.22) invariant,

(18.29)

$$\Pi H - H \Pi = \begin{pmatrix} 0 & -[\Pi_D(JP) + (JP)\,\Pi_D] \\ [\Pi_D(JP) + (JP)\,\Pi_D] & 0 \end{pmatrix} = 0,$$

as is required by the inversion symmetry of the photon.

The operation of reversal of motion, on the other hand, will make a photon of momentum $\varkappa$ and polarization S_3 appear to look like a photon of momentum $-\varkappa$ and polarization S_3, because the handedness remains unchanged if both **P** and **J** transform under Θ_D as

(18.30) $$\Theta_D \mathbf{J} \Theta_D^{-1} = -\mathbf{J}; \qquad \Theta_D \mathbf{P} \Theta_D^{-1} = -\mathbf{P}.$$

In polarization space the operator Θ can accordingly be represented by

(18.31) $$\Theta_S = \begin{pmatrix} 1 & 0 \\ 0 & 1 \end{pmatrix}$$

so that altogether

$$(18.32) \qquad \Theta = \Theta_D \times \Theta_S = \begin{pmatrix} \Theta_D & 0 \\ 0 & \Theta_D \end{pmatrix}; \qquad \Theta_D = TK;$$

where T has in the spin space spanned by $\mathbf{J}$, if they are represented by (18.21), according to (15.29) the form

$$(18.33) \qquad T = \begin{pmatrix} 0 & 0 & 1 \\ 0 & -1 & 0 \\ 1 & 0 & 0 \end{pmatrix}.$$

The Hamiltonian (18.22) is evidently invariant under Θ, and the symmetry of the photon under reversal of motion is thus guaranteed. The question of the transformation properties of photon states under particle-antiparticle conjugation cannot be answered except in conjunction with the transformation properties of fermions that are coupled to the electromagnetic field. This will be taken up in Section 28. For the purpose of the present section, it suffices to record that the operator of particle conjugation carried out twice, Γ^2, applied to a photon state is equivalent to the identity, $\Gamma^2 = I$. Since also $\Theta^2 = +I$ for photons, the photon is of "type" $|b_{++}\rangle$, in the sense explained at the end of Section 15, and no superselection rules are involved in the creation and annihilation of photons.

The solution of Eq. (18.24) is most conveniently carried out in momentum representation. Writing the six-component state vector in the combined spin-polarization space as

$$(18.34) \qquad |\varkappa, S\rangle = \begin{pmatrix} \alpha(S)\, \boldsymbol{\chi}^R(\varkappa) \\ \beta(S)\, \boldsymbol{\chi}^L(\varkappa) \end{pmatrix}$$

the two three-component vectors

$$\boldsymbol{\chi}^R = \begin{pmatrix} \chi_1^R \\ \chi_2^R \\ \chi_3^R \end{pmatrix} \qquad \text{and} \qquad \boldsymbol{\chi}^L = \begin{pmatrix} \chi_1^L \\ \chi_2^L \\ \chi_3^L \end{pmatrix}$$

will be subject to the normalization conditions

$$(18.35)$$
$$\boldsymbol{\chi}^{R*}\boldsymbol{\chi}^R = 1; \qquad \boldsymbol{\chi}^{L*}\boldsymbol{\chi}^L = 1 \qquad \text{so that} \qquad \langle \varkappa, S|\varkappa, S\rangle = |\alpha|^2 + |\beta|^2 = 1.$$

Since $\boldsymbol{\chi}^R$ and $\boldsymbol{\chi}^L$ are eigenstates of S_3 with different eigenvalues, they will satisfy the orthogonality relation

$$(18.36)$$
$$\langle \varkappa, R|\varkappa, L\rangle = \boldsymbol{\chi}^{R*}\boldsymbol{\chi}^L = \boldsymbol{\chi}^{L*}\boldsymbol{\chi}^R = \langle \varkappa, L|\varkappa, R\rangle = 0.$$

Equation (18.24) reads now, in components, with the representations (18.21) and (18.22), and $\mathbf{P} = \mathbf{\varkappa}$,

(18.37)

$$\kappa_3 \chi_1^R + (1/\sqrt{2})(\kappa_1 - i\kappa_2)\chi_2^R \qquad\qquad = \omega\chi_1^R$$

$$(1/\sqrt{2})(\kappa_1 + i\kappa_2)\chi_1^R \qquad\qquad\qquad + (1/\sqrt{2})(\kappa_1 - i\kappa_2)\chi_3^R = \omega\chi_2^R$$

$$(1/\sqrt{2})(\kappa_1 + i\kappa_2)\chi_2^R - \kappa_3\chi_3^R \qquad = \omega\chi_3^R$$

and a similar set for χ_1^L, χ_2^L, χ_3^L in which the signs of all terms on the left-hand side are reversed. In order that there exist nontrivial solutions the coefficient determinant must vanish,

(18.38)

$$\begin{vmatrix} \pm\kappa_3 - \omega & \pm(1/\sqrt{2})(\kappa_1 - i\kappa_2) & 0 \\ \pm(1/\sqrt{2})(\kappa_1 + i\kappa_2) & -\omega & \pm(1/\sqrt{2})(\kappa_1 - i\kappa_2) \\ 0 & \pm(1/\sqrt{2})(\kappa_1 + i\kappa_2) & \mp\kappa_3 - \omega \end{vmatrix}$$

$$= \omega(\kappa_1^2 + \kappa_2^2 + \kappa_3^2 - \omega^2) = 0$$

which is obviously satisfied for any $\omega \neq 0$ because of (18.18). One finds without difficulty the solutions

(18.39R)

$$\chi_1^R = -i\frac{(\kappa_1 - i\kappa_2)}{2\omega}\sqrt{\frac{\omega + \kappa_3}{\omega - \kappa_3}}; \qquad \chi_2^R = -i\sqrt{\frac{\omega^2 - \kappa_3^2}{2\omega^2}};$$

$$\chi_3^R = -i\frac{(\kappa_1 + i\kappa_2)}{2\omega}\sqrt{\frac{\omega - \kappa_3}{\omega + \kappa_3}}$$

and

(18.39L)

$$\chi_1^L = -i\frac{(\kappa_1 - i\kappa_2)}{2\omega}\sqrt{\frac{\omega - \kappa_3}{\omega + \kappa_3}}; \qquad \chi_2^L = i\sqrt{\frac{\omega^2 - \kappa_3^2}{2\omega^2}};$$

$$\chi_3^L = -i\frac{(\kappa_1 + i\kappa_2)}{2\omega}\sqrt{\frac{\omega + \kappa_3}{\omega - \kappa_3}}.$$

In terms of polar coordinates ϑ, φ characterizing the direction of propagation $\mathbf{\varkappa}$, so that $\kappa_1 = \omega\sin\vartheta\cos\varphi$, $\kappa_2 = \omega\sin\vartheta\sin\varphi$, $\kappa_3 = \omega\cos\vartheta$, one finds, as expected, that $\mathbf{\chi}^R$ represents a spin state

(18.40)
$$\mathbf{\chi}^R = -\frac{i}{2}\begin{pmatrix} (1 + \cos\vartheta)e^{-i\varphi} \\ \sqrt{2}\sin\vartheta \\ (1 - \cos\vartheta)e^{i\varphi} \end{pmatrix}$$

describing the spin $j = 1$ aligned parallel to $\varkappa$, and χ^L represents a spin state

(18.41)
$$\chi^L = \frac{i}{2}\begin{pmatrix} -(1-\cos\vartheta)\,e^{-i\varphi} \\ \sqrt{2}\sin\vartheta \\ -(1+\cos\vartheta)\,e^{i\varphi} \end{pmatrix}$$

describing the spin $j = 1$ aligned antiparallel to $\varkappa$, because these states are eigenstates of

(18.42) $\mathbf{J}_{\vartheta,\varphi} = J_1\sin\vartheta\cos\varphi + J_2\sin\vartheta\sin\varphi + J_3\cos\vartheta$

$$= \frac{1}{\sqrt{2}}\begin{pmatrix} \sqrt{2}\cos\vartheta & \sin\vartheta\,e^{-i\varphi} & 0 \\ \sin\vartheta\,e^{i\varphi} & 0 & \sin\vartheta\,e^{-i\varphi} \\ 0 & \sin\vartheta\,e^{i\varphi} & -\sqrt{2}\cos\vartheta \end{pmatrix}$$

with eigenvalues $+1$ and -1, respectively, as seen by straightforward computation. The operator $\mathbf{J}_{\vartheta,\varphi}$ has another normalized eigenstate, namely

(18.43)
$$\chi^0 = \frac{1}{\sqrt{2}}\begin{pmatrix} -\sin\vartheta\,e^{-i\varphi} \\ \sqrt{2}\cos\vartheta \\ \sin\vartheta\,e^{i\varphi} \end{pmatrix}$$

belonging to eigenvalue 0. Although this state cannot represent the spin state of an actual photon, since this would require existence of longitudinal polarization states $|\varkappa,0\rangle$ (say), it can be used to express the transversality of actual photons in the compact form

(18.44)
$$\chi^{0*}\chi^R = 0; \qquad \chi^{0*}\chi^L = 0 \qquad \text{so that} \qquad \langle\varkappa,0|\varkappa,S\rangle = 0.$$

Changing the polarization S of a photon state of given momentum $\varkappa$ means changing the amplitudes $\alpha(S)$ and $\beta(S)$ introduced in (18.23). One can abstractly describe such changes in polarization as rotations in the polarization space spanned by the eigenvectors of S_3. The generators of such rotations are, for given $\varkappa$, the operators

(18.45)
$$\begin{aligned} S_1 &= b^+(R)\,b(L) + b^+(L)\,b(R) \\ S_2 &= -ib^+(R)\,b(L) + ib^+(L)\,b(R) \\ S_3 &= b^+(R)\,b(R) - b^+(L)\,b(L) = n(R) - n(L) \\ S_0 &= b^+(R)\,b(R) + b^+(L)\,b(L) = n(R) + n(L) \end{aligned}$$

which satisfy, as a consequence of the C.R.s (18.12) for the b^+ and b, the C.R.s

(18.46)

$$[S_1 S_2] = 2iS_3; \qquad [S_2 S_3] = 2iS_1; \qquad [S_3 S_1] = 2iS_2; \qquad [S_k S_0] = 0$$

$$(k = 1, 2, 3)$$

and also

(18.47)
$$S_1^2 + S_2^2 + S_3^2 = S_0(S_0 + 2).$$

They are thus, according to (2.46), isomorphic to the Pauli matrices (2.43) and may be represented in polarization space by

(18.48)

$$S_1 = \begin{pmatrix} 0 & 1 \\ 1 & 0 \end{pmatrix}; \quad S_2 = \begin{pmatrix} 0 & -i \\ i & 0 \end{pmatrix}; \quad S_3 = \begin{pmatrix} 1 & 0 \\ 0 & -1 \end{pmatrix}; \quad S_0 = \begin{pmatrix} 1 & 0 \\ 0 & 1 \end{pmatrix}.$$

This explains, after the event, the notation S_3 for the polarization operator.

Generalization to include states containing $n_{\varkappa,S}$ photons, each of momentum $\varkappa$ and polarization S, proceeds without difficulty. The most general photon state $|\ldots n_{\varkappa,S}\ldots\rangle$ will have to be a simultaneous eigenstate of the Hamiltonian

(18.49)
$$H = \sum_{\varkappa} \sum_{S} H(\varkappa, S)\, b^+(\varkappa, S)\, b(\varkappa, S)$$

and the polarization operator

(18.50)
$$S_3 = \sum_{\varkappa} S_3(\varkappa) = \sum_{\varkappa} [b^+(\varkappa, R)\, b(\varkappa, R) - b^+(\varkappa, L)\, b(\varkappa, L)]$$

having the eigenvalues

(18.51)
$$W = \sum_{\varkappa} \sum_{S} \omega n_{\varkappa,S}$$

and

(18.52)
$$S_3 = \sum_{\varkappa} (n_{\varkappa, R} - n_{\varkappa, L}).$$

Because of the isomorphism of the polarization operators (18.48) with the operators representing an angular momentum $j = \frac{1}{2}$, the construction of many-photon states from single-photon states in polarization space has to be carried out under the observance of the rules governing the addition of angular momenta, as given in Appendix 2. Specific examples will be treated in Sections 27 and 28.

Instead of using the quantum numbers $\varkappa, S_3$ to describe single photons, as has been done here, one can alternatively use a representation in terms of the quantum numbers ω, j, m, P (energy, angular momentum, parity). The ensuing formalism, although widely employed, is less transparent, and has therefore been relegated to Appendix 3.

At this juncture the correspondence of the quantum-mechanical description of photons with the classical description of the same phenomenon in terms of transverse electromagnetic fields $\mathbf{E}$, $\mathbf{B}$ derivable from a vector potential $\mathbf{A}$ by differentiations,

$$(18.53) \qquad \mathbf{E} = -(\partial \mathbf{A}/\partial t); \qquad \mathbf{B} = \nabla \times \mathbf{A}$$

can be established in a rather straightforward manner, if one stipulates that the energy W should correspond to the value of the integral

$$(18.54) \qquad W = \tfrac{1}{2} \int (\mathbf{E}^2 + \mathbf{B}^2)\, d\mathbf{q}$$

in the sense that the expectation value of H in a given state should be numerically equal to the value of W computed with the classical fields describing the same physical situation.

This correspondence was traced by the originators of quantum electrodynamics in opposite direction, when they were groping for suitable expressions which might serve as operators representing the classical fields of electrodynamics. As it turns out, the fields $\mathbf{E}$ and $\mathbf{B}$, which prove so useful in macroscopic situations involving averages over many photons, are singularly unsuited for grasping elementary processes involving single photons. The purpose of the following considerations is thus not aimed at drawing comfort from establishing contact with classical electrodynamics. The theory of photons, laid out in this section, can stand on its own phenomenological feet. The aim is rather to exhibit the profound differences between classical fields and the quantum mechanical operators that must serve in their place.

The polarization of transverse fields propagating in direction $\varkappa$ can be conveniently specified by introduction of two real orthogonal unit vectors $\boldsymbol{\epsilon}(1)$ and $\boldsymbol{\epsilon}(2)$ which point in the direction of the intersections of the horizon of $\varkappa$ with the equatorial (x, y) plane and the $(z, \varkappa)$ plane, respectively, as indicated in Fig. 18.1. Since the horizon for any zenith intersects the equator at the east-west line,* their components are

$$(18.55) \qquad \epsilon_1(1) = \sin \varphi; \qquad \epsilon_2(1) = -\cos \varphi; \qquad \epsilon_3(1) = 0$$

(18.56)

$$\epsilon_1(2) = \cos \vartheta \cos \varphi; \qquad \epsilon_2(2) = \cos \vartheta \sin \varphi; \qquad \epsilon_3(2) = -\sin \vartheta$$

---

* The author is indebted to Dr. Luis de Sobrino, former Lieutenant in the Spanish Navy, for illumination on this point.

where ϑ and φ are the polar angles describing the direction of the propagation vector according to

(18.57)
$$n_1 = (\kappa_1/\omega) = \sin\vartheta\cos\varphi; \qquad n_2 = (\kappa_2/\omega) = \sin\vartheta\sin\varphi;$$
$$n_3 = (\kappa_3/\omega) = \cos\vartheta.$$

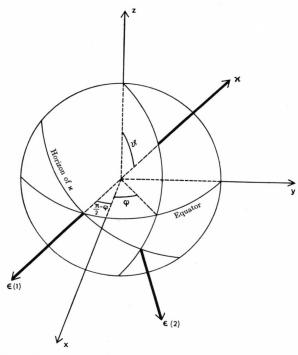

FIG. 18.1. The polarization vectors $\boldsymbol{\epsilon}(1)$ and $\boldsymbol{\epsilon}(2)$ characterizing transverse fields propagating in direction $\boldsymbol{\varkappa}$.

For the purpose of specifying circular polarization, one may introduce alternatively the complex orthogonal unit vectors

(18.58) $$\boldsymbol{\epsilon}(R) = (1/\sqrt{2})\,[\boldsymbol{\epsilon}(1)+i\boldsymbol{\epsilon}(2)]$$

(18.59) $$\boldsymbol{\epsilon}(L) = (1/\sqrt{2})\,[\boldsymbol{\epsilon}(1)-i\boldsymbol{\epsilon}(2)] = \boldsymbol{\epsilon}^*(R)$$

satisfying the orthonormality relations

(18.60) $$\boldsymbol{\epsilon}^*(R)\,\boldsymbol{\epsilon}(R) = \boldsymbol{\epsilon}^*(L)\,\boldsymbol{\epsilon}(L) = 1$$

and

(18.61) $$\boldsymbol{\epsilon}^*(R)\,\boldsymbol{\epsilon}(L) = \boldsymbol{\epsilon}^*(L)\,\boldsymbol{\epsilon}(R) = 0$$

as well as the transversality conditions

(18.62) $$\mathbf{n}\boldsymbol{\epsilon}(R) = 0; \qquad \mathbf{n}\boldsymbol{\epsilon}(L) = 0.$$

Equations (18.60)–(18.62) are formally reminiscent of the ortho-normality relations (18.35), (18.36), and the transversality condition (18.44). This suggests strongly that one *identify* the components of the complex state vectors χ^R, χ^L, and χ^0 with appropriately chosen linear combinations of the vectors $\epsilon(R)$, $\epsilon(L)$, and $\mathbf{n}$, respectively. By the substi-tution of (18.55), (18.56), and (18.57) for the components in (18.40), (18.41), and (18.43) one finds without difficulty

(18.63)

$$
\chi^R = \frac{1}{\sqrt{2}} \begin{pmatrix} -\epsilon_1(R) + i\epsilon_2(R) \\ \sqrt{2}\epsilon_3(R) \\ \epsilon_1(R) + i\epsilon_2(R) \end{pmatrix}; \qquad
\chi^L = \frac{1}{\sqrt{2}} \begin{pmatrix} -\epsilon_1(L) + i\epsilon_2(L) \\ \sqrt{2}\epsilon_3(L) \\ \epsilon_1(L) + i\epsilon_2(L) \end{pmatrix};
$$

$$
\chi^0 = \frac{1}{\sqrt{2}} \begin{pmatrix} -n_1 + in_2 \\ \sqrt{2}n_3 \\ n_1 + in_2 \end{pmatrix}
$$

With some hindsight, this correspondence between spin states and polarization vectors can be made more obvious, if one performs in spin space the unitary transformation

(18.64)

$$
U = \frac{1}{\sqrt{2}} \begin{pmatrix} -1 & 0 & 1 \\ -i & 0 & -i \\ 0 & \sqrt{2} & 0 \end{pmatrix}; \qquad
U^+ = \frac{1}{\sqrt{2}} \begin{pmatrix} -1 & i & 0 \\ 0 & 0 & \sqrt{2} \\ 1 & i & 0 \end{pmatrix}
$$

leading, for the spin operators, instead of to (18.21), to the representation

(18.65)

$$
s_1 = UJ_1 U^+ = \begin{pmatrix} 0 & 0 & 0 \\ 0 & 0 & -i \\ 0 & i & 0 \end{pmatrix}; \qquad
s_2 = UJ_2 U^+ = \begin{pmatrix} 0 & 0 & i \\ 0 & 0 & 0 \\ -i & 0 & 0 \end{pmatrix};
$$

$$
s_3 = UJ_3 U^+ = \begin{pmatrix} 0 & -i & 0 \\ i & 0 & 0 \\ 0 & 0 & 0 \end{pmatrix}
$$

and to a rearrangement of the components of the spin states, so that

$$
(18.66) \quad \eta^R = U\chi^R = \begin{pmatrix} \epsilon_1(R) \\ \epsilon_2(R) \\ \epsilon_3(R) \end{pmatrix} = \epsilon(R); \qquad \eta^L = U\chi^L = \epsilon(L);
$$

$$
\eta^0 = U\chi^0 = \mathbf{n}
$$

and the orthonormality relations

(18.67) $\eta^{R*}\eta^{R} = \eta^{L*}\eta^{L} = 1$; $\eta^{R*}\eta^{L} = \eta^{R*}\eta^{L} = 0$;

$$\eta^{0*}\eta^{R} = \eta^{0*}\eta^{L} = 0$$

are now *identical* with the relations (18.60), (18.61), and (18.62).

The eigenstates of s_3, namely

(18.68) $\boldsymbol{\xi}^1 = -\dfrac{1}{\sqrt{2}}\begin{pmatrix}1\\i\\0\end{pmatrix}$; $\boldsymbol{\xi}^0 = \begin{pmatrix}0\\0\\1\end{pmatrix}$; $\boldsymbol{\xi}^{-1} = \dfrac{1}{\sqrt{2}}\begin{pmatrix}1\\-i\\0\end{pmatrix}$

can be looked upon as a system of three orthonormal vectors which can be used to decompose any arbitrary vector **f**. This is particularly useful when one wishes to use the angular momentum representation, leading to a description of photon states in terms of so-called vector spherical harmonics, which are given in Appendix 3.

The description of the state vector (18.34) in coordinate representation requires a six-component ψ function

(18.69) $\psi(x) = \begin{pmatrix}\alpha(S)\,\psi^R(x)\\ \beta(S)\,\psi^L(x)\end{pmatrix}$; $x = (\mathbf{q}, t)$,

where $\psi^R(x)$ and $\psi^L(x)$ are three-component ψ functions satisfying Schroedinger equations which follow from (18.19) and (18.20) upon substitution of the representations $\mathbf{P} = -i\boldsymbol{\nabla}$ and $H = i(\partial/\partial t)$, and which, in the representation (18.65) of the spin matrices, read

(18.70) $-i\mathbf{s}\cdot\boldsymbol{\nabla}\psi^R = i(\partial/\partial t)\,\psi^R$; $+i\mathbf{s}\cdot\boldsymbol{\nabla}\psi^L = i(\partial/\partial t)\,\psi^L$.

By writing ψ^R and ψ^L as vectors $\boldsymbol{\psi}^R$ and $\boldsymbol{\psi}^L$, each having the components ψ_1, ψ_2, ψ_3, these equations can be written in vector notation, because of the representations (18.65),

(18.71) $\boldsymbol{\nabla}\times\boldsymbol{\psi}^R = i(\partial\boldsymbol{\psi}^R/\partial t)$; $-\boldsymbol{\nabla}\times\boldsymbol{\psi}^L = i(\partial\boldsymbol{\psi}^L/\partial t)$

and can be solved under observance of the transversality conditions

(18.72) $\boldsymbol{\nabla}\cdot\boldsymbol{\psi}^R = \boldsymbol{\nabla}\cdot\boldsymbol{\psi}^L = 0$.

The formal resemblance of these equations to Maxwell's vacuum equations for the complex field vectors $\mathbf{E} + i\mathbf{B}$ and $\mathbf{E} - i\mathbf{B}$ suggests the introduction of the hermitean *operators*

(18.73) $\mathbf{E}(x) = \dfrac{i}{\sqrt{V}}\sum_{\varkappa}\sum_{S}\sqrt{\dfrac{\omega}{2}}\left[\boldsymbol{\epsilon}(S)\,b(\boldsymbol{\varkappa}, S)\,e^{-i\kappa x} - \boldsymbol{\epsilon}^*(S)\,b^+(\boldsymbol{\varkappa}, S)\,e^{+i\kappa x}\right]$

(18.74)

$$\boldsymbol{\nabla}\times\mathbf{B}(x) = \dfrac{1}{\sqrt{V}}\sum_{\varkappa}\sum_{S}\omega\sqrt{\dfrac{\omega}{2}}\left[\boldsymbol{\epsilon}(S)\,b(\boldsymbol{\varkappa}, S)\,e^{-i\kappa x} + \boldsymbol{\epsilon}^*(S)\,b^+(\boldsymbol{\varkappa}, S)\,e^{+i\kappa x}\right]$$

where $\kappa x = \omega t - \varkappa q$ and V is a normalization volume, which can be derived by differentiations according to the prescription (18.53) from the vector potential *operator*

$$(18.75) \quad \mathbf{A}(x) = \frac{1}{\sqrt{V}} \sum_{\varkappa} \sum_{S} \frac{1}{\sqrt{2\omega}} [\boldsymbol{\epsilon}(S)\, b(\varkappa, S)\, e^{-i\kappa x} + \boldsymbol{\epsilon}^*(S)\, b^+(\varkappa, S)\, e^{+i\kappa x}].$$

One has then the decompositions

$$(18.76) \quad \frac{1}{\sqrt{V}} \sum_{\varkappa} \sum_{S} \sqrt{\omega}\, e^{-i\kappa x}\, \boldsymbol{\epsilon}(S)\, b(\varkappa, S) = \frac{1}{\sqrt{2}}\left[-i\mathbf{E} + \frac{1}{\sqrt{-\nabla^2}}(\boldsymbol{\nabla} \times \mathbf{B}) \right]$$

$$(18.77)$$
$$\frac{1}{\sqrt{V}} \sum_{\varkappa} \sum_{S} \sqrt{\omega}\, e^{+i\kappa x}\, \boldsymbol{\epsilon}^*(S)\, b^+(\varkappa, S) = \frac{1}{\sqrt{2}}\left[i\mathbf{E} + \frac{1}{\sqrt{-\nabla^2}}(\boldsymbol{\nabla} \times \mathbf{B}) \right]$$

where $\dfrac{1}{\sqrt{-\nabla^2}}$ is an operator defined by

$$(18.78) \quad \frac{1}{\sqrt{-\nabla^2}} e^{\pm i\varkappa q} = \frac{e^{\pm i\varkappa q}}{|\varkappa|} = \frac{e^{\pm i\varkappa q}}{\omega}.$$

Multiplication of (18.76) with (18.77) from the left and integration over $d\mathbf{q}$ gives, upon utilization of the normalization conditions on the left-hand side,

$$(18.79) \quad \sum_{\varkappa} \sum_{S} \omega b^+(\varkappa, S)\, b(\varkappa, S) = \tfrac{1}{2} \int \{ \mathbf{E}^2 + \mathbf{B}^2 + (i/\sqrt{-\nabla^2})[\mathbf{E}(\boldsymbol{\nabla} \times \mathbf{B})$$
$$- (\boldsymbol{\nabla} \times \mathbf{B})\mathbf{E}]\} d\mathbf{q}.$$

If one could ignore the operator nature of $\mathbf{E}$ and $\mathbf{B}$, the term containing $\mathbf{E}(\boldsymbol{\nabla} \times \mathbf{B}) - (\boldsymbol{\nabla} \times \mathbf{B})\mathbf{E}$ would obviously vanish and the desired correspondence to classical fields is established. In quantum mechanics, the presence of these terms is indispensable, however, because without them the expectation value of the integral on the right-hand side of (18.79) would become infinite for any photon state, including the vacuum state. This peculiarity stems from the operator nature of $\mathbf{E}$ and $\mathbf{B}$ which contain both creation and annihilation operators, so that the field intensity operators $\mathbf{E}^2$ and $\mathbf{B}^2$ each include terms of the form $b(\varkappa, S)\, b^+(\varkappa, S)$ whose vacuum expectation values do not vanish, leading to the infinite result

$$(18.80) \quad \langle 0| \tfrac{1}{2} \int (\mathbf{E}^2 + \mathbf{B}^2)\, d\mathbf{q} |0\rangle = \tfrac{1}{2} \sum_{\varkappa} \omega.$$

The presence of the term $\int (i/2\sqrt{-\nabla^2})[\mathbf{E}(\boldsymbol{\nabla} \times \mathbf{B}) - (\boldsymbol{\nabla} \times \mathbf{B})\mathbf{E}] d\mathbf{q}$ is needed to precisely compensate this so-called zero-point energy, as a simple calculation will show. This matter will not be pursued here beyond noting that the concept of zero-point energy does not enter the theory of photons if one adheres to the description which has led to the adoption of $\sum_{\varkappa} \sum_{S} \omega b^{+} b$ as energy operator, because the vacuum expectation value of this operator is obviously zero, and remains finite for any state containing a finite number of photons.

The conceptual differences between quantum mechanical and classical descriptions of electromagnetic polarization phenomena are equally radical. Taking the most general case of elliptical polarization as an example, one infers, in classical electrodynamics, the possibility of determining simultaneously the values of three parameters fixing the polarization, for example, the parameters e_1, e_2, γ, so that at any instant of time the electric field vector in the $[\epsilon(1), \epsilon(2)]$ plane has the components

$$(18.81) \qquad E_1 = e_1 \cos(\omega t); \qquad E_2 = e_2 \cos(\omega t + \gamma),$$

the endpoint of $\mathbf{E}$ tracing the polarization ellipse in that plane.

In quantum mechanics no such statement can meaningfully be made about the expectation values of the operators (18.73) representing these field components, because the operators b and b^+ have vanishing diagonal elements only, and therefore the expectation values of the field operators themselves will vanish in any state of definite photon number. From an operational point of view, this circumstance is a very satisfactory feature of the quantum mechanical formalism, because one cannot talk meaningfully about the measurement of electromagnetic fields without invoking, in principle, an apparatus interacting with the field so that emission and absorption acts take place making the photon number variable. One can, however, conceive measurements of field intensities, i.e. quadratic functions of the fields, in stationary situations in which the total number of photons present does not change. The parameters e_1, e_2, γ, should therefore be set in correspondence with operators containing the creation and annihilation operators at least bilinearly. The operators S_1, S_2, S_3 introduced earlier in (18.45) satisfy just that requirement, and their expectation values can be used to describe the state of polarization even in case of partially polarized light beams.

Since the polarization of the photon is a dichotomic variable, giving rise to the isomorphism of the abstract polarization operators (18.45) with the Pauli matrices, the development of Section 4 can be applied and the polarization of a photon beam described in terms if the density matrix

$$(18.82) \qquad\qquad M = \tfrac{1}{2}(I + \overline{\mathbf{P}}\mathbf{S})$$

where $|\overline{\mathbf{P}}|$ represents the degree of polarization of the photon beam.

In particular, the entire information about the polarization of the single photon state (18.23), which is a state of definite polarization so that $|\overline{\mathbf{P}}| = 1$, resides in the density matrix

(18.83) $$M(S) = \begin{pmatrix} \alpha(S)\,\alpha^*(S) & \alpha(S)\,\beta^*(S) \\ \alpha^*(S)\,\beta(S) & \beta(S)\,\beta^*(S) \end{pmatrix}$$

where the numbers $\alpha(S)$ and $\beta(S)$ are subject to the normalization constraint $|\alpha|^2 + |\beta|^2 = \operatorname{trace} M(S) = 1$ and are determined up to a common arbitrary phase factor. Using for S_1, S_2, S_3, S_0 their representations (18.60) in polarization space, one obtains for their expectation values

(18.84)
$$\overline{S}_1 = \operatorname{trace}[S_1 M(S)] = \alpha^*\beta + \alpha\beta^*$$
$$\overline{S}_2 = -i(\alpha^*\beta - \alpha\beta^*)$$
$$\overline{S}_3 = \alpha\alpha^* - \beta\beta^*$$
$$\overline{S}_0 = \alpha\alpha^* + \beta\beta^*.$$

These quantities can now meaningfully be set equal to the classical Stokes' parameters

(18.85) $S_1 = \tfrac{1}{2}e_1 e_2 \cos\gamma; \qquad S_2 = \tfrac{1}{2}e_1 e_2 \sin\gamma; \qquad S_3 = \tfrac{1}{4}(e_1^2 - e_2^2)$

which are quadratic functions of the field amplitudes e_1, e_2, establishing thus the correspondence between the parameters e_1, e_2, γ and the ψ functions α, β in polarization space.

Thus far, the discussion has been restricted to transverse photons which can be described in terms of a three-component vector potential $\mathbf{A}(\mathbf{q})$. This treatment has the aesthetic shortcoming of not being manifestly covariant under Lorentz transformations. A little thought shows, however, that this is no serious defect, because the electromagnetic potentials have the property of "gauge invariance," permitting reduction of the formally relativistic expression

(18.86)

$$A_\mu^{\mathrm{tr}}(x) = \frac{1}{\sqrt{V}} \sum_{\varkappa} \sum_{S=1}^{2} \frac{1}{\sqrt{2\omega}} [\epsilon_\mu(S)\, b(\varkappa, S)\, e^{-i\varkappa x} + \epsilon_\mu^*(S)\, b^+(\varkappa, S)\, e^{+i\varkappa x}]$$

$$(\mu = 1, 2, 3, 4)$$

to the form (18.75) by a gauge transformation

(18.87) $$A_\mu(x) \rightarrow A'_\mu(x) = A_\mu(x) + (\partial\Lambda/\partial x_\mu)$$

where Λ is a scalar function, which, for given fourth component of

$\epsilon_\mu(S)$, is constructed so that Eq. (18.87) is equivalent to the replacement

(18.88)
$$\epsilon_\mu(S) \to \epsilon'_\mu(S) = \epsilon_\mu(S) - [\epsilon_4(S)/\omega]\kappa_\mu$$

so that

(18.89)
$$\epsilon'_4(S) = 0 \quad \text{and} \quad \epsilon'_\mu \kappa_\mu = -\epsilon'\varkappa = 0 = \epsilon_\mu \kappa_\mu \quad \text{for} \quad S = 1,2.$$

The vanishing rest mass of the photon, i.e. $\kappa^2 = \kappa_\mu \kappa_\mu = \omega^2 - \varkappa^2 = 0$ is obviously the root of this particular invariance of the transversality condition (18.89).

In presence of sources of the electromagnetic field, the description (18.75) acquires another shortcoming far more serious than the aesthetic flaw just mentioned, namely it becomes incomplete: transverse photons are not sufficient to grasp, for example, the Coulomb field surrounding an electric charge. One requires an additional two polarizations, corresponding to "longitudinal" and "time-like" photons, to accommodate the entire range of electromagnetic phenomena.

The last word has almost certainly not been said about this matter. The purpose of these paragraphs is to summarize briefly the formalism invented, and widely accepted, for the description of these exotic photons, *which do not exist as free particles*.

One can easily enough extend the definition of the operator (18.86) in a relativistically covariant manner to the case of *four* independent polarizations by simply writing

(18.90)
$$A_\mu(x) = (1/\sqrt{V}) \sum_\varkappa \sum_{S=1}^4 (1/\sqrt{2\omega})[\epsilon_\mu(S)b(\varkappa,S)e^{-i\kappa x} + \epsilon_\mu^*(S)b^+(\varkappa,S)e^{i\kappa x}]$$

where now $\epsilon_\mu(S)$ are a set of *four* orthogonal unit vectors satisfying

(18.91)
$$\epsilon_\mu(S)\epsilon_\mu^*(S') = \delta_{SS'}.$$

A possible and popular choice of the polarization vectors $\epsilon_\mu(S)$ containing for $S = 1, 2$ the transverse polarizations as before is

(18.92)
$$\epsilon(1)\varkappa = 0, \quad \epsilon_4(1) = 0; \quad \epsilon(2)\varkappa = 0, \quad \epsilon_4(2) = 0;$$
$$\epsilon(3) = \varkappa/\omega, \quad \epsilon_4(3) = 0; \quad \epsilon(4) = 0, \quad \epsilon_4(4) = 1;$$

allowing one to refer meaningfully to $\epsilon_\mu(3)$ as the longitudinal and to $\epsilon_\mu(4)$ as the time-like polarization vector. With this choice the polarization vectors satisfy the formally covariant relation

(18.93)
$$\sum_{S=1}^4 \epsilon_\mu(S)\epsilon_\nu^*(S) = \delta_{\mu\nu}.$$

However, the choice (18.92) is neither necessary, nor is it necessarily the most convenient. On grounds of relativistic covariance alone it is sufficient to require instead of (18.93)

$$(18.94) \qquad \sum_{S=1}^{4} \epsilon_\mu(S)\, \epsilon_\nu^*(S) \;=\; \delta_{\mu\nu} + \kappa_\mu \kappa_\nu f(\kappa^2)$$

with arbitrary $f(\kappa^2)$. In fact, the various choices of $\epsilon_\mu(S)$, belonging to different $f(\kappa^2)$, are equivalent to different gauges adopted for the electromagnetic potentials, and keeping the function $f(\kappa^2)$ open in all calculations is a convenient way of making the gauge invariance of the description manifest.

To see this, consider a gauge transformation (18.87) where $\Lambda(x)$ is now the most general scalar function *linear* in the $A_\mu(x)$,

$$(18.95) \qquad \Lambda(x) \;=\; -F(-\Box)\,(\partial A_\nu/\partial x_\nu).$$

Here F is an *arbitrary* function of the scalar operator $\Box = \partial^2/\partial x_\mu \partial x_\mu$, and the signs have been chosen purely for convenience. With $A_\mu(x)$ given by (18.90), one finds explicitly

$$(18.96) \quad \Lambda(x) = (1/\sqrt{V}) \sum_{\boldsymbol{\varkappa}} \sum_{S} [F(\kappa^2)/\sqrt{2\omega}]\,[\kappa_\nu\, \epsilon_\nu(S)\, b(\boldsymbol{\varkappa},S)\, e^{-i\kappa x}$$
$$- \kappa_\nu\, \epsilon_\nu^*(S)\, b^+(\boldsymbol{\varkappa},S)\, e^{i\kappa x}]$$

so that

$$(18.97) \quad A'_\mu(x) = (1/\sqrt{V}) \sum_{\boldsymbol{\varkappa}} \sum_{S} (1/\sqrt{2\omega})\,\{[\epsilon_\mu(S)$$
$$+ F(\kappa^2)\, \kappa_\mu \kappa_\nu\, \epsilon_\nu(S)]\, b(\boldsymbol{\varkappa},S)\, e^{-i\kappa x} + \text{c.c.}\}.$$

The gauge transformation generated by (18.96) is thus equivalent to replacing the polarization vectors by

$$(18.98) \qquad \epsilon_\mu(S) \to \epsilon'_\mu(S) \;=\; \epsilon_\mu(S) + F(\kappa^2)\, \kappa_\mu \kappa_\nu\, \epsilon_\nu(S)$$

giving

$$(18.99) \qquad \sum_{S=1}^{4} \epsilon'_\mu(S)\, \epsilon_\nu'^*(S) \;=\; \delta_{\mu\nu} + \kappa_\mu \kappa_\nu [2F(\kappa^2) + \kappa^2 F^2(\kappa^2)],$$

which is identical with (18.94) if one drops the primes on the left and sets

$$(18.100) \qquad f(\kappa^2) \;=\; 2F(\kappa^2) + \kappa^2 F^2(\kappa^2).$$

For many applications it is more convenient to introduce a scalar function $d_l(\kappa^2)$ by

$$(18.101) \qquad f(\kappa^2) \;=\; (d_l - 1)/\kappa^2$$

which splits (18.94) into two terms, namely

$$(18.102) \qquad \sum_{S=1}^{4} \epsilon_\mu(S)\,\epsilon_\nu^*(S) = \left(\delta_{\mu\nu} - \frac{\kappa_\mu\,\kappa_\nu}{\kappa^2}\right) + d_l\,\frac{\kappa_\mu\,\kappa_\nu}{\kappa^2}.$$

The first term can be called the "transverse part" in a four dimensional sense, because it satisfies

$$(18.103) \qquad \kappa_\mu \left(\sum_{S=1}^{4} \epsilon_\mu(S)\,\epsilon_\nu^*(S)\right)^{\mathrm{tr}} = 0.$$

The arbitrariness in the gauge of the vector potentials lodges now in the arbitrariness of choice for the factor $d_l(\kappa^2)$. If one can show that observable effects do not depend on d_l, one has a "manifestly gauge-invariant" theory.

The conventions (18.92) leading to (18.93) are obviously equivalent to putting $d_l = 1$. An alternative choice, which is at least as convenient, is $d_l = 0$. For many applications it is advisable, however, to carry the unspecified factor d_l in all calculations, thus retaining and exhibiting freedom of gauge, and to settle for a particular value of d_l only when this results in an overwhelming computational advantage.

NOTES

Jordan and Klein [1] introduced the concept of creation and annihilation operators for bosons. See also Dirac [2].

Jauch and Rohrlich [3] treat transverse photon polarization and give complete references to earlier work on this subject.

Archibald [4] noticed that one can write Maxwell's vacuum equations as Schroedinger equations for two 3-component ψ functions. See also Akhiezer and Berestetskii [5].

Landau and Peierls [6] gave the decomposition of photon annihilation and creation operators in terms of electromagnetic field operators.

Rose [7] gives the unitary transformation (18.64).

For a discussion of some of the perplexities introduced through longitudinal and time-like photons see Källén [8].

Bogoliubov and Shirkov [9] use consistently a gauge convention in which the function d_l is left arbitrary.

REFERENCES

[1] P. Jordan and O. Klein, *Z. Physik* **45**, 751 (1927).
[2] P. A. M. Dirac, *Proc. Roy. Soc.* **A114**, 243 (1927).
[3] J. M. Jauch and F. Rohrlich, "The Theory of Photons and Electrons," pp. 41–46, Addison-Wesley, Reading, Massachusetts, 1955.

[4] W. J. Archibald, *Can. J. Phys.* **33**, 565 (1955).

[5] A. I. Akhiezer and V. B. Berestetskii, "Quantum Electrodynamics." The USAEC translation of the first Russian edition of this book was published at Oak Ridge, Tennessee, in 1957; an abbreviated version of the second Russian edition has been published in English by Oldbourne Press, London, 1963. A complete, new English translation of the second Russian edition is about to be published by Wiley, New York, 1964.

[6] L. Landau and R. Peierls, *Z. Physik* **62**, 188 (1930).

[7] M. E. Rose, "Multipole Fields," Chapter II. Wiley, New York, 1955.

[8] G. Källén, Quantenelektrodynamik, *in* "Encyclopedia of Physics" (S. Flügge, ed.), Vol. 5, Pt. I, Chapter II. Springer, Berlin, 1958.

[9] N. N. Bogoliubov and D. V. Shirkov, "Introduction to the Theory of Quantized Fields," Wiley (Interscience), New York, 1959.

Electrons and Positrons

Particles having a rest mass m, so that the relation between energy Ω and momentum $\mathbf{k}$ is

(19.1) $$\Omega^2 = \mathbf{k}^2 + m^2$$

need not have their spin aligned either parallel or antiparallel to the direction of $\mathbf{k}$. Even if in a given coordinate frame spin and momentum happen to be aligned, this alignment will be destroyed by a Lorentz transformation if the particle travels at a speed less than the speed of light, because spin, being a skew tensor, and momentum, being a vector, transform differently under such transformations.

Electrons and positrons, in particular, are particles belonging to spin $j = \frac{1}{2}$ which differ from neutrinos and antineutrinos in that their spin does not exhibit the firm correlation with direction of momentum $\mathbf{k}$, which makes the ψ function of massless particles obey either an equation of the form $[H - (\boldsymbol{\sigma} \cdot \mathbf{P})]u = 0$ if they are right-handed, or $[H + (\boldsymbol{\sigma} \cdot \mathbf{P})]v = 0$ if they are left-handed, as was explained in detail in Section 17. One cannot include the effect of a rest mass by simply adding a term mu or mv to the equations for u and v respectively, because if m is a scalar the resulting equations would not be invariant under inversions Π. A way of accommodating the mass without violating any invariance requirements, including invariance under Lorentz transformations, was invented by Dirac, who showed one can describe electrons by four-component ψ functions obtained by coupling the two two-component ψ functions u and v through the mass term,

(19.2)
$$(H - \boldsymbol{\sigma}\mathbf{P})u = mv$$
$$(H + \boldsymbol{\sigma}\mathbf{P})v = mu.$$

In the limit $m \to 0$ they describe uncoupled right- and left-handed massless particles of spin $j = \frac{1}{2}$, but for $m \neq 0$ these first order equations are equivalent, by iteration, to the second order two-component equations

(19.3)
$$(H + \boldsymbol{\sigma}\mathbf{P})(H - \boldsymbol{\sigma}\mathbf{P})u = m^2 u$$
$$(H - \boldsymbol{\sigma}\mathbf{P})(H + \boldsymbol{\sigma}\mathbf{P})v = m^2 v.$$

151

For free particles these equations are equivalent, because of the C.R.s of the $\boldsymbol{\sigma}$, to

$$(19.4) \qquad (H^2 - P^2 - m^2)\, u = 0; \qquad (H^2 - P^2 - m^2)\, v = 0.$$

Introducing the 4×4 matrices

$$(19.5) \qquad \boldsymbol{\gamma} = \begin{pmatrix} 0 & -\boldsymbol{\sigma} \\ \boldsymbol{\sigma} & 0 \end{pmatrix}; \qquad \gamma_4 = \begin{pmatrix} 0 & 1 \\ 1 & 0 \end{pmatrix}$$

and the four-component ψ function

$$(19.6) \qquad \psi = \begin{pmatrix} u \\ v \end{pmatrix}$$

one can, because of

$$(19.7) \qquad \boldsymbol{\gamma} \begin{pmatrix} u \\ v \end{pmatrix} = \begin{pmatrix} -\boldsymbol{\sigma}v \\ \boldsymbol{\sigma}u \end{pmatrix}; \qquad \gamma_4 \begin{pmatrix} u \\ v \end{pmatrix} = \begin{pmatrix} v \\ u \end{pmatrix}$$

write Eqs. (19.2) in the compact form

$$(19.8) \qquad (\gamma_4 H - \boldsymbol{\gamma}\mathbf{P})\psi = m\psi$$

This equation, due to Dirac, can alternatively be written

$$(19.9) \qquad (\boldsymbol{\alpha}\mathbf{P} + \beta m)\psi = H\psi$$

with

$$(19.10) \qquad \boldsymbol{\alpha} = \gamma_4 \boldsymbol{\gamma}; \qquad \beta = \gamma_4.$$

The matrices $(\boldsymbol{\gamma}, \gamma_4) \equiv \gamma_\mu$ introduced here are identical with the ones used by Feynman. They differ from the γ's used in most conventional texts by a unitary transformation. They satisfy the anti-C.R.s

$$(19.11) \quad \gamma_\mu \gamma_\nu + \gamma_\nu \gamma_\mu = 2\delta_{\mu\nu} \qquad \text{with} \qquad \delta_{\mu\nu} = \begin{pmatrix} -1 & & & \\ & -1 & & \\ & & -1 & \\ & & & +1 \end{pmatrix}.$$

A very important matrix in this abstract four-dimensional space is

$$(19.12) \qquad \gamma_5 = \gamma_1 \gamma_2 \gamma_3 \gamma_4 = \begin{pmatrix} i & 0 \\ 0 & -i \end{pmatrix}$$

which satisfies

$$(19.13) \qquad \gamma_5^2 = -1 \qquad \text{and} \qquad \gamma_5 \gamma_\mu + \gamma_\mu \gamma_5 = 0$$

The operators

$$(19.14) \qquad C_L = \tfrac{1}{2}(I + i\gamma_5) = \begin{pmatrix} 0 & 0 \\ 0 & 1 \end{pmatrix}$$

and

(19.15)
$$C_R = \tfrac{1}{2}(I - i\gamma_5) = \begin{pmatrix} 1 & 0 \\ 0 & 0 \end{pmatrix}$$

obviously project the components u and v out of ψ, respectively,

(19.16)
$$C_R\psi = \begin{pmatrix} u \\ 0 \end{pmatrix} = \psi_R; \qquad C_L\psi = \begin{pmatrix} 0 \\ v \end{pmatrix} = \psi_L.$$

Accordingly, the operator

(19.17)
$$C = -i\gamma_5 = \begin{pmatrix} 1 & 0 \\ 0 & -1 \end{pmatrix}$$

can be called the "chirality" or "handedness" operator, because ψ_R and ψ_L are eigenstates of C with eigenvalues $+1$ and -1, respectively,

(19.18)
$$C\psi_R = +\psi_R; \qquad C\psi_L = -\psi_L$$

In terms of the γ matrices, the spin operators can be represented as

(19.19)
$$\hat{\sigma}_1 = \begin{pmatrix} \sigma_1 & 0 \\ 0 & \sigma_1 \end{pmatrix} = i\gamma_2\gamma_3 \qquad \text{(cyclically)}.$$

Since $\hat{\sigma}$ and $(\gamma_4 H - \gamma\mathbf{P})$ do not commute, the solutions of the Dirac equation (19.8) are not, in general, eigenstates of $\hat{\sigma}$ for arbitrary direction of the spin. Only if the particle is at rest, namely when $P\psi = 0$ so that

(19.20)
$$\gamma_4 H\psi_{\text{rest}} = m\psi_{\text{rest}} \qquad \text{i.e.} \qquad \begin{cases} Hv = mu \\ Hu = mv \end{cases}$$

which requires

(19.21)
$$u = v \qquad \text{i.e.} \qquad \psi_{\text{rest}} = \begin{pmatrix} u \\ u \end{pmatrix}$$

can ψ be an eigenstate of $\hat{\sigma}$. By convention, the representation

(19.22)
$$\hat{\sigma}_3\psi_{\text{rest}} = \pm\psi_{\text{rest}}$$

will always be chosen.

To be quite explicit, the state vectors $\binom{1}{0}_C$; $\binom{0}{1}_C$ spanning the chirality space defined by the eigenvalues ± 1 of C, and the statevectors $\binom{1}{0}_S$; $\binom{0}{1}_S$ spanning the spin space defined by the eigenvalues ± 1 of σ_3 *in the rest frame* of the particle can be introduced. A general electron state can then be characterized by the quantum numbers $\mathbf{k}$, S, C, where C stands for the dichotomic label R, L of chirality (right-handedness, left-handedness), and S stands for the dichotomic label $\uparrow$, $\downarrow$, of spin in z direction (spin up, spin down, *in the rest frame*), and such a state is generated from vacuum $|0\rangle$ by the appropriate creation operator,

(19.23)
$$|\mathbf{k},S,C\rangle = a^+(\mathbf{k},S,C)|0\rangle$$

where

$$|\mathbf{k},S,C\rangle = |\mathbf{k},S,R\rangle \times \begin{pmatrix}1\\0\end{pmatrix}_C + |\mathbf{k},S,L\rangle \times \begin{pmatrix}0\\1\end{pmatrix}_C$$

(19.24)
$$= |\mathbf{k},\uparrow,R\rangle \times \begin{pmatrix}1\\0\end{pmatrix}_S \times \begin{pmatrix}1\\0\end{pmatrix}_C + |\mathbf{k},\downarrow,R\rangle \times \begin{pmatrix}0\\1\end{pmatrix}_S \times \begin{pmatrix}1\\0\end{pmatrix}_C$$

$$+ |\mathbf{k},\uparrow,L\rangle \times \begin{pmatrix}1\\0\end{pmatrix}_S \times \begin{pmatrix}0\\1\end{pmatrix}_C + |\mathbf{k},\downarrow,L\rangle \times \begin{pmatrix}0\\1\end{pmatrix}_S \times \begin{pmatrix}0\\1\end{pmatrix}_C.$$

In the coordinate representation

(19.25)
$$|\mathbf{k},S,C\rangle = \int |\mathbf{q}\rangle\, d\mathbf{q}\, \psi_{\mathbf{k},s,c}(\mathbf{q})$$

the ψ functions characterizing $|\mathbf{k},S,R\rangle$; $|\mathbf{k},S,L\rangle$; $|\mathbf{k},\uparrow,R\rangle$; $|\mathbf{k},\downarrow,R\rangle$; $|\mathbf{k},\uparrow,L\rangle$; $|\mathbf{k},\downarrow,L\rangle$; are u; v; u_1; u_2; v_1; v_2; respectively.

In the combined spin-chirality space (which has four dimensions), the operator of inversion of coordinates can be represented by (0 and 1 stand for 2×2 matrices in spin space)

(19.26)
$$\hat{\Pi} = \begin{pmatrix}0 & 1\\1 & 0\end{pmatrix} = \gamma_4$$

because chirality $C = -i\gamma_5$ must be odd under inversions, whereas the spin $\hat{\sigma}_j = i\gamma_k\gamma_l$ (cyclically) must be even, and this is guaranteed by (19.26) because of the anti-C.R.s (19.11) and (19.13). One expects, therefore, the operator of coordinate inversion to be entirely represented by

(19.27)
$$\Pi = \eta_\Pi \gamma_4 \Pi_D = \begin{pmatrix}0 & \Pi_D\\\Pi_D & 0\end{pmatrix}\eta_\Pi$$

where Π_D acts on the dynamical variables such as H and $\mathbf{P}$ and represents the transformation $\mathbf{q} \to -\mathbf{q}$ in the ψ functions, and η_Π is a phase factor, which in accordance with the conventions adopted in Section 14 will be chosen so that $\eta_\Pi^2 = +1$.

The operator of time reversal, on the other hand, should leave the chirality unchanged and must therefore be diagonal in chirality space. This is accomplished by putting

(19.28)
$$\hat{\Theta} = \hat{T}K = \begin{pmatrix}T & 0\\0 & T\end{pmatrix}K$$

where T is in spin space according to (15.42) of the form

$$(19.29) \qquad T = \begin{pmatrix} 0 & 1 \\ -1 & 0 \end{pmatrix} = i \begin{pmatrix} 0 & -i \\ i & 0 \end{pmatrix} = i\sigma_2$$

and K is the operator of complex conjugation, thus guaranteeing the correct transformation properties of the spin operator $\hat{\sigma}$, namely $\hat{\Theta}\hat{\sigma}\hat{\Theta}^{-1} = -\hat{\sigma}$. In terms of the matrices γ the operator $\hat{T}$ may be written

$$(19.30) \qquad \hat{T} = \gamma_1 \gamma_3 = \begin{pmatrix} i\sigma_2 & 0 \\ 0 & i\sigma_2 \end{pmatrix}$$

so that altogether the operator of time reversal can be represented by

$$(19.31) \qquad \Theta = \eta_\Theta \gamma_1 \gamma_3 \times T_D K$$

where T_D acts on the dynamical variables and represents the transformation $t \rightarrow -t$ in any time-dependent ψ function, and η_Θ is an as yet undetermined phase factor, subject only to the condition $\eta_\Theta \eta_\Theta^* = 1$.

Eigensolutions of (19.8) will now be sought which satisfy

$$(19.32)$$

$$H\psi = \Omega\psi \qquad \text{and} \qquad \mathbf{P}\psi = \mathbf{k}\psi \qquad \text{with} \qquad \Omega = +\sqrt{k^2 + m^2}.$$

It is understood that the energy Ω is never a negative number. Using $\mathbf{P} = -i(\partial/\partial\mathbf{Q})$ one has in coordinate representation

$$(19.33) \qquad (\gamma_4 \Omega + i\gamma[\partial/\partial\mathbf{q}])\psi = m\psi.$$

By writing

$$(19.34) \qquad \psi_{\mathbf{k},S,c}(\mathbf{q}) = A(\mathbf{k}, S, C) e^{i\mathbf{k}\mathbf{q}} = \begin{pmatrix} A_1 \\ A_2 \\ A_3 \\ A_4 \end{pmatrix} e^{i\mathbf{k}\mathbf{q}}$$

one obtains four linear homogeneous equations for the $A_n(\mathbf{k}, \ldots)$, namely

$$(19.35) \qquad (\gamma_4 \Omega - \gamma\mathbf{k}) A = mA$$

which read explicitly, with representation (19.5),

$$(19.36) \qquad \begin{aligned} (\Omega + k_3) A_3 + (k_1 - ik_2) A_4 &= mA_1 \\ (\Omega - k_3) A_4 + (k_1 + ik_2) A_3 &= mA_2 \\ (\Omega - k_3) A_1 - (k_1 - ik_2) A_2 &= mA_3 \\ (\Omega + k_3) A_2 - (k_1 + ik_2) A_1 &= mA_4 \end{aligned}$$

The necessary and sufficient condition for existence of a nontrivial solution is the vanishing of the coefficient determinant

(19.37)

$$\begin{vmatrix} -m & 0 & (\Omega+k_3) & (k_1-ik_2) \\ 0 & -m & (k_1+ik_2) & (\Omega-k_3) \\ (\Omega-k_3) & -(k_1-ik_2) & -m & 0 \\ -(k_1+ik_2) & (\Omega+k_3) & 0 & -m \end{vmatrix} = (\Omega^2-k^2-m^2)^2 = 0$$

which is obviously true because of (19.1).

For given energy $\Omega = +\sqrt{k^2+m^2} > 0$, Eq. (19.34) has two linearly independent solutions, namely

(19.38)

$$A(1) = \frac{1}{\sqrt{2V\Omega}} \begin{pmatrix} \dfrac{m}{\sqrt{\Omega-k_3}} \\[2mm] 0 \\[2mm] \sqrt{\Omega-k_3} \\[2mm] \dfrac{-(k_1+ik_2)}{\sqrt{\Omega-k_3}} \end{pmatrix} \quad \text{and} \quad A(2) = \frac{1}{\sqrt{2V\Omega}} \begin{pmatrix} \dfrac{(k_1-ik_2)}{\sqrt{\Omega-k_3}} \\[2mm] \sqrt{\Omega-k_3} \\[2mm] 0 \\[2mm] \dfrac{m}{\sqrt{\Omega-k_3}} \end{pmatrix}$$

which have been normalized so that

(19.39) $A^*(1)A(1) = A^*(2)A(2) = 1/V$ i.e. $\displaystyle\int_V \psi^*\psi\,d\mathbf{q} = 1$

and which satisfy the orthogonality relation

(19.40) $A^*(1)A(2) = A^*(2)A(1) = 0.$

The phase conventions have been arranged so that in the rest frame $(\mathbf{k}=0)$ the states

(19.41)

$$A(1;\mathbf{k}=0) = \frac{1}{\sqrt{2V}} \begin{pmatrix} 1 \\ 0 \\ 1 \\ 0 \end{pmatrix} \quad \text{and} \quad A(2;\mathbf{k}=0) = \frac{1}{\sqrt{2V}} \begin{pmatrix} 0 \\ 1 \\ 0 \\ 1 \end{pmatrix}$$

result, which are eigenstates of $\hat{\sigma}_3$ with eigenvalues $+1$ and -1, respectively.

Formally, Eqs. (19.36) have another pair of solutions if the energy E is taken to be $E = -\Omega = -\sqrt{k^2+m^2} < 0$, so that there are altogether *four* linearly independent solutions, labeled $A_\alpha(r)$ with $r = 1, 2, 3, 4$, which are summarized in Table 19.1. They are normalized so that

(19.42)

$A_\alpha^*(r)A_\alpha(r') = (1/V)\delta_{rr'}$ (sum over Greek subscripts appearing twice)

In analogy to the development of the theory of photons in Section 18, resulting in the possibility of representing all observables as operators bilinear in creation and annihilation operators, it is often convenient to

TABLE 19.1

THE FOUR LINEARLY INDEPENDENT SOLUTIONS OF EQ. (19.36)

$$(\Omega = +\sqrt{k^2+m^2})$$

r $\diagdown$ α	1	2	3	4	
1	$\dfrac{m}{\sqrt{\Omega-k_3}}$	$\dfrac{k_1-ik_2}{\sqrt{\Omega-k_3}}$	$-\sqrt{\Omega-k_3}$	0	
2	0	$\sqrt{\Omega-k_3}$	$\dfrac{k_1+ik_2}{\sqrt{\Omega-k_3}}$	$\dfrac{-m}{\sqrt{\Omega-k_3}}$	$\times \dfrac{1}{\sqrt{2V\Omega}}$
3	$\sqrt{\Omega-k_3}$	0	$\dfrac{m}{\sqrt{\Omega-k_3}}$	$\dfrac{k_1-ik_2}{\sqrt{\Omega-k_3}}$	
4	$\dfrac{-(k_1+ik_2)}{\sqrt{\Omega-k_3}}$	$\dfrac{m}{\sqrt{\Omega-k_3}}$	0	$\sqrt{\Omega-k_3}$	

work with space and time dependent Dirac field *operators*, which may be constructed in terms of the $A_\alpha(r)$ as

$$(19.43) \quad \psi(x) = \sum_{\mathbf{k}} \left[e^{i(\mathbf{kq}-\Omega t)} \sum_{r=1}^{2} A(r,\mathbf{k})\, a(r,\mathbf{k}) + e^{i(\mathbf{kq}+\Omega t)} \sum_{r=3}^{4} A(r,\mathbf{k})\, a(r,\mathbf{k}) \right]$$

where $a(r,\mathbf{k}) = a(\mathbf{k},r)$ is the annihilation operator of an electron with positive energy if $r = 1, 2$, and of an electron with negative energy if $r = 3, 4$. In terms of this operator and its adjoint $\bar{\psi}(x) = \psi^*(x)\gamma_4$, the energy operator for any many electron state takes the form

$$(19.44) \quad H = \int \mathscr{H}(x)\, d\mathbf{q} = \int \bar{\psi}(-i\boldsymbol{\gamma}\boldsymbol{\nabla} + m)\psi\, d\mathbf{q}$$

$$= \sum_{\mathbf{k}} \Omega \left[\sum_{r=1}^{2} a^+(r,\mathbf{k})\, a(r,\mathbf{k}) - \sum_{r=3}^{4} a^+(r,\mathbf{k})\, a(r,\mathbf{k}) \right].$$

Not surprisingly, this expression is not positive definite, because solutions of negative energy have been admitted into the theory. This difficulty of interpretation is compounded when one considers the operator of electric current density

$$(19.45) \qquad\qquad j_\mu(x) = e\bar{\psi}\gamma_\mu\psi$$

which owes its definition to the observation that it satisfies a conservation law

(19.46) $$[\partial j_\mu(x)/\partial x_\mu] = 0$$

as a consequence of the Dirac equation and its adjoint,

(19.47) $-i(\partial/\partial x_\mu)\gamma_\mu \psi + m\psi = 0;$ $i(\partial/\partial x_\mu)\bar\psi\gamma_\mu + m\bar\psi = 0.$

If one writes down the electric charge operator

(19.48) $$Q = e \int j_4(x)\,d\mathbf{q} = e \sum_{\mathbf{k}} \sum_{r=1}^{4} a^+(r,\mathbf{k})\,a(r,\mathbf{k})$$

one has now an expression which *is* positive definite as it stands.

Wanted, of course, is just the opposite, namely a positive definite energy density and an electric charge which may have negative as well as positive expectation values.

One famous way out of this dilemma is the hole theory of positrons by Dirac, who essentially availed himself of the possibility of performing a unitary transformation interchanging full states and holes as far as states of negative energy are concerned, in accordance with the procedure explained in Section 17. With the sign conventions implied by Eq. (17.25), this transformation amounts to the introduction of new operators

(19.49)
$$a_-^+(2, -\mathbf{k}) \equiv -a^+(3,\mathbf{k}) \qquad \text{replacing} \qquad a(3,\mathbf{k})$$
$$a_-^+(1, -\mathbf{k}) \equiv a^+(4,\mathbf{k}) \qquad \text{replacing} \qquad a(4,\mathbf{k})$$

so that, with the notation

(19.50)
$$a_+(r,\mathbf{k}) \equiv a(r,\mathbf{k}) \qquad \text{for} \qquad r = 1,2$$
$$A_+(r,\mathbf{k}) \equiv A(r,\mathbf{k}) \qquad \text{for} \qquad r = 1,2$$
$$A_-(2, -\mathbf{k}) \equiv -A(3,\mathbf{k})$$
$$A_-(1, -\mathbf{k}) \equiv A(4,\mathbf{k})$$

the operator (19.43) may be written

(19.51)
$$\psi(x) = \sum_{\mathbf{k}} \sum_{r=1}^{2} [e^{i(\mathbf{kq}-\Omega t)} A_+(r,\mathbf{k})\,a_+(r,\mathbf{k}) + e^{i(\mathbf{kq}+\Omega t)} A_-(r,-\mathbf{k})\,a_-^+(r,-\mathbf{k})]$$
$$= \sum_{\mathbf{k}} \sum_{r=1}^{2} [e^{i(\mathbf{kq}-\Omega t)} A_+(r,\mathbf{k})\,a_+(r,\mathbf{k}) + e^{-i(\mathbf{kq}-\Omega t)} A_-(r,\mathbf{k})\,a_-^+(r,\mathbf{k})].$$

The subscripts $(+)$ and $(-)$ have been affixed to the operators a and the amplitudes A in anticipation of an interpretation associating the subscript $(+)$ with lepton number $L = +1$ and the subscript $(-)$ with lepton number $L = -1$. The amplitudes $A_L(r, \mathbf{k})$ according to (19.50) are collected in Table 19.2. For $\overline{A}_L(r, \mathbf{k}) = A_L^*(r, \mathbf{k})\gamma_4$ one has the corresponding Table 19.3.

TABLE 19.2

THE AMPLITUDES $A_L(r, \mathbf{k})$

$$(\Omega = +\sqrt{k^2 + m^2})$$

α \ r	1	2	1	2	
1	$\dfrac{m}{\sqrt{\Omega - k_3}}$	$\dfrac{k_1 - ik_2}{\sqrt{\Omega - k_3}}$	0	$\sqrt{\Omega + k_3}$	
2	0	$\sqrt{\Omega - k_3}$	$\dfrac{-m}{\sqrt{\Omega + k_3}}$	$\dfrac{k_1 + ik_2}{\sqrt{\Omega + k_3}}$	$\times \dfrac{1}{\sqrt{2V\Omega}}$
3	$\sqrt{\Omega - k_3}$	0	$\dfrac{-(k_1 - ik_2)}{\sqrt{\Omega + k_3}}$	$\dfrac{-m}{\sqrt{\Omega + k_3}}$	
4	$\dfrac{-(k_1 + ik_2)}{\sqrt{\Omega - k_3}}$	$\dfrac{m}{\sqrt{\Omega - k_3}}$	$\sqrt{\Omega + k_3}$	0	
	$L = +1$		$L = -1$		

TABLE 19.3

THE AMPLITUDES $\overline{A}_L(r, \mathbf{k}) = A_L^*(r, \mathbf{k})\gamma_4$

$$(\Omega = +\sqrt{k^2 + m^2})$$

α \ r	1	2	1	2	
1	$\sqrt{\Omega - k_3}$	0	$\dfrac{-(k_1 + ik_2)}{\sqrt{\Omega + k_3}}$	$\dfrac{-m}{\sqrt{\Omega + k_3}}$	
2	$\dfrac{-(k_1 - ik_2)}{\sqrt{\Omega - k_3}}$	$\dfrac{m}{\sqrt{\Omega - k_3}}$	$\sqrt{\Omega + k_3}$	0	$\times \dfrac{1}{\sqrt{2V\Omega}}$
3	$\dfrac{m}{\sqrt{\Omega - k_3}}$	$\dfrac{(k_1 + ik_2)}{\sqrt{\Omega - k_3}}$	0	$\sqrt{\Omega + k_3}$	
4	0	$\sqrt{\Omega - k_3}$	$\dfrac{-m}{\sqrt{\Omega + k_3}}$	$\dfrac{(k_1 - ik_2)}{\sqrt{\Omega + k_3}}$	
	$L = +1$		$L = -1$		

From these tables one can construct immediately the matrices, needed later,

(19.52) $\sum_{r} [A_+(\mathbf{k},r)]_\alpha [\bar{A}_+(\mathbf{k},r)]_\beta =$

$$\begin{pmatrix} m & 0 & \Omega+k_3 & k_1-ik_2 \\ 0 & m & k_1+ik_2 & \Omega-k_3 \\ \Omega-k_3 & -(k_1-ik_2) & m & 0 \\ -(k_1+ik_2) & \Omega+k_3 & 0 & m \end{pmatrix} \times (1/2V\Omega)$$

and

(19.53) $\sum_{r} [\bar{A}_-(\mathbf{k},r)]_\alpha [A_-(\mathbf{k},r)]_\beta =$

$$\begin{pmatrix} -m & 0 & \Omega-k_3 & -(k_1+ik_2) \\ 0 & -m & -(k_1-ik_2) & \Omega+k_3 \\ \Omega+k_3 & k_1+ik_2 & -m & 0 \\ k_1-ik_2 & \Omega-k_3 & 0 & -m \end{pmatrix} \times (1/2V\Omega)$$

which in terms of γ matrices may be expressed as

(19.52) $\sum_{r} [A_+(\mathbf{k},r)]_\alpha [\bar{A}_+(\mathbf{k},r)]_\beta = (1/2V\Omega)[\gamma_4\Omega - \mathbf{k}\gamma + mI]_{\alpha\beta}$

> or

$\sum_{r} A_+(\mathbf{k},r)\,\bar{A}_+(\mathbf{k},r) = (1/2V\Omega)(\hat{k}+m)$

(19.53) $\sum_{r} [\bar{A}_-(\mathbf{k},r)]_\beta [A_-(\mathbf{k},r)]_\alpha = (1/2V\Omega)[\gamma_4\Omega - \mathbf{k}\gamma - mI]_{\alpha\beta}$

> or

$\sum_{r} A_-(\mathbf{k},r)\,\bar{A}_-(\mathbf{k},r) = (1/2V\Omega)(\hat{k}-m)$

where $\hat{k} = k_\nu\gamma_\nu = \gamma_4\Omega - \mathbf{k}\gamma.$

With this relabeling, the energy operator takes the form

(19.54) $H = \sum_{\mathbf{k}} \sum_{r=1}^{2} \Omega[N_+(\mathbf{k},r) + N_-(\mathbf{k},r) - 2]$

where

(19.55) $N_\pm(\mathbf{k},r) = a_\pm^+(\mathbf{k},r)\,a_\pm(\mathbf{k},r)$

are now interpreted as the number operators of electrons and positrons, respectively. The expression (19.54) is obviously invariant under interchanges $\mathbf{k} \leftrightarrow -\mathbf{k}$ and/or $r = 1, 2 \leftrightarrow r = 2, 1$. The conventions inherent in the notations (19.50) are such, however, that any positron state labeled $(\mathbf{k},r)$ differs mechanically from the corresponding electron state labeled $(\mathbf{k},r)$ only by its handedness.

   In this hole theory of positrons, the infinite vacuum energy $-2\sum_{\mathbf{k}}\Omega$

is disregarded since it is the same for all states and thus, in principle, not observable. A modified definition of the electric current density which eliminates from the outset any nonvanishing vacuum charge, without invalidating the conservation law (19.46), is

$$(19.56) \qquad j_\mu(x) = (e/2)[\bar{\psi}, \gamma_\mu \psi] = (e/2)[\gamma_\mu]_{\alpha\beta}[\bar{\psi}_\alpha \psi_\beta - \psi_\beta \bar{\psi}_\alpha]$$

giving in particular for the total electric charge

$$(19.57) \qquad Q = e \sum_{\mathbf{k}} \sum_{r=1}^{2} [N_+(\mathbf{k}, r) - N_-(\mathbf{k}, r)]$$

which is consistent with the interpretation of N_+ and N_- as number operators for electrons and positrons, respectively.*

It is instructive and useful for later applications to write out the operator (19.56) for the current density in terms of the operators $a_\pm^+(\mathbf{k}, r)$ and $a_\pm(\mathbf{k}, r)$; using the anti-C.R.s of these operators one finds

(19.56)

$$
\begin{aligned}
(e/2)[\bar{\psi}, \gamma_\mu \psi] = e &\sum_{\mathbf{k}} \sum_{\mathbf{k'}} \sum_{r} \sum_{r'} \{\exp(-i[(\mathbf{k'} - \mathbf{k})\mathbf{q} - (\Omega' - \Omega)t]) \\
&\times \bar{A}_+(\mathbf{k'}, r') \gamma_\mu A_+(\mathbf{k}, r) a_+^+(\mathbf{k'}, r') a_+(\mathbf{k}, r) \\
&- \exp(i[(\mathbf{k'} - \mathbf{k})\mathbf{q} - (\Omega' - \Omega)t]) \bar{A}_-(\mathbf{k'}, r') \gamma_\mu \\
&\times A_-(\mathbf{k}, r) a_-^+(\mathbf{k}, r) a_-(\mathbf{k'}, r') \\
&+ \exp(-i[(\mathbf{k'} + \mathbf{k})\mathbf{q} - (\Omega' + \Omega)t]) \bar{A}_+(\mathbf{k'}, r') \gamma_\mu \\
&\times A_-(\mathbf{k}, r) a_+^+(\mathbf{k'}, r') a_-^+(\mathbf{k}, r) \\
&- \exp(i[(\mathbf{k'} + \mathbf{k})\mathbf{q} - (\Omega' + \Omega)t]) \bar{A}_-(\mathbf{k'}, r') \gamma_\mu \\
&\times A_+(\mathbf{k}, r) a_+(\mathbf{k}, r) a_-(\mathbf{k'}, r')\} \\
&- e \sum_{\mathbf{k}} \sum_{r} [\bar{A}_+(\mathbf{k}, r) \gamma_\mu A_+(\mathbf{k}, r) - \bar{A}_-(\mathbf{k}, r) \gamma_\mu A_-(\mathbf{k}, r)].
\end{aligned}
$$

The last term, not containing any operators a or a^+, vanishes for all μ, as can be seen by straightforward computation from Table 19.3, and (19.57) follows as a special case for $\mu = 4$.

The ingenious hole theory, necessitated by the admittance of negative energy solutions into the theory, can be avoided altogether from the outset, if one exploits a peculiar symmetry property of the Dirac equation which is known in the literature as charge-conjugation symmetry. By this is meant that if one interchanges the number operators N_+ and N_-

$$(19.58) \qquad N_+ \rightleftarrows N_-$$

---

* If one uses instead of (19.42) a relativistically covariant normalization to Ω/m particles per unit volume, then, in the definition (19.51), a factor $\sqrt{m/\Omega}$ is required.

then all mechanical operators, such as P_μ, are not affected, whereas the operator j_μ, *with the understanding that* e *is a numerically fixed parameter*, changes sign. This property is already manifestly incorporated in expressions (19.54) and (19.56).

By a curious and rather remarkable property of the Dirac equation and its solutions, not shared, for example, by the equation governing the neutrino, the effect of (19.58) can be represented by an operator $\hat{\Gamma}$ which affects only the spin-chirality space, so that one may write instead of (19.58)

$$(19.58') \quad \psi \to \psi' = \eta_\Gamma \hat{\Gamma}\psi = \eta_\Gamma \psi \tilde{\hat{\Gamma}}; \qquad \bar{\psi} \to \bar{\psi}' = \eta_\Gamma^* \hat{\Gamma}^{-1}\psi = \eta_\Gamma^* \psi \tilde{\hat{\Gamma}}^{-1}$$

where $\hat{\Gamma}$ does not affect the creation and annihilation operators, and where η_Γ is an as yet undetermined phase factor subject only to the condition $\eta_\Gamma \eta_\Gamma^* = 1$. A representation of $\hat{\Gamma}$ in terms of the matrices γ_μ may be obtained by requiring ψ' to satisfy the Dirac equation

$$(19.59) \qquad [-i\gamma_\mu(\partial/\partial x_\mu) + m]\psi' = 0$$

provided ψ and $\bar{\psi}$ satisfy the same equation, i.e. (19.47), respectively. It turns out that this condition is sufficient to ensure the invariance of P_μ and the change in sign in j_μ. Substituting in (19.59) for ψ' one finds

$$(19.60) \qquad \hat{\Gamma}[-i\hat{\Gamma}^{-1}\gamma_\mu \hat{\Gamma}(\partial\bar{\psi}/\partial x_\mu) + m\bar{\psi}] = 0.$$

This is a consequence of the second equation (19.47) provided

(19.61)

$$\hat{\Gamma}^{-1}\gamma_\mu \Gamma = -\tilde{\gamma}_\mu, \quad \text{in components} \quad [\hat{\Gamma}^{-1}\gamma_\mu \hat{\Gamma}]_{\alpha\beta} = -[\gamma_\mu]_{\beta\alpha}.$$

With the representation (19.5) one has

$$(19.62) \qquad [\gamma_\mu]_{\alpha\beta} = \begin{cases} +[\gamma_\mu]_{\beta\alpha} & \text{for} \quad \mu = 2,4 \\ -[\gamma_\mu]_{\beta\alpha} & \text{for} \quad \mu = 1,3. \end{cases}$$

One can thus satisfy (19.61) by choosing

$$(19.63) \qquad \hat{\Gamma} = i\gamma_2\gamma_4 = i\begin{pmatrix} -\sigma_2 & 0 \\ 0 & \sigma_2 \end{pmatrix} = \begin{pmatrix} 0 & 1 & 0 & 0 \\ -1 & 0 & 0 & 0 \\ 0 & 0 & 0 & -1 \\ 0 & 0 & 1 & 0 \end{pmatrix}$$

so that

$$(19.64) \qquad \tilde{\hat{\Gamma}} = -\hat{\Gamma}, \quad \text{i.e.} \quad \tilde{\hat{\Gamma}}\hat{\Gamma}^{-1} = -I$$

and $\hat{\Gamma}$ becomes unitary,

$$(19.65) \qquad \hat{\Gamma}^{-1} = -\hat{\Gamma}^*, \quad \text{i.e.} \quad \hat{\Gamma}\hat{\Gamma}^+ = I.$$

The proof that j_μ changes sign under the operation $\psi \to \psi'$ is now straight-forward. Using (19.64) and (19.61) one finds

$$
\begin{aligned}
j''_\mu &= (e/2)\,[\bar{\psi}', \gamma_\mu \psi'] = (e/2)\,[\gamma_\mu]_{\alpha\beta}[\bar{\psi}'_\alpha \psi'_\beta - \psi'_\beta \bar{\psi}'_\alpha] \\
&= -(e/2)\,[\gamma_\mu]_{\alpha\beta}[\hat{\Gamma}^{-1}]_{\alpha\gamma}[\hat{\Gamma}]_{\beta\delta}[\psi_\gamma \bar{\psi}_\delta - \bar{\psi}_\delta \psi_\gamma] \\
&= (e/2)\,[\hat{\Gamma}^{-1}]_{\gamma\alpha}[\gamma_\mu]_{\alpha\beta}[\hat{\Gamma}]_{\beta\delta}[\bar{\psi}_\delta \psi_\gamma - \psi_\gamma \bar{\psi}_\delta] \\
&= -(e/2)\,[\gamma_\mu]_{\delta\gamma}[\bar{\psi}_\delta \psi_\gamma - \psi_\gamma \bar{\psi}_\delta] = -j_\mu.
\end{aligned}
$$

(19.66)

It should be stressed that if one wishes to replace the arrow in Eq. (19.58), i.e. if one wants to formally define an operator Γ connecting ψ' with ψ by

(19.67) $$ \psi' = \Gamma\psi = \eta_\Gamma \hat{\Gamma}\bar{\psi} = \eta_\Gamma \hat{\Gamma}\psi^* \gamma_4 = \eta_\Gamma \hat{\Gamma}\gamma_4 K\psi $$

then this operator *must* be antiunitary,

(19.68) $$ \Gamma = \eta_\Gamma \hat{\Gamma}\gamma_4 K = \eta_\Gamma i\gamma_2 K. $$

Since γ_2 is pure imaginary and $\gamma_2 \gamma_2 = -I$, and because $\eta_\Gamma \eta_\Gamma^* = 1$, one has

(19.69) $$ \Gamma^2 = i\gamma_2 K i\gamma_2 K = -\gamma_2 \gamma_2 = +I $$

and no superselection rule is generated in the electron-positron system by pure charge-conjugation symmetry. However, if one considers now the operation of combined inversion, using (19.27),

(19.70) $$ \Sigma = \Pi\Gamma = \eta_\Pi \eta_\Gamma \gamma_4 \Pi_D\, i\gamma_2 K $$

one notices that Σ too is antiunitary, and has the additional property*

(19.71) $$ \Sigma^2 = i\gamma_4 \gamma_2 K i\gamma_4 \gamma_2 K = -\gamma_4 \gamma_2 \gamma_4 \gamma_2 = \gamma_2^2 \gamma_4^2 = -I. $$

Thus the spaces spanned by the state vector $|\ldots N_L(\mathbf{k}, r)\ldots\rangle$ and its spatial inverse $\Sigma|\ldots N_L(\mathbf{k}, r)\ldots\rangle$ are separated by a superselection rule, which in view of the fact that Γ interchanges $L \rightleftarrows -L$ may be interpreted as the reason for the conservation of lepton number L as far as electrons and positrons are concerned.

One may thus from the outset circumvent the introduction of negative energy states entirely, by considering only positive energy solutions of the Dirac equation and defining positron solutions as obtained by the operation of combined inversion Σ from corresponding electron solutions. The consistency of this procedure is borne out if one computes with

---

* Note that this would be true even if one had chosen the convention $\eta_\Pi^2 = -1$, because η_Π enters Σ^2 only in the combination $\eta_\Pi \eta_\Pi^*$.

the representation (19.70) the effect of Σ on the first two columns of Table 19.2, namely

$$(19.72) \qquad \Sigma A_{+1}(\mathbf{k}, r) = \eta_\Pi \eta_\Gamma A_{-1}(-\mathbf{k}, r).$$

The effect of combined inversion is thus a change in sign of the momentum $\mathbf{k}$ and the replacement of lepton number $L = +1$ by lepton number $L = -1$. The operator (19.51), in particular, may be written

$$(19.73) \qquad \psi(x) = \sum_{\mathbf{k}} \sum_{r=1}^{2} (I + \Sigma) e^{i(\mathbf{k}\mathbf{q} - \Omega t)} A_{+1}(\mathbf{k}, r) a_{+1}(\mathbf{k}, r)$$

with the understanding that the operator of combined inversion applied to an annihilation operator has the effect

$$(19.74) \qquad \Sigma a_{+1}(\mathbf{k}, r) \Sigma^{-1} = \eta_\Pi \eta_\Gamma a_{-1}^+(-\mathbf{k}, r).$$

These considerations strengthen the point of view, already expressed on the occasion of the corresponding development for neutrino and antineutrino, that an antilepton should be considered as the *spatial* inverse of the lepton, and not as its time reverse as had been suggested by Feynman.

The operation of time reversal, when applied to electrons, is also an antiunitary operation, satisfying, since γ_1 and γ_3 are real and satisfy $\gamma_1^2 = \gamma_3^2 = -I$,

$$(19.75) \qquad \Theta^2 = \gamma_1 \gamma_3 K \gamma_1 \gamma_3 K = \gamma_1 \gamma_3 \gamma_1 \gamma_3 = -\gamma_3^2 \gamma_1^2 = -1$$

so that time-reversed states of electron states are also separated from electron states by a superselection rule. One is tempted to speculatively identify such states with the corresponding muon states, so that conservation of muon number L_μ, already mentioned in Section 17, flows from that particular superselection rule. Such speculation intensifies the riddle posed by the muon's mass which, except for the number L_μ, seems to be the only attribute by which a muon can be distinguished dynamically from an electron. Since at present there exists no satisfactory dynamical theory of the masses of elementary particles, one cannot dismiss the possibility that in the actual physical world the masses of electron and muon are connected by the transformation property $m_\mu = \Theta m_e \Theta^{-1}$.

Dirac's equation in four-component form (19.8) looks deceptively simple. The wealth of information contained in it is brought out more transparently when one attempts to write the equation in two-component form. To this end, consider again Eq. (19.2) and remember that, according to (19.21), the two two-component functions u and v are equal in the rest frame, i.e. whenever the eigenvalue of $\mathbf{P}$ vanishes. Knowledge of the two-component ψ function u is thus sufficient in the rest frame, and it ought to be possible to find a two-component w, say, from which all information contained in the four-component ψ can be extracted.

To be precise, one wants to find a two-component ψ function w, taking the place of the four-component function ψ, and 2×2 matrices representing any operator F, taking the place of any given 4×4 matrix representing an operator $\hat{F}$, so that the expectation value of F in the state w and of $\hat{F}$ in the state ψ are equal,

$$(19.76) \qquad \langle \psi | \hat{F} | \psi \rangle = \langle w | F | w \rangle$$

provided the normalizations

$$(19.77) \qquad \langle \psi | \psi \rangle = 1$$

and

$$(19.78) \qquad \langle w | w \rangle = 1$$

are adopted. Strict adherence to the condition (19.78) is decisive for the consistency of the procedure.

To actually carry out the transition from a four-component to a two-component description, remember that finite momentum **P** means v and u are not equal and write

$$(19.79) \qquad v = (I + W) u$$

where W is a 2×2 operator that can, in principle, be derived from Eq. (19.2), as will be shown by successive approximations below. The desired two-component function w likewise must be obtainable by some operation G, say, applied to u,

$$(19.80) \qquad w = Gu; \qquad u = G^{-1} w.$$

The connection between G and W is culled immediately from the normalization conditions (19.77) and (19.78), which read

$$(19.81) \quad \langle \psi | \psi \rangle = \langle u | u \rangle + \langle v | v \rangle = \langle u | I + (I + W^+)(I + W) | u \rangle = I$$

and

$$(19.82) \qquad \langle w | w \rangle = \langle u | G^+ G | u \rangle = 1.$$

This can be true only if, up to some phase,

$$(19.83) \qquad G^+ G = I + (I + W^+)(I + W).$$

Now the most general 4×4 operator is of the form

$$(19.84) \qquad \hat{F} = \begin{pmatrix} F_{11} & F_{12} \\ F_{21} & F_{22} \end{pmatrix} \qquad (F_{jk} \text{ are } 2 \times 2 \text{ matrices})$$

so that its expectation value can be written in terms of two-component functions

$$(19.85)$$
$$\begin{aligned} \langle \psi | \hat{F} | \psi \rangle &= \langle u | F_{11} | u \rangle + \langle u | F_{12} | v \rangle + \langle v | F_{21} | u \rangle + \langle v | F_{22} | v \rangle \\ &= \langle u | F_{11} + F_{12}(I + W) + (I + W^+) F_{21} + (I + W^+) F_{22}(I + W) | u \rangle \\ &= \langle w | (G^+)^{-1} [F_{11} + F_{12}(I + W) + (I + W^+) F_{21} \\ &\quad + (I + W^+) F_{22}(I + W)] G^{-1} | w \rangle. \end{aligned}$$

The condition (19.76) is therefore satisfied provided one uses as 2×2 operator F the expression

$$(19.86)$$
$$F = (G^+)^{-1} [F_{11} + F_{12}(I + W) + (I + W^+) F_{21} + (I + W^+) F_{22}(I + W)] G^{-1}.$$

The remaining task is to find expressions for G and W from Eq. (19.2). By substitution of (19.79) one finds the operator equations

$$(19.87) \qquad \begin{aligned} H - \boldsymbol{\sigma}\mathbf{P} &= (I + W) m \\ (H + \boldsymbol{\sigma}\mathbf{P})(I + W) &= mI \end{aligned}$$

so that one has simply

$$(19.88) \qquad I + W = I + W^+ = (1/m)(H - \boldsymbol{\sigma}\mathbf{P})$$

and therefore by (19.83)

$$(19.89) \qquad G^+ G = [I - i(I + W)][I + i(I + W)].$$

Thus, up to some arbitrary phase,

$$(19.90) \qquad G = I + i(I + W) = \sqrt{2}\, e^{i(\pi/4)} [I + (1/\sqrt{2})\, e^{i(\pi/4)} W]$$

$$(19.91) \qquad G^+ = I - i(I + W) = \sqrt{2}\, e^{-i(\pi/4)} [I + (1/\sqrt{2})\, e^{-i(\pi/4)} W].$$

The expressions for G^{-1} and $(G^+)^{-1}$ needed in (19.86) can now be obtained as a series in powers of W. Keeping terms up to order P^3/m^3 one has the approximations

$$(19.92) \qquad H = \sqrt{m^2 + P^2} = m + (P^2/2m) + \dots$$

$$(19.93) \qquad W = (1/m)(H - \boldsymbol{\sigma}\mathbf{P} - m) = -(\boldsymbol{\sigma}\mathbf{P}/m) + (P^2/2m^2) + \dots$$

(19.94)
$$W^2 = (P^2/m^2) - (P^2 \boldsymbol{\sigma}\mathbf{P}/m^3) + \ldots$$

(19.95)
$$W^3 = -(P^2 \boldsymbol{\sigma}\mathbf{P}/m^3) + \ldots$$

and from these follow

(19.96)
$$G^{-1} = [(1-i)/2] I - (1/2) W + [(1+i)/4] W^2 - (i/4) W^3 + \ldots$$
$$= [(1-i)/2] I + (\boldsymbol{\sigma}\mathbf{P}/2m) + i(P^2/4m^2) - (P^2 \boldsymbol{\sigma}\mathbf{P}/4m^3) + \ldots$$

(19.97)
$$(G^+)^{-1} = [(1+i)/2] I + (\boldsymbol{\sigma}\mathbf{P}/2m) - i(P^2/4m^2) - (P^2 \boldsymbol{\sigma}\mathbf{P}/4m^3) + \ldots$$

Collecting terms one finds for (19.86)

(19.98)
$$\begin{aligned}
F = {}& (1/2)\,(F_{11} + F_{12} + F_{21} + F_{22}) \\
& + (1/4m)\,\{[(F_{11} - F_{22})\,\boldsymbol{\sigma}\mathbf{P} + \boldsymbol{\sigma}\mathbf{P}(F_{11} - F_{22})] \\
& + i[(F_{11} + F_{22})\,\boldsymbol{\sigma}\mathbf{P} - \boldsymbol{\sigma}\mathbf{P}(F_{11} + F_{22})] \\
& + [(F_{21} - F_{12})\,\boldsymbol{\sigma}\mathbf{P} - \boldsymbol{\sigma}\mathbf{P}(F_{21} - F_{12})] \\
& + i[(F_{21} + F_{12})\,\boldsymbol{\sigma}\mathbf{P} - \boldsymbol{\sigma}\mathbf{P}(F_{21} + F_{12})]\} \\
& + (1/8m^2)\,\{-[(F_{11} + F_{22})\,P^2 + P^2(F_{11} + F_{22})] \\
& + i[(F_{11} - F_{22})\,P^2 - P^2(F_{11} - F_{22})] \\
& - [(F_{21} + F_{12})\,P^2 + P^2(F_{21} + F_{12})] \\
& + i[(F_{21} - F_{12})\,P^2 + P^2(F_{21} - F_{12})] \\
& + 2[\boldsymbol{\sigma}\mathbf{P}(F_{11} + F_{22})\,\boldsymbol{\sigma}\mathbf{P} - i\boldsymbol{\sigma}\mathbf{P}(F_{21} - F_{12})\,\boldsymbol{\sigma}\mathbf{P}]\} \\
& + (1/8m^3)\{-[(F_{11} - F_{22})\,P^2\,\boldsymbol{\sigma}\mathbf{P} + P^2\,\boldsymbol{\sigma}\mathbf{P}(F_{11} - F_{22})] \\
& - i[(F_{11} + F_{22})\,P^2\,\boldsymbol{\sigma}\mathbf{P} - P^2\,\boldsymbol{\sigma}\mathbf{P}(F_{11} + F_{22})] \\
& + [(F_{21} - F_{21})\,P^2\,\boldsymbol{\sigma}\mathbf{P} - P^2\,\boldsymbol{\sigma}\mathbf{P}(F_{12} - F_{21})] \\
& - i[(F_{12} + F_{21})\,P^2\,\boldsymbol{\sigma}\mathbf{P} - P^2\,\boldsymbol{\sigma}\mathbf{P}(F_{12} + F_{21})] \\
& - [\boldsymbol{\sigma}\mathbf{P}(F_{12} - F_{21})\,P^2 - P^2(F_{12} - F_{21})\,\boldsymbol{\sigma}\mathbf{P}] \\
& + i[\boldsymbol{\sigma}\mathbf{P}(F_{11} + F_{22})\,P^2 - P^2(F_{11} + F_{22})\,\boldsymbol{\sigma}\mathbf{P}]\} \\
& + \text{terms of order } (P^4/m^4) \text{ and higher.}
\end{aligned}$$

This formula ought to be sufficient for any practical purposes. It is valid, incidentally, even if $\mathbf{P}$ is the operator of momentum of a Dirac particle in a fixed external field, to be explained in Section 20.

As a simple example the matrices γ are given in two-component form in Table 19.4 for the field free case and under expansion up to and including the power P^2/m^2.

TABLE 19.4

REPRESENTATION OF THE MATRICES γ IN TWO-COMPONENT FORM, UP TO TERMS OF ORDER P^2/m^2 INCLUSIVE

(It will be noticed that the operators α can be interpreted as representing the vector of kinematic velocity. This completes the emancipation of the momentum from the velocity, which goes like a red thread through the development of classical mechanics in the 19th century.)

$\hat{F}$	F_{11}	F_{22}	F_{12}	F_{21}	F
γ	0	0	$-\sigma$	σ	$(1/2m)[\sigma(\sigma\mathbf{P})-(\sigma\mathbf{P})\sigma] = -(i/2m)(\sigma\times\mathbf{P})$
γ_4	0	0	1	1	$[1-(P^2/2m^2)]I$
$\alpha = \gamma_4\gamma$	σ	$-\sigma$	0	0	$(1/2m)[(\sigma\mathbf{P})\sigma+\sigma(\sigma\mathbf{P})] = (\mathbf{P}/m)I$
γ_5	1	-1	0	0	$(\sigma\mathbf{P})/m$

NOTES

Dirac [1] found the equation which bears his name.

The conventions used regarding Dirac's equation agree with the ones given by Feynman [2].

Dirac [3] proposed the hole theory of positively charged electrons to circumvent difficulties arising from the states of negative energy, which had been examined earlier by Oppenheimer [4].

Charge conjugation as a symmetry operation was apparently first proposed by Kramers [5].

Becker [6] gave the first correct reduction of Dirac's equation to two-component form. See also Chraplyvy [7].

Earlier work by various authors following a procedure by Darwin [8] is incorrect, because the two-component ψ functions used by Darwin are not properly normalized, giving rise to nonhermitean terms in the two-component Hamiltonian.

REFERENCES

[1] P. A. M. Dirac, *Proc. Roy. Soc.* **A117**, 610 (1928); **A118**, 351 (1928).
[2] R. P. Feynman, "Theory of Fundamental Processes," Chapter 23. Benjamin, New York, 1961.
[3] P. A. M. Dirac, *Proc. Roy. Soc.* **A126**, 360 (1931).
[4] J. R. Oppenheimer, *Phys. Rev.* **35**, 562 (1930).
[5] H. A. Kramers, *Proc. Acad. Sci. Amsterdam* **A40**, 814 (1937).
[6] R. Becker, *Nachr. Akad. Wiss. Goettingen Math.-Phys. Kl.* p. 39 (1945).
[7] Z. V. Chraplyvy, *Phys. Rev.* **91**, 388 (1953); **92**, 1310 (1953); **99**, 317 (1955).
[8] C. G. Darwin, *Proc. Roy. Soc.* **A118**, 654 (1928).

The Lack of Sufficient Reason for Actually Existing Interactions

The apparent capriciousness of nature which provides just four supposedly basic interactions of widely disparate strengths is matched perhaps only by the equally enigmatic apparent arbitrariness exhibited in the mass spectrum of the so-called elementary particles.

The various classical responses to the baffling mystery of interaction have resulted in tenable edifices, such as Einstein's theory of gravitation, and in failures, such as Einstein's unified field theory. This type of approach to the problem of interaction has now generally been abandoned, partly because gravitation and electromagnetism turned out to comprise only a fraction of observable interaction phenomena.

The experimental exploration of the so-called strong and weak interactions, which appear empirically to be as disconnected as are gravitation and electromagnetism, has gathered evidence on a vast scale since the advent of quantum mechanics, lending urgency to all those attempts which cast among the tenets of quantum mechanics for a vehicle to which a theory of interactions might be attached.

The arbitrariness in the phase of a state vector, in particular, has been a favorite starting point for efforts aimed at deriving the specific form of actually existing interactions from invariance arguments. A typical line of reasoning runs as follows.

Consider a single particle state $|1\rangle = a^+|0\rangle$ characterized by a set of ψ functions $\phi_\tau(\mathbf{q})$ so that

$$(20.1) \qquad W_\tau = \int \phi_\tau^*(\mathbf{q})\,\phi_\tau(\mathbf{q})\,d\mathbf{q} = 1$$

and

$$(20.2) \qquad R_{\tau\tau'}\,e^{i\gamma_{\tau\tau'}} = \int \phi_\tau^*(\mathbf{q})\,\phi_\tau(\mathbf{q})\,d\mathbf{q} = \text{complex number}$$

so that

(20.3) $P_{\tau\tau'} = R_{\tau\tau'}^2 =$ Probability for finding the values τ' of the observables if the particle is known to be in the state characterized by the values τ.

169

In terms of the eigenstates of momentum $\mathbf{k}$, for example, and their ψ functions $\psi_\mathbf{k}(\mathbf{q})$, any such state can be obtained by a linear superposition

$$(20.4) \qquad \phi_\tau(q) = \sum_\mathbf{k} c_\mathbf{k}(\tau)\, \psi_\mathbf{k}(\mathbf{q})$$

and one can write in terms of the expansion coefficients

$$(20.5) \qquad W_\tau = \sum_\mathbf{k} |c_\mathbf{k}(\tau)|^2 ; \qquad R_{\tau\tau'}\, e^{i\gamma_{\tau\tau'}} = \sum_\mathbf{k} c_\mathbf{k}^*(\tau')\, c_\mathbf{k}(\tau).$$

Now the following question arises: If one writes

$$(20.6) \quad \phi_\tau(\mathbf{q}) = A_\tau(\mathbf{q})\, e^{i\alpha_\tau(\mathbf{q})} ; \qquad A_\tau(\mathbf{q}) \text{ and } \alpha_\tau(\mathbf{q}) \; real \text{ functions,}$$

to what extent is the phase $\alpha_\tau(\mathbf{q})$ determined by the two quantities which have an observable meaning, namely W_τ and $R_{\tau\tau'}$?

Clearly, W_τ is independent of $\alpha_\tau(\mathbf{q})$, and the phase remains undetermined by W_τ. The only requirement imposed by given $R_{\tau\tau'}$ is that the integral (20.2) must have a definite modulus. Consequently, the integrand, although it need not have a definite phase at each point, must have a definite phase difference between any two points in space, whether neighboring or not. This follows, by generalization, from the simple rules governing the addition of complex numbers. Suppose one wants to have, in the sum $R(\cos\gamma + i\sin\gamma) = R_1(\cos\gamma_1 + i\sin\gamma_1) + R_2(\cos\gamma_2 + i\sin\gamma_2)$, only R determined, but to allow γ to remain arbitrary. Using the elementary formula $R^2 = R_1^2 + R_2^2 + 2R_1 R_2 \cos(\gamma_1 - \gamma_2)$ one sees that, for given R, only the phase difference $\gamma_1 - \gamma_2$ is determined.

Thus, the change in phase of $\phi_\tau^*(\mathbf{q})\phi_\tau(\mathbf{q})$ along a closed curve must vanish. This requires then that the change in phase of $\phi_\tau(\mathbf{q})$ along a closed curve shall be opposite and equal to that in $\phi_\tau^*(\mathbf{q})$ and hence the same in all $\phi_\tau(\mathbf{q})$. Result:

The change in phase of a ψ function along a closed curve must be the same for all ψ functions, independent of τ.

In other words, the change in phase along a closed curve must be something determined by the dynamical system itself, independent of the particular state considered. This *suggests* exploiting the nonintegrability of phase to accommodate features of the environment, such as provided by some external field in which the particle moves.

To investigate this possibility write, generalizing to dependence in space and time,

$$(20.7) \qquad \Phi_\tau(x) = \tilde{\Phi}_\tau^0(x)\, e^{i\beta(x)}$$

where $\Phi_\tau^0(x)$ is an "ordinary" ψ function, i.e. one with a definite phase at each point $x = (\mathbf{q}, t)$ in space-time, and the indeterminacy in the phase is put into the factor $e^{i\beta(x)}$. It will be noted from the foregoing that $\beta(x)$

is not required to be a function of x having definite values at each point, but $\beta(x)$ must have definite derivatives,

$$(20.8) \qquad\qquad \kappa_\nu = \partial\beta/\partial x_\nu$$

at each point, which do not, in general, satisfy the integrability condition $\partial\kappa_\nu/\partial x_\mu = \partial\kappa_\mu/\partial x_\nu$. Now the change in phase, around a closed curve, should be observable because $R_{\tau'\tau}$ depends on it. In four dimensions this change in phase is, by Stokes' theorem,

$$(20.9) \qquad \oint \kappa_\mu \, dx_\mu = \int\int [(\partial\kappa_\nu/\partial x_\mu) - (\partial\kappa_\mu/\partial x_\nu)] \, dS_{\mu\nu}$$

where $dS_{\mu\nu}$ is the skew tensor element of the surface bounded by the curve.

It is now very tempting to identify the derivatives of the phase β with the electromagnetic potentials, so that

$$(20.10) \qquad\qquad \kappa_\nu = -eA_\nu$$

and

$$(20.11) \qquad (\partial\kappa_\nu/\partial x_\mu) - (\partial\kappa_\mu/\partial x_\nu) = -eF_{\mu\nu}$$

can be identified with the electromagnetic field tensor, if e stands for the numerical value of the electric charge of the particle under consideration. The homogeneous Maxwell equations

$$(20.12) \qquad\qquad \epsilon_{\kappa\lambda\mu\nu}(\partial F_{\mu\nu}/\partial x_\lambda) = 0$$

are then automatically satisfied, and are equivalent to requiring that the right-hand side of (20.9) must not depend on which surface bounded by the curve given on the left-hand side is taken. Indeed, if one has two such surfaces, then the difference in the integral $\int F_{\mu\nu} dS_{\mu\nu}$ over them will be given, by Gauss' theorem, as

$$(20.13) \qquad \Delta \int F_{\mu\nu} dS_{\mu\nu} = \int \epsilon_{\kappa\lambda\mu\nu}(\partial F_{\mu\nu}/\partial x_\lambda) \, d^4 x$$

where the integral on the right is taken over the volume between the two surfaces, and (20.12) is necessary to guarantee the vanishing of this expression.

The identification (20.10) gives rise to some observable effects whose importance for the quantum mechanical concept of interaction was first realized by Aharonov and Bohm, and which are bound to startle anyone who has been brought up with classical electrodynamics. For example, if a coherent beam of electrons is taken around both sides of a solenoid, an interference pattern is observed which will shift continuously with continuously varied flux F through the solenoid. This is predicted by the identification (20.10), because if one considers the part of the ψ function at the point of interference which is a linear superposition of two ψ functions corresponding to path 1 taken around one side and path 2 taken around the other side of the solenoid,

(20.14)

$$\phi = \phi_1 e^{i\beta_1} + \phi_2 e^{i\beta_2}; \qquad \beta_1 = -e \int_1 \mathbf{A}\,d\mathbf{q}; \qquad \beta_2 = -e \int_2 \mathbf{A}\,d\mathbf{q}$$

then the interference between the two beams will depend on

(20.15) $$\beta_2 - \beta_1 = e \oint \mathbf{A}\,d\mathbf{q} = eF$$

where the integral follows the closed path formed by paths 1 and 2 around the solenoid, and F is consequently the entire flux through the solenoid.

Since this effect will occur even though the electron beam may never enter any region in which the electromagnetic field is unequal to zero, Bohm and Aharonov have argued that in quantum mechanics potentials acquire the status of observables which they do not have in classical electrodynamics, pointing out that this conclusion is unavoidable if one wishes to adhere to the concept of *local* interactions as a basic requirement.

One can, however, formulate quantum electrodynamics without the use of potentials, if one admits the kind of nonlocality inherent in the very notion of a path-dependent phase, as has been shown by Mandelstam. It would appear, then, more reasonable to accept the Bohm-Aharonov experiment as an indication of a profoundly nonlocal feature acquired by ψ functions in an electromagnetic field. This feature can be extracted from the ψ function (20.7) in yet another fashion. By differentiation one obtains

(20.16) $$(\partial \Phi_\tau / \partial x_\nu) = e^{i\beta}[(\partial/\partial x_\nu) + i\kappa_\nu]\Phi_\tau^0 = e^{i\beta}[(\partial/\partial x_\nu) - ieA_\nu]\Phi_\tau^0.$$

It follows that if Φ_τ satisfies any equation involving the operator of momentum-energy $P_\mu = i(\partial/\partial x_\mu)$, then Φ_τ^0 will satisfy the corresponding equation in which P_μ has been replaced by $P_\mu + eA_\mu$. On the basis of the identification (20.10), one would then have to conclude that *the ψ function Φ always satisfies the same equation, whether there is a field or not, and the whole effect of the field is in making the phase nonintegrable.* This is equivalent to having the "ordinary" part Φ^0 of the ψ function, namely the part having a definite phase, satisfy the equation with P_μ replaced by $P_\mu + eA_\mu$. In particular, the Dirac equation for an electron in an electromagnetic field now reads

(20.17) $$\gamma_\mu(P_\mu + eA_\mu)\psi = m\psi$$

where ψ is now meant to have a definite phase at each point in space and time.

It will be noticed that this equation is invariant under phase transformations involving *single-valued* functions $\lambda(x)$

(20.18)
$$\psi \to \psi \, e^{ie\lambda(x)}$$

provided transformation (20.18) is accompanied among the vector potentials by a gauge transformation

(20.19)
$$A_\nu \to A_\nu + (\partial\lambda/\partial x_\nu).$$

This invariance property can be made manifest if one introduces the explicit path dependence of Φ

(20.20)
$$\Phi(x,P) = \Phi^0(x) \exp\left[-ih \int_{-\infty}^{x} A_\mu(\xi)\,d\xi_\mu\right]$$

and the gauge-invariant derivative

(20.21)
$$\partial_\mu \Phi(x,P) = \lim_{dx_\mu \to 0} \{[\Phi(x+dx_\mu, P') - \Phi(x,P)]/dx_\mu\}$$

where the path P' is obtained from P by giving it an extension dx_μ in x_μ direction, i.e. P' passes through the end point x of the path P. The Dirac equation can then be written

(20.22)
$$i\gamma_\mu \partial_\mu \Phi = m\Phi.$$

It should be kept in mind that the operators ∂_μ, ∂_ν do not commute if a field is present,

(20.23)
$$(\partial_\mu \partial_\nu - \partial_\nu \partial_\mu)\Phi = -ie\Phi F_{\mu\nu}.$$

As a calculational aid, the electromagnetic potentials are, of course, always extremely convenient when one seeks to actually solve Eq. (20.22).

The rule of replacing P_μ by $P_\mu + eA_\mu$ in presence of an electromagnetic field has been known for a long time, and its success in giving right answers in empirical situations where the representation of the electromagnetic field by a *classical* potential is meaningful, is well known. The comparison of the energy levels in a Coulomb field resulting from (20.17) with observation can be found in practically all texts on quantum mechanics. The good approximations to reality provided by this empirical rule are the more astonishing in view of the number of arbitrary features characterizing this "derivation" of that rule by the line of reasoning employed above.

Perhaps the most serious shortcoming of the argument leading to the identification (20.10) is the complete arbitrariness of the value of the parameter e, identified as the electric charge of the particle considered, which could be set equal to zero, for example, thus making the entire discussion up to this point an empty exercise. In other words, no sufficient reason for the existence of either charged or uncharged particles has thus far been advanced. It is, in particular, not at all clear why one could not have, in analogy to electrons satisfying (20.17), neutrinos coupled to the electromagnetic field satisfying the correspondingly modified neutrino equation $\sigma_\mu(P_\mu + eA_\mu)\psi = 0$.

In this connection, it seems worth recalling a curious argument by Dirac based on the fundamental indeterminacy modulo $n2\pi$ (n an integer) in the phase of any complex number. Considering the single-valued part Φ^0 of any ψ function, Dirac argues that the change in phase around a small closed curve must be small and cannot therefore be a nonvanishing multiple of 2π, because in the limit of an infinitesimal circuit this would conflict with the continuity of Φ. There is an exceptional case, however, when Φ^0 vanishes, since then its phase does not have a meaning. Since Φ^0 is complex its vanishing will require two conditions, so that in general the points at which $\Phi(x)$ vanishes will lie along a "nodal line." From continuity, one can now only infer that the change in phase along a circuit around a nodal line must go over into $n2\pi$ in the limit. This integer n will thus be a characteristic of the nodal line, and its sign can be associated with the direction of the circuit, which in turn may be associated with a direction along the nodal line. If one considers now a large circuit in space with a number of nodal lines passing through it, then the total change in phase along the curve is

$$(20.24) \qquad \Delta\beta = 2\pi \sum_i n_i + e \iint \mathbf{B} \cdot \mathbf{dS}.$$

Applied to a closed surface, (20.24) must vanish,

$$(20.25) \quad \sum_{c.s.} n_i = -(e/2\pi) \iint \mathbf{B} \cdot \mathbf{dS} \qquad \text{(c.s. means closed surface)}$$

If $\sum_{c.s.} n_i \neq 0$, some nodal lines must have end points inside the closed surface. Thus the endpoints of nodal lines, if they exist, must be the same for all ψ functions, and represent sources of magnetic flux $4\pi f = (2\pi n/e)$, where f is the strength of the magnetic monopole at the end point,

$$(20.26) \qquad\qquad f = (n/2e).$$

These considerations do not, of course, show that such nodal lines with or without end points must exist in nature. As in case of electric charge e, the argument leading to the prediction of f does not contain sufficient reason for the existence or nonexistence of any finite value of n. All experimental evidence available to date indicates that magnetic monopoles do *not* exist in nature.

Equation (20.26) is, incidentally, invariant under inversion of coordinates and under reversal of motion only if the nodal characteristic n changes sign under either transformation, independent of whether e does or does not change sign under either transformation, because of the opposite transformation character of electric and magnetic fields.

NOTES

Probably the first attempt to find the reason for the electromagnetic interaction in the invariance of the Schroedinger equation under phase transformations is due to London [1].

Dirac [2] used the indeterminacy modulo 2π in the phase of any single-valued ψ function to speculate on the possible existence of magnetic monopoles.

Aharonov and Bohm [3] drew attention to experiments which demonstrate directly the nonintegrability of phase in presence of electromagnetic fields. The experiment of Aharonov and Bohm with magnetic flux enclosed by a split electron beam was actually performed by Chambers [4].

Mandelstam [5] has given a formulation of electrodynamics without potentials, and shown in which sense the path dependence of phase implies a basic nonlocality of the ψ function in presence of electromagnetic fields.

Readers who wish to review at this point the comparison with experiment of the solutions of Eq. (20.17) as applied to atomic hydrogen may find the slim booklet by Series [6] particularly concise and comprehensive.

REFERENCES

[1] F. London, *Z. Physik* **42**, 375 (1927).
[2] P. A. M. Dirac, *Proc. Roy. Soc.* **A133**, 60 (1931).
[3] Y. Aharonov and D. Bohm, *Phys. Rev.* **115**, 485 (1959).
[4] R. G. Chambers, *Phys. Rev. Letters* **5**, 3 (1960).
[5] S. Mandelstam, *Ann. Phys. N.Y.* **19**, 1 (1962).
[6] G. W. Series, "Spectrum of Atomic Hydrogen," Oxford Univ. Press, London and New York, 1957.

The Idea of the Compensating Field

Since the requirement of invariance under phase transformations (20.18) *necessitates* the presence of *some* field which compensates by a gauge transformation (20.19) the effect of (20.18) on the equation governing the ψ function, some authors have attempted to elevate a generalized form of gauge invariance to a fundamental principle, and have sought to find in the idea of the "compensating field" the raison d'être for actually occurring interactions. The formulation of this approach is greatly aided if one uses as starting point a so-called action principle

(21.1) $$\int L[\psi, (\partial\psi/\partial x_\nu)]\, d^4x \;=\; \text{Extremum}$$

from which flow the equations governing the ψ functions as Euler-Lagrange equations of this variational principle,

(21.2) $$(\delta L/\delta\psi) \equiv (\partial L/\partial\psi) - (\partial/\partial x_\nu)[\partial L/\partial(\partial\psi/\partial x_\nu)] \;=\; 0.$$

For example, the "Lagrangian" L giving rise to the Dirac equation and its adjoint by the recipe (21.2) is

(21.3) $$L_D = i\bar\psi\gamma_\mu(\partial\psi/\partial x_\mu) - m\bar\psi\psi.$$

It should be understood that introduction of (21.1) is a purely formal device, and the step leading from (21.3) by (21.2) to the Dirac equation can in no sense be considered as a "derivation" of this equation.

Suppose now one wants to insist on the invariance of L under unitary transformations of the type

(21.4) $$U = \exp[i\epsilon_\alpha(x)S_\alpha]$$

where S_α ($\alpha = 1, 2, \ldots, n$) are n hermitean operators and $\epsilon_\alpha(x)$ n real functions of space and time. The idea is to associate the operators S_α with internal properties of elementary particles, such as electric charge, isospin, etc. For example, one may visualize two spatially separate observers with *different* conventions about the labeling of nucleons as neutrons and protons looking at the *same* event involving strong interactions which are invariant under changes in these conventions, i.e.

invariant under rotations in the abstract isospin space. *The generators of space time transformations* associated with dynamical rather than internal properties of elementary particles *are expressly excluded from consideration at this point.* The coordinates x refer to a flat space-time continuum in which displacements, in particular, are integrable. If one makes the parameters characterizing space-time transformations, such as translations, rotations and Lorentz transformations, coordinate dependent, then the generators S_α will, in general, not commute with the parameters $\epsilon_\alpha(x)$, and one is forced to consider nonintegrable, curved spaces. This question will be taken up in some detail in Section 22.

The transformation (20.18) is the special case with $n = 1$, $S = eI$ and $\epsilon(x) = \lambda(x)$. The Lagrangian (21.3) is then not invariant as it stands, because

$$(21.5) \qquad U^{-1}(\partial/\partial x_\mu)\,U \;=\; (\partial/\partial x_\mu) + U^{-1}(\partial U/\partial x_\mu).$$

This deficiency of L_D can be repaired, however, if one introduces a compensating field B_μ which transforms as

$$(21.6) \qquad B'_\mu \;=\; U^{-1} B_\mu\, U + (i/g)\, U^{-1}(\partial U/\partial x_\mu)$$

so that, if ψ transforms as

$$(21.7) \qquad \psi' \;=\; U^{-1}\psi; \qquad \bar\psi' \;=\; U\bar\psi$$

then

$$(21.8)$$
$$\tilde{L}_D = i\bar\psi\gamma_\mu[(\partial/\partial x_\mu) - igB_\mu]\psi - m\bar\psi\psi = i\bar\psi'\,\gamma_\mu[(\partial/\partial x_\mu) - igB'_\mu]\psi' - m\bar\psi'\,\psi'$$

is now invariant under the transformation (21.4). In the special case mentioned above, the still arbitrary coupling parameter g will then have to be identified with the numerical value of the electronic charge e, and B_μ with the vector potentials A_μ of electrodynamics.

It is evident that one cannot introduce scalar fields through this requirement of gauge invariance. Any compensating field must, by necessity, be a vector field. Therefore, if compensating fields are accepted as the primary agents of interaction between fermions, the pion field, which for a long time was thought to be the "glue" which holds the nucleons together, will have to be demoted from the ranks of fields whose quanta are considered as elementary particles. Any gauge theory of strong interaction requires, in principle, vector mesons as agents of interaction, and in such a theory one must seek to obtain pions as composite particles, for example as bound states made up out of nucleons and antinucleons as had originally been surmised by Fermi and Yang.

The transformation formulae (21.5) and (21.6) are, in general, rather complicated, since they depend on the C.R.s between the operators S_α. One has the expansions

(21.9)

$$U^{-1}(\partial U/\partial x_\mu) = i(\partial\epsilon_\alpha/\partial x_\mu)S_\alpha - (i^2/2!)\,\epsilon_\alpha(\partial\epsilon_\beta/\partial x_\mu)\,[S_\alpha, S_\beta]$$
$$+ (i^3/3!)\,e_\alpha\,e_\beta(\partial\epsilon_\gamma/\partial x_\mu)\,[S_\alpha, [S_\beta, S_\gamma]] - + \ldots$$

(21.10)

$$U^{-1}B_\mu\,U = B_\mu - i\epsilon_\alpha[S_\alpha, B_\mu] + (i^2/2!)\,\epsilon_\alpha\,\epsilon_\beta[S_\alpha, [S_\beta, B_\mu]] - + \ldots$$

If one were to regard *local* gauge invariance as embodied in the transformations (21.4) for *any* operators S_α as compulsory, one would in fact be faced with an embarrassment of riches regarding possible compensating fields. Some physical reasons are needed to restrict the large variety of possible formal choices for compensating fields. With an eye on such reasons, to be given in the form of some examples later, one can attain a substantial reduction in complexity if one restricts consideration to operators S_α which satisfy C.R.s of the type

(21.11) $$[S_\alpha, S_\beta] = C_{\alpha\beta\gamma}S_\gamma.$$

The "structure constants" $C_{\alpha\beta\gamma}$ of the set of transformations S_α are then independent of the representations used for the S_α, and satisfy the relations

(21.12) $$C_{\alpha\beta\gamma} + C_{\beta\alpha\gamma} = 0$$

and

(21.13) $$C_{\alpha\beta\epsilon}\,C_{\epsilon\gamma\delta} + C_{\beta\gamma\epsilon}\,C_{\epsilon\alpha\delta} + C_{\gamma\alpha\epsilon}\,C_{\epsilon\beta\delta} = 0$$

where (21.12) is a consequence of definition (21.11), and (21.13) follows from the Jacobi identity satisfied by triple commutators. Transformations that are isomorphic to the Euclidean rotations in three dimensions and four dimensions, and Lorentz transformations are special cases of (21.11).

Introducing without restriction of generality new field variables $B_{\alpha,\mu}$. which commute with all S_α, by

(21.14) $$B_\mu = S_\alpha B_{\alpha,\mu}$$

one can consolidate the two expressions (21.9) and (21.10) into

(21.15) $$U^{-1}(\partial U/\partial x_\mu) = S_\alpha[i(\partial\epsilon_\alpha/\partial x_\mu) - (i^2/2!)\,\epsilon_\beta(\partial\epsilon_\gamma/\partial x_\mu)\,C_{\beta\gamma\alpha}$$
$$+ (i^3/3!)\,\epsilon_\beta\,\epsilon_\gamma(\partial\epsilon_\delta/\partial x_\mu)\,C_{\gamma\delta\epsilon}\,C_{\beta\epsilon\alpha} - + \ldots]$$

(21.16) $$U^{-1}B_\mu\,U = S_\alpha[B_{\alpha,\mu} - i\epsilon_\beta\,B_{\gamma,\mu}\,C_{\beta\gamma\alpha}$$
$$+ (i^2/2!)\,\epsilon_\beta\,\epsilon_\gamma\,B_{\delta,\mu}\,C_{\gamma\delta\epsilon}\,C_{\beta\epsilon\alpha} - + \ldots].$$

Therefore, (21.6) is satisfied for *infinitesimal* gauge transformations if the $B_{\alpha,\mu}$ transform according to

(21.17)
$$\delta B_{\alpha,\mu} = B_{\alpha,\mu} - B'_{\alpha,\mu} = i\epsilon_\beta B_{\gamma,\mu} C_{\beta\gamma\alpha} + (1/g)\,(\partial\epsilon_\alpha/\partial x_\mu).$$

This formula gives a clue to the construction of the gauge-invariant Lagrangian L_B for the field $B_{\alpha,\mu}$ which will give the field equations satisfied by the compensating field, and which should in the special case of the electromagnetic field coincide with the well known Lagrangian

(21.18)
$$L_A = -\tfrac{1}{4} F_{\mu\nu} F_{\mu\nu}$$

containing only the gauge-invariant fields

(21.19)
$$F_{\mu\nu} = (\partial A_\nu/\partial x_\mu) - (\partial A_\mu/\partial x_\nu).$$

Denoting this unknown Lagrangian with

(21.20)
$$L_B = L_B(B_{\alpha,\mu}; B_{\alpha,\mu|\nu}); \qquad B_{\alpha,\mu|\nu} \equiv (\partial B_{\alpha,\mu}/\partial x_\nu)$$

one requires as a consequence of the invariance under transformations (21.17)

(21.21)
$$\delta L_B = (\partial L_B/\partial B_{\alpha,\mu})\,\delta B_{\alpha,\mu} + (\partial L_B/\partial B_{\alpha,\mu|\nu})\,\delta B_{\alpha,\mu|\nu} = 0.$$

Substituting (21.17) and (note that δ and $\partial/\partial x_\nu$ commute)

(21.22)
$$\delta B_{\alpha,\mu|\nu} = i\epsilon_\beta B_{\gamma,\mu|\nu} C_{\beta\gamma\alpha} + i(\partial\epsilon_\beta/\partial x_\nu)\,B_{\gamma,\mu} C_{\beta\gamma\alpha} + (1/g)\,(\partial^2 \epsilon_\alpha/\partial x_\nu\,\partial x_\mu)$$

and collecting coefficients of ϵ_α, $\partial\epsilon_\alpha/\partial x_\mu$ and $\partial^2 \epsilon_\alpha/\partial x_\nu\,\partial x_\mu$ one obtains

(21.23) $C_{\alpha\beta\gamma}[B_{\beta,\mu}(\partial L_B/\partial B_{\gamma,\mu}) + B_{\beta,\mu|\nu}(\partial L_B/\partial B_{\gamma,\mu|\nu})] = 0$

(21.24) $(\partial L_B/\partial B_{\alpha,\mu}) + ig C_{\alpha\beta\gamma} B_{\beta,\nu}(\partial L_B/\partial B_{\gamma,\nu|\mu}) = 0$

(21.25) $(\partial L_B/\partial B_{\alpha,\mu|\nu}) + (\partial L_B/\partial B_{\alpha,\nu|\mu}) = 0$

Equation (21.25) can be satisfied only if the derivatives of B enter L_B in the combination $B_{\alpha,[\mu|\nu]} = B_{\alpha,\nu|\mu} - B_{\alpha,\mu|\nu}$, and, from (21.24), it follows further that $B_{\alpha,[\mu|\nu]}$ enters L_B only in the combination

(21.26)
$$G_{\alpha,\mu\nu} = B_{\alpha,[\mu|\nu]} - (ig/2)\,C_{\beta\gamma\alpha}(B_{\beta,\mu} B_{\gamma,\nu} - B_{\beta,\nu} B_{\gamma,\mu}) = -G_{\alpha,\nu\mu}.$$

Equation (21.23) is not sufficient to determine L_B uniquely. However, if one wishes to obtain a general expression which contains (21.18) as special case, one must introduce the "field strengths"

(21.27)
$$G_{\mu\nu} = (\partial B_\nu/\partial x_\mu) - (\partial B_\mu/\partial x_\nu) - ig(B_\mu B_\nu - B_\nu B_\mu) = S_\alpha G_{\alpha,\mu\nu}$$

and write

(21.28)
$$L_B = -\tfrac{1}{4} \text{trace } G_{\mu\nu} G_{\mu\nu} = -\tfrac{1}{4} G_{\alpha,\mu\nu} G_{\beta,\mu\nu} \text{ trace } S_\alpha S_\beta$$
$$= -\tfrac{1}{4} G_{\alpha,\mu\nu} G_{\beta,\mu\nu} C_{\alpha\gamma\delta} C_{\delta\gamma\beta}.$$

The proof that L_B does indeed satisfy (21.23) is cumbersome and will be omitted here.

The entire Lagrangian describing the mutually interacting ψ fields and B-fields is now

(21.29)
$$L = \tilde{L}_D + L_B$$

where the derivatives of ψ occur in the combination $[(\partial/\partial x_\mu) - igS_\alpha B_{\alpha,\mu}]\psi$ so that

(21.30)
$$[\partial \tilde{L}_D/\partial(\partial\psi/\partial x_\mu)]S_\alpha\psi = (i/g)[\partial\tilde{L}_D/\partial B_{\alpha,\mu}].$$

Incidentally, the arbitrary coupling parameter g can alternatively be introduced as a factor with which L_B may be multiplied without changing the conclusions of this section. Formally, this is equivalent to introducing, instead of B_μ, the fields $B'_\mu = gB_\mu$ as variables, so that

$$\partial_\mu\psi \equiv [(\partial/\partial x_\mu) - iS_\alpha B'_{\alpha,\mu}]\psi \qquad \text{and} \qquad L_B = g^{-1} L_B(B').$$

Now, the invariance of L under an infinitesimal gauge transformation, transforming $B_{\alpha,\mu}$ according to (21.17) and ψ according to

(21.31)
$$\delta\psi = i\epsilon_\alpha S_\alpha \psi$$

(21.32)
$$\delta\bar{\psi} = -i\epsilon_\alpha S_\alpha \bar{\psi}$$

(21.33)
$$\delta(\partial\psi/\partial x_\mu) = i(\partial\epsilon_\alpha/\partial x_\mu)S_\alpha\psi + i\epsilon_\alpha S_\alpha(\partial\psi/\partial x_\mu)$$

leads to a conservation law of "current"

(21.34)
$$(\partial J_{\alpha,\mu}/\partial x_\mu) = 0,$$

which can be derived as follows.

Since L_B had already been chosen so that $\delta L_B = 0$ under that transformation, one need write down only

(21.35)
$$\delta\tilde{L}_D = (\partial\tilde{L}_D/\partial\bar{\psi})\,\delta\bar{\psi} + (\partial\tilde{L}_D/\partial\psi)\,\delta\psi + [\partial\tilde{L}_D/\partial(\partial\psi/\partial x_\mu)]\delta(\partial\psi/\partial x)$$
$$+ (\partial\tilde{L}_D/\partial B_{\alpha,\mu})\,\delta B_{\alpha,\mu} = 0.$$

Use has been made here of the fact that $\tilde{L}_D$ contains neither $(\partial\tilde{\psi}/\partial x_\mu)$ nor $B_{\alpha,\,\mu|\nu}$. The first term vanishes because of the field equation

(21.36) $$(\delta L/\delta\tilde{\psi}) = (\partial\tilde{L}_D/\partial\tilde{\psi}) = 0$$

and the remainder reads upon substitution of (21.17), (21.31), and (21.33):

(21.37)
$$\epsilon_\alpha\{(\partial\tilde{L}_D/\partial\psi)\,S_\alpha\psi + [\partial\tilde{L}_D/\partial(\partial\psi/\partial x_\mu)]\,S_\alpha(\partial\psi/\partial x_\mu) + C_{\alpha\beta\gamma}\,B_{\beta,\,\mu}(\partial\tilde{L}_D/\partial B_{\gamma,\,\mu})\}$$
$$+ (\partial\epsilon_\alpha/\partial x_\mu)\{[\partial\tilde{L}_D/\partial(\partial\psi/\partial x_\mu)]\,S_\alpha\psi - (i/g)\,(\partial\tilde{L}_D/\partial B_{\alpha,\,\mu})\} = 0.$$

The second bracket vanishes on account of (21.30) and for the first bracket one can exploit the field equations

(21.38)
$$(\delta L/\delta\psi) = (\partial\tilde{L}_D/\partial\psi) - (\partial/\partial x_\mu)\,[\partial\tilde{L}_D/\partial(\partial\psi/\partial x_\mu)] = 0$$

and

(21.39)
$$(\delta L/\delta B_{\gamma,\,\mu}) = (\partial\tilde{L}_D/\partial B_{\gamma,\,\mu}) + (\partial L_B/\partial B_{\gamma,\,\mu}) - (\partial/\partial x_\nu)\,(\partial L_B/\partial B_{\gamma,\,\mu|\nu}) = 0$$

to yield

(21.40)
$$(\partial/\partial x_\mu)\{[\partial\tilde{L}_D/\partial(\partial\psi/\partial x_\mu)]\,S_\alpha\psi\} + C_{\alpha\beta\gamma}[-B_{\beta,\,\mu}(\partial L_B/\partial B_{\gamma,\,\mu})$$
$$+ B_{\beta,\,\mu}(\partial/\partial x_\nu)\,(\partial L_B/\partial B_{\gamma,\,\mu|\nu})] = 0$$

At this point, Eq. (21.33) allows one to substitute for $C_{\alpha\beta\gamma}\,B_{\beta,\,\mu}(\partial L_B/\partial B_{\gamma,\,\mu})$ and write (21.40)

(21.41)
$$(\partial/\partial x_\mu)\{[\partial\tilde{L}_D/\partial(\partial\psi/\partial x_\mu)]\,S_\alpha\psi + C_{\alpha\beta\gamma}B_{\beta,\,\nu}(\partial L_B/\partial B_{\gamma,\,\nu|\mu})\} = 0$$

which is of the form (21.34) with the current density

(21.42)
$$J_{\alpha,\,\mu} = [\partial\tilde{L}_D/\partial(\partial\psi/\partial x_\mu)]\,S_\alpha\psi + C_{\alpha\beta\gamma}\,B_{\beta,\,\nu}(\partial L_B/\partial B_{\beta,\,\nu|\mu}) = j_{\alpha,\,\mu}(\psi) + j_{\alpha,\,\mu}(B).$$

Equations (21.39), governing the field B, can be cast in the form

(21.43) $$(\partial L_B/\partial B_{\gamma,\,\mu}) - (\partial/\partial x_\nu)\,(\partial L_B/\partial B_{\gamma,\,\mu|\nu}) = igj_{\alpha,\,\mu}(\psi).$$

These equations are, except in case $C_{\alpha\beta\gamma} = 0$ which corresponds to a type of interaction resembling electromagnetism, essentially nonlinear, because the compensating field B acts as its own source in a manner indicated by the presence of the second term on the right-hand side of the current (21.42). The vector particles which in a full quantum theory must appear as the carriers of interaction are, therefore, except in case $C_{\alpha\beta\gamma} = 0$, expected to share with the primary source particles the symmetry property which gives rise to their existence. Thus any vector mesons which mediate the interaction caused by the isospin symmetry of nucleons must themselves carry isospin. Particles associated with a one parameter symmetry are exceptional in that they do not possess the relevant source property. Thus photons do not carry an electric charge.

If one specializes to invariance under rotations in isospace, so that

(21.44)
$$S_\alpha = T_\alpha \quad (\alpha = 1, 2, 3); \qquad C_{\alpha\beta\gamma} = i\epsilon_{\alpha\beta\gamma}; \qquad \mathrm{trace}\, S_\alpha S_\beta = 2\delta_{\alpha\beta};$$

one obtains the equations characterizing a vector field $\mathbf{B}_\mu$ first introduced by Yang and Mills in an attempt to account for some of the facts of strong interactions in terms of vector mesons coupled to the isospin of baryons,

(21.45)
$$G_{\alpha,\mu\nu} \equiv \mathbf{G}_{\mu\nu} = \mathbf{B}_{[\mu|\nu]} + g(\mathbf{B}_\mu \times \mathbf{B}_\nu)$$

(21.46)
$$\partial_\mu \psi \equiv [(\partial/\partial x_\mu) - ig(\mathbf{T}\cdot\mathbf{B}_\mu)]\psi; \qquad (\partial_\mu \partial_\nu - \partial_\nu \partial_\mu)\psi = -ig(\mathbf{G}_{\mu\nu}\cdot\mathbf{T})\psi$$

(21.47)
$$L_B = -\tfrac{1}{2}(\mathbf{G}_{\mu\nu}\cdot\mathbf{G}_{\mu\nu})$$

(21.48)
$$J_{\alpha,\mu} \equiv \mathbf{J}_\mu = i\bar{\psi}\gamma_\mu \mathbf{T}\psi + i(\mathbf{G}_{\mu\nu} \times \mathbf{B}_\nu) = i\bar{\psi}\gamma_\mu \mathbf{T}\psi + i(\mathbf{B}_{[\mu|\nu]} \times \mathbf{B}_\nu)$$

(21.49)
$$(\delta L_B/\delta \mathbf{B}_\mu) = (\partial \mathbf{G}_{\mu\nu}/\partial x_\nu) - ig\mathbf{J}_\mu.$$

The compensating field has the generally attractive feature of independence from the particular form taken by the operators S_α which generate the symmetry property attached to the ψ field. It depends only on the structure constants $C_{\alpha\beta\gamma}$ which are the same for all representations of the operators S_α. Physically, this means the compensating field B is the same for all ψ fields that possess the particular kind of symmetry represented in terms of the operators S_α. Thus there is only one universal electromagnetic field which compensates by gauge transformation the effects of the one parameter phase transformation generated by the operator of electric charge.

A number of schemes have been proposed linking the empirical symmetries of the strong interactions which reveal themselves through conservation of various attributes such as baryonic charge, hypercharge, and isospin (see Section 29), with the existence of various types of vector meson fields, each field being generated in the manner described above as a compensating field B from the requirement of gauge invariance, so that the mentioned conservation laws are identified with equations of the type of (21.41). Notably, a scheme by Sakurai postulates the existence of three types of vector mesons which, respectively, are the dynamical agents attached by the idea of the compensating field to the two one-parameter attributes of baryonic charge and hypercharge, and the one three-parameter attribute of isospin.

Similarly, there exist a number of proposals to account for the weak interactions by the intermediary of some vector meson generated as a compensating field by the particular gauge symmetries characteristic of the particles engaging in these interactions.

However, all attempts to identify actually observed vector mesons with any of the possible compensating fields B that may mediate other than electromagnetic interactions encounter a disappointing feature of this theory. The formalism developed above cannot accommodate compensating fields B which contain in the Lagrangian terms of the type

$$(21.50) \qquad -(\mu_0^2/2)\, B_{\alpha,\mu} B_{\alpha,\mu} \qquad (\mu_0 \text{ a constant})$$

without destroying the general pauge invariance which has been the very motivation for this approach. On the other hand, such terms are needed if one wishes to describe any of the actually observed vector mesons other than the photon, because only by inclusion of terms (21.50) will the field equations be of the form

$$(21.51) \qquad (\partial G_{\mu\nu}/\partial x_\nu) - \mu_0^2 B_\mu = ig j_\mu$$

needed if the interaction is to give rise to short-range forces and, therefore, to mesons with finite rest masses in accordance with observation.

Now, some authors have used the fact that the derivation of the expression (21.42) for the current and the conservation law (21.41) are not affected by the addition of a term (21.50) to L_B as an excuse to consider "partially gauge-invariant" theories in which a field B, not longer deserving the name of "compensating field," is coupled to a conserved current after the fashion of (21.51). Such an approach loses its aspect of complete meaninglessness if one keeps in mind the hybrid nature of the theory developed above. The field B up to this point has been conceived as a classical field. In a consistent quantum theory both the fields ψ and B should always appear in conjunction with the creation and annihilation operators which give rise, in particular, to so-called vacuum polarization effects. One can then argue with Schwinger that, in case of sufficiently strong coupling, the vector mesons may acquire, even in a strictly gauge invariant theory, the propagation characteristics of particles with finite rest mass.

From an intuitive point of view this can be made plausible by the following line of reasoning. In actual electrodynamics the comparative weakness of the coupling constant ($e^2 = 1/137$) guarantees the stability of the photon against possible disintegration into two or more atoms of positronium, because the binding energy of positronium is of order $-e^2 m$ as compared to the restmasses m of its constituents. Similarly, any external charge Q brought into the electrodynamic vacuum will cause a polarization into virtual positroniums only in its vicinity, and will (apart from a possible renormalization of its numerical value) retain an uncompensated amount which acts as a source of a long range coulomb field. If the strength of the coupling were now allowed to increase (and it should be possible to do this analytically without destroying the gauge invariance, which is a structural property of the theory independent of the parameter e^2), at a certain critical value of order $e^2 \approx 1$ an entirely different situation would arise. The vector meson would become unstable against disintegration into various bound states of the source field and acquire propagation characteristics usually associated with massive vector mesons that are unstable against decay into two or more pions, which in turn may be considered as bound states of nucleons and antinucleons. Any external "charge" introduced into this kind of vacuum would induce a chain of polarization events which would effectively transport the original charge to spatial infinity, leading to complete compensation of the original charge in any finite volume. Consequently, no long range field of the "charge" could be maintained, and this amounts again to the absence of vector mesons with vanishing rest mass in this case.

The analytic penetration of this attractive idea, which has been promulgated and shown to be feasible in some simplified models by Schwinger, has not been completed at the time of writing. Therefore, all gauge theories of strong and weak interactions remain stalled in a state of animated suspension until this crucial point, of whether massive vector mesons can be consistently accommodated in a strictly gauge-invariant theory, has been settled.

NOTES

Yang and Mills [1] first attempted a theory of strong interactions in terms of a vector meson field which is a compensating field needed to guarantee coordinate dependent invariance under rotations in isospace.

The idea to consider pions as bound nucleon-antinucleon states is due to Fermi and Yang [2].

Sakurai [3] has attempted a theory of strong interactions by invoking three types of compensating fields, corresponding to three types of vector mesons, one generated by a three-parameter group of transformations and the other two each generated by a one-parameter transformation group. See also Schwinger [4] and Lee and Yang [5].

Roman [6] has proposed a scheme of compensating fields which incorporates the weak interactions. See also Salam and Ward [7].

Glashow and Gell-Mann [8] have considered some consequences of so-called "partially gauge-invariant" theories of interaction.

Schwinger [9] has given a formal, but somewhat unrealistic, example demonstrating the consistency of compensating fields having finite rest-mass with strict gauge-invariance in case of sufficiently strong coupling.

REFERENCES

[1] C. N. Yang and R. L. Mills, *Phys. Rev.* **96**, 191 (1954).

[2] E. Fermi and C. N. Yang, *Phys. Rev.* **76**, 1739 (1949).

[3] J. Sakurai, *Ann. Phys. N.Y.* **11**, 1 (1960).

[4] J. Schwinger, *Ann. Phys. N.Y.* **2**, 407 (1957).

[5] T. D. Lee and C. N. Yang, *Phys. Rev.* **98**, 1501 (1955).

[6] P. Roman, *Nuovo Cimento* **21**, 747 (1961).

[7] A. Salam and J. C. Ward, *Nuovo Cimento* **11**, 568 (1959).

[8] S. L. Glashow and M. Gell-Mann, *Ann. Phys. N.Y.* **15**, 437 (1961).

[9] J. Schwinger, *Phys. Rev.* **128**, 2425 (1962).

Gravitation as a Compensating Field

The only other interaction besides electromagnetism which does not require field equations corresponding to particles of finite rest mass, and is, therefore, already on the level of classical field theory free from any obvious inconsistency with a gauge-invariance principle, is gravitation.

There is, of course, one fundamental difference between electromagnetism and gravitation. Whereas electromagnetic theory can be formulated so that it becomes globally invariant under Lorentz transformations, in the presence of gravitation one can, in principle, only require *local* Lorentz invariance, for the simple reason that, in the presence of gravitation, inertial frames of reference are, in general, accelerated with respect to each other if they are some finite distance apart. The description of an event in space-time requires, therefore, *two* labels, namely the distance x of the event from the origin of the inertial frame in which the event is described, and the label u which tells where the origin of that inertial frame is located in an underlying curvilinear coordinate system.

For the purpose of illustration consider two observers (0) and $(\bar{0})$ located at the origins of two inertial frames, respectively. The observer (0) is placed at the center of the earth, and the other observer $(\bar{0})$ is oscillating in a tunnel drilled through the center of the earth, as drawn in Fig. 22.1. In the underlying curvilinear continuum the two observers will describe worldlines which may be rendered graphically as indicated

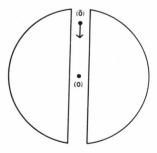

Fig. 22.1. Example of two inertial frames accelerated with respect to each other.

187

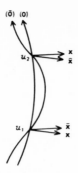

FIG. 22.2. The world lines of observers (0) and ($\bar{0}$) described in an underlying curvilinear continuum.

in Fig. 22.2. It is intuitively clear that the two frames carried along by the two observers cannot be connected by a Lorentz transformation, except *locally* whenever the worldlines intersect, because only at the instant when the two origins coincide are the two frames in unaccelerated motion with respect to each other. In fact, the displacement operation becomes nonintegrable under these conditions, because the two inertial coordinate frames x and $\bar{x}$ carried along without rotation by each observer will be rotated with respect to each other between successive meetings at u_1 and u_2 (say), as indicated in Fig. 22.2.

These considerations can be made formally precise through the introduction of a set of 16 functions $h_\mu^k(u)$ which allow, at each continuum point u_μ, the transformation to a local inertial coordinate frame x_k by

(22.1) $\qquad (\partial x^k / \partial u^\mu) = h_\mu^k(u) \qquad$ and $\qquad (\partial u^\mu / \partial x^k) = h_k^\mu(u)$

so that $\qquad h_\mu^k h_k^\nu = \delta_\mu^\nu.$

(By convention, Latin indices refer to components in local inertial frames, Greek indices to components in the underlying continuum.) Since the h_μ^k are not required to satisfy any integrability condition, i.e.

(22.2) $\qquad\qquad h_{\mu|\nu}^k - h_{\nu|\mu}^k \neq 0; \qquad h_{\mu|\nu}^k \equiv \partial h_\mu^k / \partial u^\nu,$

they may be considered as a set of 16 independent *given* functions representing the properties of the *given* gravitational field in which observers tied to inertial frames may find themselves.

The purpose of this section is then to show how the requirement of *local* Lorentz invariance induces dynamical restrictions on the h_μ^k in the form of field equations which the h_μ^k must satisfy, in analogy to the restrictions imposed on the vector potentials of electrodynamics by the requirement of gauge invariance leading to Maxwell's field equations as a possible set of such equations.

Contact with the standard formalism in the theory of gravitation can be made by noting that if the metric of any inertial frame x is taken to be

(22.3) $\qquad g_{ik} = \delta_{ik} \qquad$ so that $\qquad ds^2 = \delta_{ik}\, dx^i\, dx^k,$

then the metric $g_{\mu\nu}$ of the continuum u, defined by

(22.4) $\qquad ds^2 = g_{\mu\nu}\, du^\mu\, du^\nu = (\partial x^i/\partial u^\mu)\,(\partial x^k/\partial u^\nu)\,\delta_{ik}\, du^\mu\, du^\nu$

can be expressed in terms of the functions h_μ^k as

(22.5) $\qquad g_{\mu\nu} = h_\mu^i\, h_\nu^k\, \delta_{ik} = h_{k\mu}\, h_\nu^k$

and therefore all other quantities characterizing the geometry of the continuum u, such as the affinities $\Gamma_{\mu\nu}^\lambda$ and the curvature tensor $R_{\mu\nu\sigma}^\lambda$, are expressible in terms of the field variables h_μ^k.

The question is now whether the functions h_μ^k representing gravitation can be related to a compensating field B whose existence will become necessary when one requires invariance under *local* Lorentz transformations which connect any *local* inertial frame x^k with another *local* frame $\bar{x}^k$ by

(22.6) $\qquad \bar{x}^k = x^k + \epsilon_l^k(u)\, x^l; \qquad \epsilon^{kl}(u) = -\epsilon^{lk}(u)$

affecting the functions h_μ^k themselves according to

(22.7) $\qquad \delta h_\mu^k = \epsilon_l^k(u) h_\mu^l$

and transforming any ψ function according to

(22.8) $\qquad \delta\psi = \tfrac{1}{2}\epsilon^{kl}(u)\, M_{kl}\psi.$

Here the generators M_{kl} representing the Lorentz transformation satisfy the C.R.s (see Appendix 4)

(22.9) $\qquad [M_{kl}, M_{mn}] = \tfrac{1}{2}C_{kl,mn}^{ab}\, M_{ab}; \qquad M_{ab} = -M_{ba}$

where

(22.10) $\qquad C_{kl,mn}^{ab}\, C_{ab,ij}^{mn} = 4(\delta_{ki}\,\delta_{lj} - \delta_{kj}\,\delta_{li}).$

Indeed, any action integral

(22.11) $\qquad I = \int L[\psi, (\partial\psi/\partial x^k)]\, d^4 x$

governing the dynamics of the ψ function in an inertial frame reads in curvilinear coordinates

(22.12)
$$I = \int L[\psi(u), h_k^\mu(u)\,(\partial\psi/\partial u^\mu)]\,h\,d^4u \equiv \int \mathscr{L}[\psi, (\partial\psi/\partial u^\mu), h_\mu^k]\,d^4u;$$
$$h = \det|h_\mu^k|; \qquad \mathscr{L} = hL;$$

and is not invariant, as it stands, under the transformation (22.6). In analogy to the development leading to (21.8) this deficiency can be repaired, however, if one postulates the existence of a compensating field

(22.13)
$$B_\mu^{kl}(u) = -B_\mu^{lk}(u)$$

transforming according to

(22.14)
$$\delta B_\mu^{kl} = \tfrac{1}{4}C_{ab,\,ij}^{kl}\,\epsilon^{ab}(u)\,B_\mu^{ij} + (\partial\epsilon^{bl}/\partial u^\mu) = \epsilon_m^k\,B_\mu^{ml} + \epsilon_m^l\,B_\mu^{km} + (\partial\epsilon^{kl}/\partial u^\mu)$$

and replaces $\mathscr{L}$ by

(22.15)
$$\tilde{\mathscr{L}} = \mathscr{L}(\psi, \partial_\mu\psi, h_\mu^k)$$

where $\partial_\mu\psi$ stands for

(22.16)
$$\partial_\mu\psi \equiv [(\partial/\partial u^\mu) - \tfrac{1}{2}B_\mu^{kl}\,M_{kl}]\,\psi.$$

The relationship between the compensating field B and the gravitational field h is now obtained when one notices the identity of the "gauge-invariant derivative" (22.16) with what is commonly known as the "covariant derivative" of ψ. To see this write down (22.16) for the special case of a tensor field ψ with components ψ^{ij} in the local Lorentz frame,

(22.17)
$$\partial_\mu\psi^{ij} = (\partial\psi^{ij}/\partial u^\mu) - B_\mu^{ik}\,\psi_k^j - B_\mu^{jk}\,\psi_k^i$$

and go over to the components $\psi^{\rho\sigma}$ of ψ in the curvilinear frame by

(22.18)
$$\psi^{\rho\sigma} = h_i^\rho h_j^\sigma\,\psi^{ij}; \qquad \psi^{ij} = h_\rho^i h_\sigma^j\,\psi^{\rho\sigma}.$$

After multiplication of (22.17) with $h_i^\rho h_j^\sigma$ and utilization of $\psi_k^j = \delta_{kl}\psi^{jl}$ as well as $h_k^\mu h_\nu^k = \delta_\nu^\mu$, one finds

(22.19)
$$h_i^\rho h_j^\sigma\,\partial_\mu\psi^{ij} = (\partial\psi^{\rho\sigma}/\partial u^\mu) + \Gamma_{\tau\mu}^\rho\,\psi^{\tau\sigma} + \Gamma_{\tau\mu}^\sigma\,\psi^{\rho\tau}$$

with

(22.20)
$$\Gamma_{\tau\mu}^\rho \equiv h_i^\rho h_{\tau|\mu}^i - B_{\tau,\,\mu}^\rho$$

where

(22.21) $$B^{\rho}_{\tau,\mu} = h^{\rho}_j h_{\tau k} B^{jk}_{\mu}.$$

If one can now show that the symbols $\varGamma^{\rho}_{\tau\mu}$ introduced in (22.20) are identical with the affinities

(22.22) $$\varGamma^{\rho}_{\tau\mu} = \tfrac{1}{2} g^{\rho\sigma}(g_{\sigma\tau|\mu} + g_{\mu\sigma|\tau} - g_{\mu\tau|\sigma})$$

then one has established the identity of (22.19) as the covariant derivative. To this end write down Eq. (22.19) for the metric tensor itself,

(22.23) $$h_i h_j \partial_\mu(\delta^{ij}) = g_{\rho\sigma|\mu} + \varGamma^{\rho}_{\tau\mu} g^{\tau\sigma} + \varGamma^{\sigma}_{\tau\mu} g^{\rho\tau}.$$

Since, by definition (22.17) and the antisymmetry of the fields B^{ik}_{μ} in the indices i and k, one has

(22.24) $$\partial_\mu(\delta^{ij}) = -B^{ij}_{\mu} - B^{ji}_{\mu} = 0$$

the right-hand side of (22.23) must vanish. One can solve then uniquely for the $\varGamma$, provided one assumes them to satisfy

(22.25) $$\varGamma^{\rho}_{\mu\nu} = \varGamma^{\rho}_{\nu\mu}$$

and obtains then Eq. (22.22). Therefore, Eq. (22.23) is nothing but the covariant derivative, denoted $g_{\rho\sigma;\mu}$, which vanishes, and (22.19) is recognized as the covariant derivative,

(22.26) $$h^{\rho}_i h^{\sigma}_j \partial_\mu \psi^{ij} \equiv \psi^{\rho\tau}_{;\mu}.$$

Equations (22.20) and (22.21) contain, therefore, the desired relationship between the compensating field B^{ij}_{μ} and the gravitational field h^k_{μ}.

As in the corresponding case treated in Section 21, the Lagrangian for the field B, which must be added to $\tilde{\mathscr{L}}$ to give the entire Lagrangian

(22.27) $$\mathscr{L} = \tilde{\mathscr{L}} + \mathscr{L}_B$$

from which the field equations for the gravitational field h are obtained by variation

(22.28) $$(\delta\mathscr{L}/\delta h^i_\mu) = 0$$

is not uniquely determined by the requirement of gauge invariance. All one can say is that $\mathscr{L}_B$ must be of the form

(22.29) $$\mathscr{L}_B = \mathscr{L}_B(h^k_\mu, F^{kl}_{\mu\nu})$$

where $F^{kl}_{\mu\nu}$ is defined in terms of the compensating field by

(22.30) $$F^{kl}_{\mu\nu} = (\partial B^{kl}_\nu/\partial u^\mu) - (\partial B^{kl}_\mu/\partial u^\nu) - \tfrac{1}{4} C^{kl}_{ab,ij}(B^{ab}_\mu B^{ij}_\nu - B^{ab}_\nu B^{ij}_\mu).$$

One can easily show by straightforward computation, using relations

(22.20) and (22.21), that $F^{kl}_{\mu\nu}$ is related to the well known curvature tensor

(22.31) $\qquad R^\kappa_{\lambda\mu\nu} = (\partial\Gamma^\kappa_{\lambda\mu}/\partial u^\nu) - (\partial\Gamma^\kappa_{\lambda\nu}/\partial u^\mu) + \Gamma^\sigma_{\lambda\mu}\,\Gamma^\kappa_{\sigma\nu} - \Gamma^\sigma_{\lambda\nu}\,\Gamma^\kappa_{\sigma\mu}$

by

(22.32) $\qquad\qquad\qquad\qquad F^{kl}_{\mu\nu} = h^{l\lambda}\,h^k_\kappa\,R^\kappa_{\lambda\mu\nu}.$

The gravitational field equations proposed by Einstein, in particular, are obtained if one chooses

(22.33) $\qquad\qquad\qquad\qquad \mathscr{L}_B = \kappa^{-1}\,hR$

with

(22.34) $\qquad R = g^{\mu\nu}\,R_{\mu\nu} = h^\mu_l h^\nu_k F^{kl}_{\mu\nu}; \qquad R_{\mu\nu} = R^\kappa_{\mu\nu\kappa};$

where the coupling parameter κ, as the electric charge in the corresponding case, remains completely undetermined by the theory at this stage.

For further details the reader is referred to the work of Utiyama.

NOTES

Utiyama [1] first developed fully the general formalism of compensating fields and applied it, in particular, to the case of the gravitational interaction. See also the review article by Adamskii [2].

REFERENCES

[1] R. Utiyama, *Phys. Rev.* **101**, 1597 (1956).
[2] V. B. Adamskii, *Soviet Phys. Usp.* (*English Transl.*) **4**, 607 (1962).

The Starting Point of Quantum Electrodynamics

The considerations of Sections 20 and 21 suggest, though do not prove, that in a complete quantum theory of fermions interacting through the medium of the electromagnetic field, the Hamiltonian of uncoupled fermions and photons, reading in the notation of (19.54) and (18.79), under omission of any zero-point energies,

$$(23.1) \qquad H^0 = H^0_\psi + H^0_A$$

$$(23.2) \qquad H^0_\psi = \sum_{\mathbf{k}} \sum_{r} \Omega[N_+(\mathbf{k}, r) + N_-(\mathbf{k}, r)]; \qquad \Omega = +\sqrt{k^2 + m^2}$$

$$(23.3) \qquad H^0_A = \sum_{\varkappa} \sum_{S} \omega n(\varkappa, S); \qquad \omega = |\varkappa|$$

should be augmented by an interaction Hamiltonian

$$(23.4) \qquad H' = -\int A_\mu j_\mu \, d\mathbf{q}$$

where now both $A_\mu(x)$ and $j_\mu(x)$ are given, as in (18.86) and (19.56), in terms of creation and annihilation operators b^+, b and a^+, a, respectively. *This interaction Hamiltonian is a function of time*, containing eight basic terms, each linear in the photon operators b^+ and b, and bilinear in the electron-positron operators a^+_L and/or a_L. The integration over $\mathbf{q}$ in (23.4) yields for each of these terms a δ function in the momenta of the involved particles, incorporating the conservation of momentum in all transitions between states caused by the interaction. With the notation (see Appendix 6)

$$(23.5) \qquad \delta(\mathbf{k}) = [1/(2\pi)^3] \int e^{i\mathbf{k}\mathbf{q}} \, d\mathbf{q}$$

one finds

$$(23.6) \quad H'(t) = -[e(2\pi)^3/\sqrt{V}] \sum_{\varkappa} \sum_{\mathbf{k}'} \sum_{\mathbf{k}} \sum_{S} \sum_{r'} \sum_{r} (1/\sqrt{2\omega})$$

$$\times [\epsilon_\mu(S) \, \overline{A}_+(\mathbf{k}', r') \gamma_\mu A_+(\mathbf{k}, r) \, b(\varkappa, S) \, a^+_+(\mathbf{k}', r') \, a_+(\mathbf{k}, r)$$

$$\times e^{i(\Omega'-\Omega-\omega)t} \delta(\mathbf{k}' - \mathbf{k} - \varkappa) + \epsilon^*_\mu(S) \, \overline{A}_+(\mathbf{k}', r') \gamma_\mu A_+(\mathbf{k}, r)$$

$$\times b^+(\varkappa, S) \, a^+_+(\mathbf{k}', r') \, a_+(\mathbf{k}, r) \, e^{i(\Omega'-\Omega+\omega)t} \delta(\mathbf{k}' - \mathbf{k} + \varkappa)$$

193

$$-\epsilon_{\mu}(S)\,\overline{A}_{-}(\mathbf{k}',r')\,\gamma_{\mu}\,A_{-}(\mathbf{k},r)\,b(\boldsymbol{\varkappa},S)\,a_{-}^{+}(\mathbf{k},r)\,a_{-}(\mathbf{k}',r')$$

$$\times e^{-i(\Omega'-\Omega+\omega)t}\,\delta(\mathbf{k}'-\mathbf{k}+\boldsymbol{\varkappa})-\epsilon_{\mu}^{*}(S)\,\overline{A}_{-}(\mathbf{k}',r')\,\gamma_{\mu}\,A_{-}(\mathbf{k},r)$$

$$\times b^{+}(\boldsymbol{\varkappa},S)\,a_{-}^{+}(\mathbf{k},r)\,a_{-}(\mathbf{k}',r')\,e^{-i(\Omega'-\Omega-\omega)t}\,\delta(\mathbf{k}'-\mathbf{k}-\boldsymbol{\varkappa})$$

$$+\epsilon_{\mu}(S)\,\overline{A}_{+}(\mathbf{k}',r')\,\gamma_{\mu}\,A_{-}(\mathbf{k},r)\,b(\boldsymbol{\varkappa},S)\,a_{+}^{+}(\mathbf{k}',r')\,a_{-}^{+}(\mathbf{k},r)$$

$$\times e^{i(\Omega'+\Omega-\omega)t}\,\delta(\mathbf{k}'+\mathbf{k}-\boldsymbol{\varkappa})+\epsilon_{\mu}^{*}(S)\,\overline{A}_{+}(\mathbf{k}',r')\,\gamma_{\mu}\,A_{-}(\mathbf{k},r)$$

$$\times b^{+}(\boldsymbol{\varkappa},S)\,a_{+}^{+}(\mathbf{k}',r')\,a_{-}^{+}(\mathbf{k},r)\,e^{i(\Omega'+\Omega+\omega)t}\,\delta(\mathbf{k}'+\mathbf{k}+\boldsymbol{\varkappa})$$

$$-\epsilon_{\mu}(S)\,\overline{A}_{-}(\mathbf{k}',r')\,\gamma_{\mu}\,A_{+}(\mathbf{k},r)\,b(\boldsymbol{\varkappa},S)\,a_{+}(\mathbf{k},r)\,a_{-}(\mathbf{k}',r')$$

$$\times e^{-i(\Omega'+\Omega+\omega)t}\,\delta(\mathbf{k}'+\mathbf{k}+\boldsymbol{\varkappa})-\epsilon_{\mu}^{*}(S)\,\overline{A}_{-}(\mathbf{k}',r')\,\gamma_{\mu}\,A_{+}(\mathbf{k},r)$$

$$\times b^{+}(\boldsymbol{\varkappa},S)\,a_{+}(\mathbf{k},r)\,a_{-}(\mathbf{k}',r')\,e^{-i(\Omega'+\Omega-\omega)t}\,\delta(\mathbf{k}'+\mathbf{k}-\boldsymbol{\varkappa})].$$

This expression for the interaction Hamiltonian of quantum electro-dynamics has been written out in all detail to impress upon the reader the formidable computational task faced whenever one tries to extract from it information about the outcome of possible experiments.

Since the inception of quantum electrodynamics two avenues of attack on the mathematical problem of how to disentangle the ramifications of the specific form (23.6) for H' have yielded results, leading in many instances to quite fabulous numerical agreement between prediction and observation, namely:

(i) the development of computational techniques making the application of so-called perturbation theory to H' tractable, and

(ii) the exploitation of symmetry properties of H', giving rise to a number of so-called selection rules and other general consequences aiding the computation of experimentally accessible quantities such as cross sections and lifetimes.

If the separation of the Hamiltonian $H = H^0 + H'$ into an uncoupled and an interaction term is at all meaningful, the results inferred from symmetry considerations ought to be independent of the numerical value of the coupling parameter e. The same cannot be said for perturbation theory whose validity appears to be dependent on the assumption of "weak coupling," meaning $e^2 < 1$, and is even in this case beset by a number of perplexities which required development of rather daring mathematical techniques known as "renormalization" procedures. The case of strong coupling, meaning $e^2 > 1$, has thus far resisted all attempts aimed at its mathematical penetration, despite massive efforts that have been brought to bear on this problem. The task of developing a strong-coupling quantum electrodynamics has become more urgent since the advent of novel points of view, touched upon in Section 21, which make it appear probable that all fundamental interactions are, in fact, mediated by vector mesons, for which the photon may serve as a model.

Because of the peculiar combinations in which the creation and annihilation operators appear in the various terms in H', each term will have nonvanishing matrix elements only between states which differ from each other appropriately in the occupation numbers. Thus a term

$$(23.7) \qquad b(\mathbf{x})\, a_+^{\pm}(\mathbf{k}')\, a_+(\mathbf{k})$$

will give a contribution unequal zero only to the matrix element

$$(23.8) \qquad \langle n(\mathbf{x}) - 1,\, 1_+(\mathbf{k}'),\, 0_+(\mathbf{k}) | H' | n(\mathbf{x}),\, 0_+(\mathbf{k}'),\, 1_+(\mathbf{k}) \rangle.$$

The classification of such contributions is greatly aided by a graphical technique due to Feynman, which consists of the following conventions.

Each term in H' is represented by a vertex from which emerge or into which enter lines, one line emerging for each photon creation, electron creation, and positron annihilation operator, and one line entering for each photon annihilation, electron annihilation, and positron creation operator. Lines representing photons are drawn dotted, and lines representing electrons or positrons are drawn solid, with the further convention that all lines representing electrons are directed upwards and lines representing positrons are directed downwards.

Thus the eight terms of H' are represented by the following "Feynman Graphs" (letting now $\mathbf{k}$ stand for the labels $\mathbf{k}, r$ and $\mathbf{x}$ for $\mathbf{x}, S$):

(I) $b(\mathbf{x})\, a_+^{\pm}(\mathbf{k}')\, a_+(\mathbf{k})$

(II) $b^+(\mathbf{x})\, a_+^{\pm}(\mathbf{k}')\, a_+(\mathbf{k})$

(III) $b(\mathbf{x})\, a_-^{\pm}(\mathbf{k})\, a_-(\mathbf{k}')$

(IV) $b^+(\mathbf{x})\, a_-^{\pm}(\mathbf{k})\, a_-(\mathbf{k}')$

(V) $b(\mathbf{x})\, a_+^{\pm}(\mathbf{k}')\, a_-^{\pm}(\mathbf{k})$

(VI) $b^+(\mathbf{x})\, a_+^{\pm}(\mathbf{k}')\, a_-^{\pm}(\mathbf{k})$

(VII) $b(\mathbf{x})\, a_+(\mathbf{k})\, a_-(\mathbf{k}')$

(VIII) $b^+(\mathbf{x})\, a_+(\mathbf{k})\, a_-(\mathbf{k}')$

One envisages *the possibility of dissecting any actual interaction process into* such elementary vertices representing *virtual elementary acts* of the following eight types:

(I) Absorption of a photon under scattering of an electron
(II) Emission of a photon under scattering of an electron
(III) Absorption of a photon under scattering of a positron
(IV) Emission of a photon under scattering of a positron
(V) Absorption of a photon under creation of an electron-positron pair
(VI) Emission of a photon under creation of an electron-positron pair
(VII) Absorption of a photon under annihilation of an electron-positron pair
(VIII) Emission of a photon under annihilation of an electron-positron pair

In accordance with the fundamental dynamical postulate developed in Section 7, any state vector $|u(t)\rangle = |b(t)\rangle_I$ describing a system of electromagnetically interacting fermions in the interaction picture, which is appropriate here, will satisfy (7.10)

$$(23.9) \qquad i(\partial/\partial t)|u(t)\rangle = H'_I(t)|u(t)\rangle.$$

Comparison of the consequences of this equation with experiment is facilitated if one performs a formal integration leading to introduction of the concept of *the scattering matrix.*

For a first orientation, this concept can be adumbrated by the following line of reasoning which might be useful as a mnemonic device, pending a more detailed treatment to be given later. In an infinitesimal time interval Δt one has

$$(23.10) \qquad |u(t+\Delta t)\rangle = \exp[-iH'(t)\,\Delta t]\,|u(t)\rangle$$

so that, by iteration

(23.11)
$$|u(t+2\Delta t)\rangle = \exp[-iH'(t+\Delta t)\,\Delta t]\exp[-iH'(t)\,\Delta t]\,|u(t)\rangle, \qquad \text{etc.}$$

Now, since in general $H'(t+\Delta t)$ and $H'(t)$ do not commute, one must take into account

$$(23.12) \qquad e^A\,e^B \neq e^{A+B} \qquad \text{if} \qquad AB \neq BA$$

preventing one from writing down immediately

$$\prod_{n=0}^{\cdots} \exp[-iH'(t+n\Delta t)\,\Delta t] = \exp\left[-i\int_{t_0}^{\cdots} H'(t)\,dt\right].$$

However, one can arrive at a closed formula by introduction of a time-ordering operator P, so that

(23.13) $|u(t+2\,\varDelta t)\rangle \,=\, P[\exp{(-i\{H'(t+\varDelta t)+H'(t)\}\,\varDelta t)]}|u(t)\rangle$

where

(23.14) $P[H(t_1)\,H(t_2)] \,=\, \begin{cases} H(t_2)\,H(t_1) & \text{if} \quad t_2 > t_1 \\ H(t_1)\,H(t_2) & \text{if} \quad t_2 < t_1 \end{cases}$

and generally

(23.15)

$P[H(t_1)\,H(t_2)\ldots H(t_n)] \,=\, H(t_i)\,H(t_j)\ldots H(t_k) \quad \text{with} \quad t_i > t_j > \ldots > t_k.$

One can thus write down, *formally*, the solution

(23.16) $|u(t)\rangle \,=\, P\left[\exp\left(-i\int_{t_0}^{t} H'(t)\,dt\right)\right]|u(t_0)\rangle,$

in which $P[\ ]$ can be looked upon as a unitary operator connecting the state vector at time t_0 with the state vector at time t. Incidentally, time ordering after the prescription (23.15) is a relativistically invariant concept, provided P is applied to operators which commute for space-like points.

When applying this solution to experimentally realizable situations one is normally interested in a comparison of the "final" state of the system $|u(t = +\infty)\rangle$ with the "initial" state in the remote past $|u(t = -\infty)\rangle$. With these limits one has then

(23.17) $|u(t = +\infty)\rangle \,=\, S|u(t = -\infty)\rangle$

where the unitary scattering operator S is defined by

(23.18) $S \,=\, P\left[\exp\left(-i\int_{-\infty}^{+\infty} H'(t)\,dt\right)\right] \,=\, P\left[\exp\left(-i\int \mathscr{H}'(x)\,d^4x\right)\right]$

with

(23.19) $\mathscr{H}'(x) \,=\, -A_\mu(x)\,j_\mu(x)$

in the case of quantum electrodynamics.

The matrix elements between some initial state, labeled by a complete set τ of quantum numbers as $|\tau\rangle = |u(t = -\infty)\rangle$, and some final state, labeled similarly $|\tau'\rangle = |u(t = +\infty)\rangle$, turn out to be generally of the form

(23.20) $\langle\tau'|S|\tau\rangle \,=\, \delta_{\tau'\tau}+\langle\tau'|R|\tau\rangle\,\delta(\mathbf{P}'-\mathbf{P})\,\delta(P_0'-P_0)$

where $\langle\tau'|R|\tau\rangle$ (the "reaction matrix") is some regular function of the momenta **p** and the energies p_0 of the particles involved in the transition. **P** and P_0 stand for the total momentum and the total energy of the system in state τ, and the δ functions express the conservation of these dynamical quantities.

When computing transition probabilities between states $\tau' \neq \tau$, so that $\delta_{\tau' \tau} = 0$, one has to keep in mind when squaring (23.20)

$$(23.21) \quad [\delta(\mathbf{P'}-\mathbf{P})\,\delta(P_0'-P_0)]^2 = [1/(2\pi)^4]\,\delta(\mathbf{P'}-\mathbf{P})\,\delta(P_0'-P_0)\int d^4x$$

(see Appendix 6), so that one can meaningfully define only the transition probability per unit volume and unit time,

(23.22)

$$w_{\tau' \tau} = \frac{|\langle\tau'|S|\tau\rangle|^2}{\int d^4x} = [1/(2\pi)^4]|\langle\tau'|R|\tau\rangle|^2\,\delta(\mathbf{P'}-\mathbf{P})\,\delta(P_0'-P_0).$$

By standard procedures one can introduce at this point the idea of the "cross section" for scattering into a given final state, essentially by dividing (23.22) by the number of incoming particles per unit time and unit area. Some authors have spent considerable effort on defining cross sections in a relativistically covariant manner. Although such definitions can be useful, especially if one wishes to make calculations in the barycentric system, they are rather cumbersome and require complicated notations when more than two particles in the initial or final state are considered. Since in practice one deals usually with a laboratory frame of reference with one of the particles in the initial state at rest, and rarely considers situations with more than three particles in the final state, it is usually most convenient to make the so-called phase space considerations as one goes along in the description of a particular experimental situation. The standard example for phase space considerations in quantum electrodynamics is the well known Compton effect, which is treated for the purpose of illustrating the use of the δ function in Appendix 6. Readers interested in other examples are referred to the work by Fermi quoted at the end of this section.

Since the scattering matrix is the concept which most closely expresses the quantum mechanical view taken of processes involving transitions from an initial to a final state, it warrants a somewhat more detailed treatment than the derivation of expression (21.18) given above.

Consider quite generally a physical object for which, by assumption, it makes sense to write the Hamiltonian

$$(23.23) \qquad\qquad H = H^0 + H'$$

where H' represents the interaction energy of the system. Assume further that H', if written in the state picture, does not contain the time t explicitly.

Following the development of Section 7, one has various possibilities of describing the object.

(i) *In the state picture* the state of the system at time t can be defined through observation at time t and satisfies, if it is denoted by $|b(t)\rangle_s$, the Schroedinger equation

(23.24) $$i(\partial|b\rangle_s/\partial t) = H|b\rangle_s.$$

Solutions of the Schroedinger equation in absence of H' will be denoted $|c(t)\rangle_s$ so that in the state picture

(23.25) $$i(\partial|c\rangle_s/\partial t) = H^0|c\rangle_s.$$

(ii) *In the operator picture* the state is a fixed vector in time and may be taken as identical with the initial value of $|b(t)\rangle_s$ in the state picture, $|b\rangle_0 = |b(0)\rangle_s$. The transformation between state and operator picture is mediated by the unitary operations [see Eq. (7.17)]

(23.26) $$|b(t)\rangle_s = e^{-iHt}|b\rangle_0; \qquad |c(t)\rangle_s = \exp(-iH^0t)|c\rangle_0.$$

(iii) *In the interaction picture* the concept of state has again a different meaning. One speaks of a "state at time t" in the interaction picture, denoted $|b(t)\rangle_I = |u(t)\rangle$, and means a description obtained from observations at time t *and then mentally reduced*, so to speak, *by interaction free motion to time* $t = 0$. The connection of any state vector $|u(t)\rangle$ in the interaction picture with the corresponding state vector $|b\rangle_0$ in the operator picture is thus

(23.27) $$|b(t)\rangle_I = |u(t)\rangle = \exp(iH^0t)\, e^{-iHt}|b\rangle_0$$

i.e. the development of the state in time is determined essentially by H'.

Introducing the notation

(23.27') $\quad U(t) = e^{iHt}\exp(-iH^0t); \qquad U^+(t) = \exp(iH^0t)\, e^{-iHt}$

one can write Eq. (23.27)

(23.28) $$|u(t)\rangle = U^+(t)|b\rangle_0$$

and has thus

(23.29) $\quad |u(t_2)\rangle = U^+(t_2)\, U(t_1)\, |u(t_1)\rangle = U(t_2, t_1)\,|u(t_1)\rangle$

with

(23.30) $$U(t_2, t_1) = U^+(t_2)\, U(t_1).$$

This transformation operator $U(t_2, t_1)$ satisfies the equation

(23.31) $i[\partial U(t_2, t_1)/\partial t_2] = [H'(t_2)]_I \, U(t_2, t_1)$

where

(23.32) $[H'(t)]_I = \exp(iH^0 t) \, H' \exp(-iH^0 t)$

is the interaction Hamiltonian in the interaction picture (the "effective" interaction) containing the time dependence as in the example (23.6) given above. To prove Eq. (23.31), note that, according to (23.27) and (23.26), the "state at time t" in the interaction picture is the "state" in the state picture reduced by interaction free motion to time $t = 0$,

(23.33) $|u(t)\rangle = \exp(iH^0 t) \, |b(t)\rangle_s$

and use Eq. (23.24) so that by differentiation one obtains

(23.34)
$$i[\partial|u(t)\rangle/\partial t] = -H^0 \exp(iH^0 t) \, |b(t)\rangle_s + i \exp(iH^0 t) \, [\partial|b(t)\rangle_I/\partial t]$$
$$= \exp(iH^0 t) \, H' |b(t)\rangle_s = \exp(iH^0 t) \, H' \exp(-iH^0 t) \, |u(t)\rangle.$$

Equation (23.31) can be integrated formally, using the boundary condition $U(t,t) = 1$, to give

(23.35) $U(t_2, t_1) = 1 - i \int_{t_1}^{t_2} [H'(t)]_I \, U(t, t_1) \, dt.$

Writing

(23.36) $U(t, t_1) = U^{(0)}(t, t_1) + U^{(1)}(t, t_1) + U^{(2)}(t, t_1) + \ldots$

one obtains by iteration, starting with $U^{(0)}(t, t_1)$,

(23.37)
$$U(t_2, t_1) = 1 - i \int_{t_1}^{t_2} [H'(t)]_I \, dt + (-i)^2 \int_{t_1}^{t_2} [H'(t)]_I \, dt \int_{t_1}^{t} [H'(t')]_I \, dt' + \ldots$$

which can be summed to the time ordered product

(23.38) $U(t_2, t_1) = P\left[\exp\left(-i \int_{t_1}^{t_2} H'(t) \, dt\right)\right].$

From this the scattering operator S, connecting initial state $|\tau\rangle = |u(t = -\infty)\rangle$ with final state $|\tau'\rangle = |u(t = +\infty)\rangle$, is obtained as the limit

(23.39) $S = U(+\infty, -\infty).$

Although the interaction picture is particularly adapted to description of scattering processes, such processes can also be grasped in the state picture as well as in the operator picture through introduction of the concepts of "final and initial configuration" and "outgoing and ingoing states."

Consider, for simplicity's sake, a system which possesses no bound states, and in it are found only so-called single channel scattering processes, which means to each solution $|b(t)\rangle_s$ of the Schroedinger equation (23.24) there should correspond *uniquely* a solution $|c_+(t)\rangle_s$ of (23.25) through the requirement

$$(23.40) \qquad \lim_{t\to+\infty} |b(t)\rangle_s = \lim_{t\to+\infty} |c_+(t)\rangle_s$$

and another solution $|c_-(t)\rangle_s$ of (23.25) through the requirement

$$(23.41) \qquad \lim_{t\to-\infty} |b(t)\rangle_s = \lim_{t\to-\infty} |c_-(t)\rangle_s.$$

These definitions of the "result of scattering" $|c_+\rangle$ and the "cause of scattering" $|c_-\rangle$ imply the notion of an interaction which is absent for large time $t > 0$ and for large time $t < 0$ and which is "turned on" only during the scattering process proper in between these times.

One can now introduce the concept of "final and initial configuration," denoted $|c_+(0)\rangle$ and $|c_-(0)\rangle$, which are obtained, respectively, by calculating the result of scattering $|c_+(t)\rangle$ for large $t > 0$ backward to $t = 0$, and the cause of scattering $|c_-(t)\rangle$ for large $t < 0$ forward to $t = 0$, assuming interaction free motion in accordance with the second equation (23.26). Having done this, one is now in a position to relate the actual state $|b(0)\rangle$ at time $t = 0$ to the final configuration $|c_+(0)\rangle$ by

$$(23.42) \qquad |b(0)\rangle = \Omega_+ |c_+(0)\rangle$$

where

$$(23.43) \qquad \Omega_+ = \lim_{t\to+\infty} e^{iHt} \exp(-iH^0 t)$$

and to the initial configuration $|c_-(0)\rangle$ by

$$(23.44) \qquad |b(0)\rangle = \Omega_- |c_-(0)\rangle$$

where

$$(23.45) \qquad \Omega_- = \lim_{t\to-\infty} e^{iHt} \exp(-iH^0 t).$$

In the operator picture one can thus characterize any scattering state $|b\rangle_0$ either by its initial or by its final configuration. Conversely, a state with final configuration $|c_+(0);\tau\rangle$, where τ stands for the set of quantum numbers characterizing that state, can be denoted $|b_\tau(\text{out})\rangle_0$ and called the "outgoing state," and a state with initial configuration $|c_-(0);\tau\rangle$ can be denoted $|b_\tau(\text{in})\rangle_0$ and called the "ingoing state."

An alternative scattering operator can now be defined in the operator picture, after a precedent by Yang and Feldman, as the projection operator which projects the outgoing states on the ingoing states,

$$(23.46) \qquad |b_\tau(\text{in})\rangle = S_Y |b_\tau(\text{out})\rangle \qquad \text{for all } \tau,$$

so that the probability amplitude for finding the outgoing state $|b_{\tau'}(\text{out})\rangle$, if the ingoing state is known to be $|b_\tau(\text{in})\rangle$, is

$$(23.47) \qquad \langle b_{\tau'}(\text{out}) | b_\tau(\text{in})\rangle = \langle b_{\tau'}(\text{out}) | S_Y | b_\tau(\text{out})\rangle.$$

The scattering operator S_Y may be expressed in terms of the operators $\Omega_\pm$, because one may write (23.46)

$$(23.48) \qquad \Omega_- |c_-(0);\tau\rangle = S_Y \Omega_+ |c_+(0);\tau\rangle$$

and since $|c_-(0);\tau\rangle \equiv |c_+(0);\tau\rangle$, one has

$$(23.49) \qquad S_Y = \Omega_- \Omega_+^\dagger$$

where use has been made of the unitarity

$$(23.50) \qquad \Omega \Omega^+ = I.$$

NOTES

Feynman's book [1] contains, *inter alia*, reprints of Feynman's early papers in which the graph technique is developed. Fermi's book [2] contains a number of examples in which the most ingenious use is made of phase space considerations. The scattering operator was invented by Wheeler [3]. The alternative definition of S_Y is due to Yang and Feldman [4].

Readers interested in how to overcome difficulties encountered through presence of bound states, and in the case of multichannel scattering, are referred to the works by Ekstein [5] and Jauch [6]. See also the review article by Brenig and Haag [7].

REFERENCES

[1] R. P. Feynman, "Quantum Electrodynamics," Benjamin, New York, 1961.
[2] E. Fermi, "Elementary Particles," Yale Univ. Press, New Haven, Connecticut, 1951.
[3] J. A. Wheeler, *Phys. Rev.* **52**, 1107 (1937).
[4] C. N. Yang and D. Feldman, *Phys. Rev.* **79**, 972 (1950).
[5] H. Ekstein, *Nuovo Cimento* **4**, 1017 (1956).
[6] J. M. Jauch, *Helv. Phys. Acta* **31** 127 (1958); **31**, 661 (1958).
[7] W. Brenig and R. Haag, *Fortschr. Physik* **7**, 183 (1959).

Perturbation Theory and the Propagator Concept

Experimentally the coupling parameter e of quantum electrodynamics is found to have the value $e^2 \simeq 1/137$, and, on the strength of this weakness in the coupling between charged fermions and photons, a so-called perturbation treatment of the interaction has been developed, consisting essentially of the hope that an expansion of the scattering operator S in the form

$$(24.1) \quad S = I + \sum_{n \geqslant 1} \int \ldots \int \mathscr{S}_n(x_1, \ldots, x_n) \, d^4 x_1 \ldots d^4 x_n = I + \sum_{n \geqslant 1} S_n$$

is meaningful even though some of its terms turn out to result in diverging matrix elements.

By comparison with (23.18), each of the terms in expansion (24.1) is obtained by the substitution

$$(24.2) \qquad \mathscr{S}_n = [(-i)^n/n!] P[\mathscr{H}'(x_1) \ldots \mathscr{H}'(x_n)]$$

and the matrix elements to be computed are

(24.3)

$$\langle \tau' | S_n | \tau \rangle = (-ie/2)^n (1/n!) \int \ldots \int \langle \tau' | P\{[\bar{\psi}(x_1), \gamma_{\mu_1} \psi(x_1)] A_{\mu_1}(x_1) \ldots$$

$$[\bar{\psi}(x_n), \gamma_{\mu_n} \psi(x_n)] A_{\mu_n}(x_n)\} | \tau \rangle \, d^4 x_1 \ldots d^4 x_n.$$

The reduction of this—on first sight—overwhelmingly complicated expression to a form more tractable for actual computation can be done in a sequence of steps, which are the fruits of laborious efforts carried out by a number of workers in the late 1940's and early 1950's.

(i) Instead of using the "time ordered" product

$$P[A(x_1) A(x_2) \ldots A(x_n)] = A(x_i) A(x_j) \ldots A(x_k) \quad \text{with} \quad t_i > t_j > \ldots > t_k$$

of a set of n boson or fermion operators as defined in (23.15), it is more convenient to work with the "chronological product" denoted $T[A(x_1) A(x_2) \ldots A(x_n)]$ which differs from P by a sign factor η

$$(24.4) \quad T[A(x_1) A(x_2) \ldots A(x_n)] = \eta A(x_i) A(x_j) \ldots A(x_k)$$

$$\text{with} \quad t_i > t_j > \ldots > t_k$$

203

where $\eta = +1$ if the conversion of the sequence $A(x_1)A(x_2)\ldots A(x_n)$ into $A(x_i)A(x_j)\ldots A(x_k)$ requires an even number of transpositions of fermion operators, and $\eta = -1$ if this conversion involves an odd number of such transpositions. Since S_n contains all fermion operators in pairs, any change in sequence $\bar{\psi}(x_i)\psi(x_i)\bar{\psi}(x_j)\psi(x_j) \rightarrow \bar{\psi}(x_j)\psi(x_j)\bar{\psi}(x_i)\psi(x_i)$ involves an even number of transpositions of fermion operators, and one has therefore simply

$$(24.5) \qquad P[\mathcal{H}'(x_1)\ldots\mathcal{H}'(x_n)] = T[\mathcal{H}'(x_1)\ldots\mathcal{H}'(x_n)]$$

(ii) All operators $A_\mu(x)$, $\psi(x)$ and $\bar{\psi}(x)$ are separated into their "negative frequency part" and their "positive frequency part":

$$A_\mu(x) = A_\mu^{(-)}(x) + A_\mu^{(+)}(x)$$

$$(24.6) \qquad \psi(x) = \psi^{(-)}(x) + \psi^{(+)}(x)$$

$$\bar{\psi}(x) = \bar{\psi}^{(-)}(x) + \bar{\psi}^{(+)}(x)$$

being essentially a separation into annihilation and creation operators, since ψ functions with phase $e^{ikx} = e^{i(\mathbf{kq}-\omega t)}$ have always been associated with annihilation operators and phases $e^{-ikx} = e^{-i(\mathbf{kq}-\omega t)}$ with creation operators. Thus, according to definitions (18.86) and (19.51)

$$A_\mu^{(-)}(x) = (1/\sqrt{V}) \sum_\varkappa \sum_S [\epsilon_\mu(S)/\sqrt{2\omega}]\, e^{i\varkappa x}\, b(\varkappa, S)$$

$$A_\mu^{(+)}(x) = (1/\sqrt{V}) \sum_\varkappa \sum_S [\epsilon_\mu^*(S)/\sqrt{2\omega}]\, e^{-i\varkappa x}\, b^+(\varkappa, S)$$

$$\psi^{(-)}(x) = \sum_\mathbf{k} \sum_r A_+(\mathbf{k}, r)\, e^{ikx}\, a_+(\mathbf{k}, r)$$

$$(24.7) \qquad \psi^{(+)}(x) = \sum_\mathbf{k} \sum_r A_-(\mathbf{k}, r)\, e^{-ikx}\, a_-^+(\mathbf{k}, r)$$

$$\bar{\psi}^{(-)}(x) = \sum_\mathbf{k} \sum_r \overline{A}_-(\mathbf{k}, r)\, e^{ikx}\, a_-(\mathbf{k}, r)$$

$$\bar{\psi}^{(+)}(x) = \sum_\mathbf{k} \sum_r \overline{A}_+(\mathbf{k}, r)\, e^{-ikx}\, a_+^+(\mathbf{k}, r).$$

(iii) One aims at reordering the operators making up S_n so that all creation operators stand to the left of all annihilation operators, and with this aim in mind one defines the "normal product" denoted $N[\ldots]$ of two boson operators $A_1(x_1)$ and $A_2(x_2)$ as

$$(24.8)$$

$$N[A_1(x_1)A_2(x_2)] = A_1^{(+)}(x_1)A_2^{(+)}(x_2) + A_1^{(+)}(x_1)A_2^{(-)}(x_2)$$
$$+ A_2^{(+)}(x_2)A_1^{(-)}(x_1) + A_1^{(-)}(x_1)A_2^{(-)}(x_2)$$

and of two fermion operators $\psi_1(x_1)$ and $\psi_2(x_2)$ as

(24.9)

$$N[\psi_1(x_1)\,\psi_2(x_2)] = \psi_1^{(+)}(x_1)\,\psi_2^{(+)}(x_2) + \psi_1^{(+)}(x_1)\,\psi_2^{(-)}(x_2) - \psi_2^{(+)}(x_2)\,\psi_1^{(-)}(x_1)$$
$$+ \psi_1^{(-)}(x_1)\,\psi_2^{(-)}(x_2).$$

The definition of the normal product of any number of boson and/or fermion operators follows from these basic definitions by induction. The minus sign in front of the third term on the right-hand side of (24.9) has been introduced to accommodate the anti-C.R.s between fermion operators consistently.

Introduction of the normal product is advantageous for three reasons:

(a) The vacuum expectation value of any normal product vanishes,

(24.10) $$\langle 0|N[\ldots]|0\rangle = 0.$$

(b) Since

(24.11) $$N[A_1(x_1)\,A_2(x_2)] = N[A_2(x_2)\,A_1(x_1)]$$

and

(24.12) $$N[\psi_1(x_1)\,\psi_2(x_2)] = -N[\psi_2(x_2)\,\psi_1(x_1)]$$

one can treat any normal product $N[\ldots]$ as if all boson operators inside the bracket $[\ldots]$ *always* commute and as if all Fermion operators inside $[\ldots]$ *always* anticommute.

(c) The operators in the interaction Hamiltonian $\mathscr{H}'(x)$ are already in ordered normal form, as is immediately obvious from (23.6) when one remembers that Boson operators and Fermion operators commute. One can thus write

(24.13) $$\mathscr{H}'(x) = -e\,N[\bar{\psi}(x)\,\gamma_\mu\,A_\mu(x)\,\psi(x)].$$

This is, of course, a consequence of the particular definition of the current density, which was arranged such that the vacuum expectation value of the current vanished.

(iv) The conversion of a chronological product into a sum of normal products is possible, because the chronological product of two operators differs as a consequence of C.R.s or anti-C.R.s from the normal product of these two operators only by a number, which is called the "chronological pairing." Denoting this "chronological pairing" number associated with two boson operators $A_1(x_1)$ and $A_2(x_2)$ by $\overline{A_1(x_1)\,A_2(x_2)}$, one has by definition

(24.14)

$$T[A_1(x_1)\,A_2(x_2)] = N[A_1(x_1)\,A_2(x_2)] + \overline{A_1(x_1)\,A_2(x_2)},$$

and denoting similarly the "chronological pairing" number associated with two Fermion operators $\psi_1(x_1)$ and $\psi_2(x_2)$ by $\overbracket{\psi_1(x_1)\psi_2}(x_2)$ one has

(24.15)

$$T[\psi_1(x_1)\psi_2(x_2)] = N[\psi_1(x_1)\psi_2(x_2)] + \overbracket{\psi_1(x_1)}\psi_2(x_2).$$

Equation (24.10) allows one to compute any chronological pairing as a vacuum expectation value:

(24.16) $$\overbracket{A_1(x_1)\,A_2}(x_2) = \langle 0|T[A_1(x_1)\,A_2(x_2)]|0\rangle$$

(24.17) $$\overbracket{\psi_1(x_1)\psi_2}(x_2) = \langle 0|T[\psi_1(x_1)\psi_2(x_2)]|0\rangle.$$

By successive use of prescriptions (24.14) and (24.15), one can now express any chronological product as a sum of terms containing only normal products and chronological pairings. More precisely, by induction one arrives at a theorem first proven by Wick: *Any chronological product is equal to the sum of all possible normal products that can be formed with all possible pairings.* The meaning of this statement is perhaps best communicated by writing out its application to the cases of two, three, and four operators:

(24.18)

$$T[AB] = N[AB] + N[\overbracket{AB}]$$

$$T[ABC] = N[ABC] + N[\overbracket{AB}C] + N[A\overbracket{BC}] + N[\overbracket{A}B\overbracket{C}]$$

$$T[ABCD] = N[ABCD] + N[\overbracket{AB}CD] + N[\overbracket{A}B\overbracket{C}D] + N[\overbracket{A}BC\overbracket{D}]$$

$$+ N[A\overbracket{BC}D] + N[A\overbracket{B}C\overbracket{D}] + N[AB\overbracket{CD}] + N[\overbracket{AB}\overbracket{CD}]$$

$$+ N[\overbracket{A\overbracket{BC}D}] + N[\overbracket{A}\overbracket{BC}\overbracket{D}].$$

The notation

(24.19)

$$N[A\overbracket{BC}D] \equiv \eta \overbracket{CD}N[AB]; \qquad N[\overbracket{A\overbracket{BC}D}] \equiv \eta \overbracket{AD}\overbracket{BC}; \qquad \text{etc.}$$

has been used, where η is again the sign factor equal to $+1$ or -1 depending on whether an even or odd number of interchanges in fermion operators are needed to convert the sequence of operators $ABCD$ into the sequences $CDAB$, $ADBC$, etc., respectively.

The ground is now prepared for the conversion of the term S_n of the scattering operator (24.1) into a sum of terms which contain only normal products and chronological pairings. For $n \geqslant 3$ this is still a rather formidable task, but for $n = 2$ one obtains, upon observation of (24.5), (24.13), and Wick's theorem, and keeping in mind that chronological pairings between a boson and a fermion operator, as well as $\overline{\psi\psi}$ and $\overline{\overline{\psi}\overline{\psi}}$, vanish, a decomposition of

(24.20)

$$S_2 = -(e^2/2) \int\int T\{N[\overline{\psi}(x_1)\gamma_{\mu_1}A_{\mu_1}(x_1)\psi(x_1)]N[\overline{\psi}(x_2)\gamma_{\mu_2}A_{\mu_2}(x_2)$$
$$\times \psi(x_2)]\}\,d^4x_1\,d^4x_2$$

into eight terms

(24.21)
$$S_2 = \sum_{C=I}^{VIII} S_2^{(C)}$$

which will now be written down:

(24.22)

$$S_2^{(I)} = -(e^2/2)\int\int N[\overline{\psi}(x_1)\gamma_{\mu_1}A_{\mu_1}(x_1)\psi(x_1)\overline{\psi}(x_2)\gamma_{\mu_2}A_{\mu_2}(x_2)\psi(x_2)]\,d^4x_1\,d^4x_2$$

$$S_2^{(II)} = -(e^2/2)\int\int N[\overline{\psi}(x_1)\gamma_{\mu_1}A_{\mu_1}(x_1)\overbrace{\psi(x_1)\,\overline{\psi}(x_2)}\gamma_{\mu_2}A_{\mu_2}(x_2)\psi(x_2)]\,d^4x_1\,d^4x_2$$

$$S_2^{(III)} = -(e^2/2)\int\int N[\overbrace{\overline{\psi}(x_1)\gamma_{\mu_1}A_{\mu_1}(x_1)\psi(x_1)\overline{\psi}(x_2)\gamma_{\mu_2}A_{\mu_2}(x_2)\psi(x_2)}]\,d^4x_1\,d^4x_2$$

$$S_2^{(IV)} = -(e^2/2)\int\int N[\overline{\psi}(x_1)\gamma_{\mu_1}\overbrace{A_{\mu_1}(x_1)\psi(x_1)\overline{\psi}(x_2)\gamma_{\mu_2}A_{\mu_2}(x_2)}\psi(x_2)]\,d^4x_1\,d^4x_2$$

$$S_2^{(V)} = -(e^2/2)\int\int N[\overline{\psi}(x_1)\gamma_{\mu_1}\overbrace{A_{\mu_1}(x_1)\,\overbrace{\psi(x_1)\,\overline{\psi}(x_2)}\gamma_{\mu_2}A_{\mu_2}(x_2)}\psi(x_2)]\,d^4x_1\,d^4x_2$$

$$S_2^{(VI)} = -(e^2/2)\int\int N[\overbrace{\overline{\psi}(x_1)\gamma_{\mu_1}\overbrace{A_{\mu_1}(x_1)\psi(x_1)\overline{\psi}(x_2)\gamma_{\mu_2}A_{\mu_2}(x_2)}\psi(x_2)}]\,d^4x_1\,d^4x_2$$

$$S_2^{(VII)} = -(e^2/2)\int\int N[\overbrace{\overline{\psi}(x_1)\gamma_{\mu_1}A_{\mu_1}(x_1)\overbrace{\psi(x_1)\,\overline{\psi}(x_2)}\gamma_{\mu_2}A_{\mu_2}(x_2)\psi(x_2)}]\,d^4x_1\,d^4x_2$$

$$S_2^{(VIII)} = -(e^2/2)\int\int N[\overbrace{\overline{\psi}(x_1)\gamma_{\mu_1}\overbrace{A_{\mu_1}(x_1)\,\overbrace{\psi(x_1)\,\overline{\psi}(x_2)}\gamma_{\mu_2}A_{\mu_2}(x_2)}\psi(x_2)}]\,d^4x_1\,d^4x_2$$

The conventions, explained in Section 23, leading to the Feynman graphs representing the various terms in the interaction Hamiltonian, allow a picturesque classification and extraordinarily suggestive interpretation of every term in the scattering matrix, after it has been ordered into normal products, if these conventions are extended to envelop chronological pairings, *which are not operators*, as follows.

(a) Draw a vertex labeled x_i for each integration variable x_i. Since every such variable appears in conjunction with one and only one factor γ_{μ_i}, this convention is equivalent to drawing a vertex labeled x_i for each factor γ_{μ_i}.

(b) Draw a solid line entering the vertex x_i for each unpaired operator $\psi(x_i)$, a solid line leaving x_i for each unpaired $\bar{\psi}(x_i)$, a dotted line without direction connected to x_i for each unpaired $A(x_i)$. These conventions are consistent with the ones adopted in Section 23, but are less detailed because they do not yet distinguish between ingoing electrons and outgoing positrons, between outgoing electrons and ingoing positrons, and between ingoing and outgoing photons.

(c) Since chronological pairings occur only as connections of one symbol ψ and one symbol $\bar{\psi}$ and never between two symbols ψ or two symbols $\bar{\psi}$, and only between symbols ψ and $\bar{\psi}$ belonging to *different* vertices x_i and x_k, the following conventions are sufficient and consistent. Draw a solid line connecting the vertices x_i and x_k in direction from x_k to x_i for each pairing $\overline{\psi(x_i)\bar{\psi}(x_k)}$, a solid line directed from x_i to x_k for each pairing $\overline{\bar{\psi}(x_i)\psi(x_k)}$, and a dotted line without direction between x_i and x_k for each pairing $\overline{A(x_i)A(x_k)}$.

With these conventions all remaining operators $\bar{\psi}$, ψ, A in S_n contribute only "external lines" and all pairings $\overline{\psi\bar{\psi}}$, $\overline{\psi\bar{\psi}}$, $\overline{AA}$ contribute only "internal lines," such that all fermion lines can be drawn by following arrows of direction from end to end without hiatus. For example, the eight terms (24.22) will correspond to the following eight graphs:

$$S_2^{(I)} \qquad\qquad S_2^{(II)} \qquad\qquad S_2^{(III)} \qquad\qquad S_2^{(IV)}$$

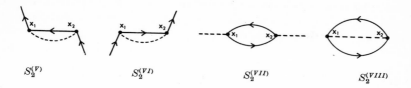

$S_2^{(V)}$ $\qquad$ $S_2^{(VI)}$ $\qquad$ $S_2^{(VII)}$ $\qquad$ $S_2^{(VIII)}$

With the exception of $S_2^{(I)}$ all graphs representing the various terms in S_2 have one or more internal lines. After some contemplation of these drawings one is inevitably tempted to adopt, with Feynman, a picturesque manner of speaking about them which conjures up for every internal line the image of a particle *propagating* in a state of virtual existence between the vertices of that line. Accordingly, the factors

$$\overline{\psi(x_i)\,\psi(x_k)}$$

and

$$\overline{A_{\mu_i}(x_i)\,A_{\mu_k}(x_k)}$$

will from now on be referred to as "fermion propagator" and "photon propagator," respectively.

One has to go back in the history of physics to Faraday's concept of the field line if one wants to find a mnemonic device which matches Feynman's graph in propagandistic persuasiveness. This historical analogy may serve here as timely warning against all too literal acceptance of mental images based mainly on a fabric of *conventions*, however consistent that fabric may appear. Thus, Maxwell was led by all too literal acceptance of Faraday's field concept to an ether theory of vacuum which ultimately turned out to be abortive. Similar temptations are lurking behind Feynman's graphs, especially the ones of the type $S_2^{(VIII)}$.

Evaluation of matrix elements requires separation of all remaining operators into negative and positive frequency parts. Accordingly, one has a decomposition of $S_2^{(I)}$ into $2^6 = 64$ terms, of $S_2^{(II)}$, $S_2^{(III)}$, and $S_2^{(IV)}$ each into $2^4 = 16$ terms, of $S_2^{(V)}$, $S_2^{(VI)}$, and $S_2^{(VII)}$ each into $2^2 = 4$ terms, and $S_2^{(VIII)}$ requires no further decomposition. In the language of Feynman's graph this amounts to sorting the external lines into electron lines, positron lines, ingoing photon lines, and outgoing photon lines. If one adopts again, as in Section 23, *for external lines* the convention of drawing electron lines pointing upward and positron lines pointing downward, as well as drawing *all* external photon lines pointing upward, one arrives at a unique decomposition of each basic graph $S_2^{(C)}$ into a sum of "*directed graphs.*" One has thus, for example,

(24.22) $$S_2^{(II)} = \sum_{d=1}^{16} S_2^{(II,d)}$$

i.e.

and a completely analogous decomposition of $S_2^{(III)}$.

It is at this stage of the development that there emerges quite trans-parently the consistency of a manner of speaking about fermion lines, which refers to positrons as "electrons running backward in time," by associating the upward direction in each directed graph with the direc-tion of time. This convention may be extended to fermion propagators, so that occurrence of a factor $\bar{\psi}(x_k)\psi(x_i)$ is represented graphically as a virtual electron propagating from x_k to x_i provided $t_k < t_i$ and as a virtual positron propagating from x_k to x_i provided $t_k > t_i$.

Thus one may think about a graph

which represents a contribution to the Compton effect on positrons as consisting of two virtual events occurring at x_1 and x_2 that are, for $t_2 > t_1$, two virtual scattering processes with photon absorption at x_2 and photon emission at x_1,

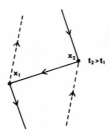

and, for $t_2 < t_1$, virtual annihilation of an electron-positron pair at x_1 and virtual creation of a pair at x_2,

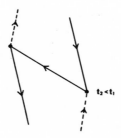

However, since an integration over x_1 and x_2 is performed in the computation of the corresponding term of $S_2^{(II)}$, the propagator will automatically take care of these possibilities, and one need not adhere to any up-down conventions as far as internal lines are concerned.

The conventional nature of the phrase "positrons are electrons running backward in time" should be abundantly clear by now, and the reader will not be misled into concluding positron *states* are necessarily obtained by the operation of time reversal from electron *states*.

From the expansion (24.22) for $S_2^{(II)}$ and an identical expression for $S_2^{(III)}$ one can immediately read which terms in S_2 will contribute, for example, to the Compton effect on electrons. The only graphs containing one ingoing electron, one ingoing photon, one outgoing electron, and one outgoing photon are

(24.23)

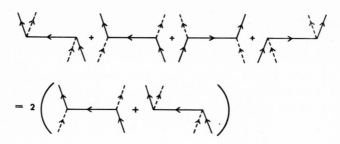

or, in analytical language,

(24.24)

$$S_2^{(\text{C.E.})} = -e^2 \int\int \bar{\psi}^{(+)}(x_1)\,\gamma_{\mu_1}\,\psi(x_1)\,\bar{\psi}(x_2)\,\gamma_{\mu_2}\,\psi^{(-)}(x_2)[A_{\mu_2}^{(+)}(x_2)\,A_{\mu_1}^{(-)}(x_1)$$
$$+ A_{\mu_1}^{(+)}(x_1)\,A_{\mu_2}^{(-)}(x_2)]\,d^4x_1\,d^4x_2.$$

Similarly, one can read, for example, from the expansion for $S_2^{(\text{IV})}$,

(24.25)
$$S_2^{(\text{IV})} = \sum_{d=1}^{16} S_2^{(\text{IV})(d)}$$

i.e.

the term which contributes to electron-electron scattering, namely the one represented by the sixth graph on the right-hand side of (24.25):

(24.26)

$$S_2^{(\text{E.E.S.})} = (e^2/2) \int\int \bar{\psi}^{(+)}(x_1)\,\gamma_{\mu_1}\,\bar{\psi}^{(+)}(x_2)\,\psi^{(-)}(x_1)\,\gamma_{\mu_2}\,\psi^{(-)}(x_2)$$

$$\times \overline{A_{\mu_1}(x_1)\,A_{\mu_2}(x_2)}\,d^4x_1\,d^4x_2.$$

One final task remains to be done before computation of matrix elements can be performed, namely evaluation of the propagators.

According to (24.17) the fermion propagator is defined as

(24.27)

$$\overline{\psi(x_1)\,\bar{\psi}(x_2)} = \langle 0|T[\psi(x_1)\,\bar{\psi}(x_2)]|0\rangle = \begin{cases} \langle 0|\psi(x_1)\,\bar{\psi}(x_2)|0\rangle & \text{for } t_1 > t_2 \\ -\langle 0|\bar{\psi}(x_2)\,\psi(x_1)|0\rangle & \text{for } t_1 < t_2 \end{cases}$$

Upon decomposition of the operators $\psi(x_1)$ and $\bar{\psi}(x_2)$ into positive and negative frequency parts, nonvanishing contributions to this vacuum expectation value are made by one term only in each case, namely

(24.28)

$$\overline{\psi(x_1)\,\bar{\psi}(x_2)} = \begin{cases} \langle 0|\psi^{(-)}(x_1)\,\bar{\psi}^{(+)}(x_2)|0\rangle & \text{for } t_1 > t_2 \\ -\langle 0|\bar{\psi}^{(-)}(x_2)\,\psi^{(+)}(x_1)|0\rangle & \text{for } t_1 < t_2 \end{cases}$$

By substitution of the expansions (24.7) one obtains

(24.29)

$$\overline{\psi(x_1)\,\bar{\psi}(x_2)} = \begin{cases} \sum_{\mathbf{k}}\sum_{r} A_+(\mathbf{k},r)\,\bar{A}_+(\mathbf{k},r)\exp[ik(x_1-x_2)] & \text{for } t_1 > t_2 \\ -\sum_{\mathbf{k}}\sum_{r} \bar{A}_-(\mathbf{k},r)\,A_-(\mathbf{k},r)\exp[-ik(x_1-x_2)] & \text{for } t_1 < t_2 \end{cases}$$

and summation over spins gives according to (19.52) and (19.53)

(24.30) $\quad \overline{\psi(x_1)\,\bar{\psi}(x_2)}$

$$= \begin{cases} \displaystyle\sum_{\mathbf{k}} \frac{(\hat{k}+m)}{2V\Omega}\exp[ik(x_1-x_2)] \\[2mm] \qquad = \displaystyle\sum_{\mathbf{k}} \frac{(\gamma_4\Omega - \boldsymbol{\gamma}\mathbf{k}+m)}{2V\Omega}\exp[-i\mathbf{k}(\mathbf{q}_1-\mathbf{q}_2)+i\Omega(t_1-t_2)], \\[4mm] -\displaystyle\sum_{\mathbf{k}} \frac{(\hat{k}-m)}{2V\Omega}\exp[-ik(x_1-x_2)] \\[2mm] \qquad = -\displaystyle\sum_{\mathbf{k}} \frac{(-\gamma_4\Omega - \boldsymbol{\gamma}\mathbf{k}+m)}{-2V\Omega}\exp[-i\mathbf{k}(\mathbf{q}_1-\mathbf{q}_2)-i\Omega(t_1-t_2)] \end{cases}$$

where Ω is always meant to be equal to $+\sqrt{\mathbf{k}^2+m^2}$, and where in the second equation a change in summation label $\mathbf{k} \to -\mathbf{k}$ has been performed.

This formula can be consolidated by utilizing the integral

(24.31) $$I = \int_F \frac{f(k_4)\, e^{ikx}}{k^2 - m^2}\, dk_4 = \int_F \frac{f(k_4)\, e^{-i\mathbf{k}\mathbf{q}}\, e^{+ik_4 t}}{(k_4 - \Omega)\,(k_4 + \Omega)}\, dk_4$$

where $k^2 = k_4^2 - \mathbf{k}^2$, and $\int_F$ stands for integration along "Feynman's

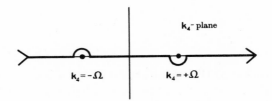

FIG. 24.1. Feynman's contour.

contour" in the complex k_4 plane, as indicated in Fig. 24.1. One has then to distinguish two cases.

(a) $t > 0$. The contour can be completed in the upper half plane, so that

(24.32a)
$$I = 2\pi i\, \mathrm{Res}\,(k_4 = +\Omega) = 2\pi i [f(\Omega)/2\Omega]\, e^{-i\mathbf{k}\mathbf{q}}\, e^{+i\Omega t} \qquad (t > 0).$$

(b) $t < 0$. The contour can be completed in the lower half plane, so that

(24.32b)
$$I = -2\pi i\, \mathrm{Res}\,(k_4 = -\Omega) = -2\pi i [f(-\Omega)/-2\Omega]\, e^{-i\mathbf{k}\mathbf{q}}\, e^{-i\Omega t} \qquad (t < 0).$$

Thus the propagator (24.30) can be written, under conversion of the sum $\sum_k (.../V)$ into the integral $[1/(2\pi)^3] \int ...\, d\mathbf{k}$

(24.33)
$$\overline{\psi(x_1)\, \bar{\psi}(x_2)} = \frac{1}{i(2\pi)^4} \int_F \frac{\hat{k} + m}{k^2 - m^2} \exp\,[ik(x_1 - x_2)]\, d^4 k = -\overline{\bar{\psi}(x_2)\, \psi(x_1)}.$$

The computation of the photon propagator proceeds in complete analogy to the one of the fermion propagator. By substituting into

(24.34)

$$\overline{A_\mu(x_1)\, A_\nu(x_2)} = \begin{cases} \langle 0|\, A_\mu(x_1)\, A_\nu(x_2)\, |0\rangle = \langle 0|\, A_\mu^{(-)}(x_1)\, A_\nu^{(+)}(x_2)\, |0\rangle \\ \qquad\qquad\qquad \text{for } \ t_1 > t_2 \\[2mm] \langle 0|\, A_\nu(x_2)\, A_\mu(x_1)\, |0\rangle = \langle 0|\, A_\nu^{(-)}(x_2)\, A_\mu^{(+)}(x_1)\, |0\rangle \\ \qquad\qquad\qquad \text{for } \ t_1 < t_2 \end{cases}$$

the expansions (24.7) one obtains

(24.35)

$$
\overline{A_\mu(x_1)\,A_\nu(x_2)} = \begin{cases}
\displaystyle\sum_\varkappa \sum_S \frac{\epsilon_\mu(S)\,\epsilon_\nu^*(S)}{2V\omega}\exp\left[i\kappa(x_1-x_2)\right] \\[2ex]
\displaystyle = \sum_\varkappa \sum_S \frac{\epsilon_\mu\,\epsilon_\nu^*}{2V\omega}\exp\left[-i\varkappa(\mathbf{q}_1-\mathbf{q}_2)+i\omega(t_1-t_2)\right], \\[2ex]
\displaystyle \sum_\varkappa \sum_S \frac{\epsilon_\nu(S)\,\epsilon_\mu^*(S)}{2V\omega}\exp\left[i\kappa(x_2-x_1)\right] \\[2ex]
\displaystyle = -\sum_\varkappa \sum_S \frac{\epsilon_\nu\,\epsilon_\mu^*}{-2V\omega}\exp\left[-i\varkappa(\mathbf{q}_1-\mathbf{q}_2)-i\omega(t_1-t_2)\right]
\end{cases}
$$

The summation over the polarizations can be carried out in the gauge-invariant manner explained in Section 18 leading to (18.102), and formula (24.35) can be consolidated once again by exploiting the integral formula

(24.36)

$$
\int_F \frac{e^{ikx}}{k^2}\,dk_4 = e^{-ik\mathbf{q}}\int_F \frac{e^{ik_4 t}}{(k_4-\omega)(k_4+\omega)}\,dk_4 = \begin{cases}
\displaystyle \frac{2\pi i}{2\omega}e^{-i\varkappa\mathbf{q}+i\omega t}\;(t>0) \\[2ex]
\displaystyle \frac{-2\pi i}{-2\omega}e^{-i\varkappa\mathbf{q}-i\omega t}\;(t<0)
\end{cases}
$$

yielding for the photon propagator finally

(24.37)

$$
\overline{A_\mu(x_1)\,A_\nu(x_2)} = \frac{1}{i(2\pi)^4}\int_F \frac{1}{\kappa^2}\left\{\left[\delta_{\mu\nu}-\frac{\kappa_\mu\kappa_\nu}{\kappa^2}\right]+d_l\frac{\kappa_\mu\kappa_\nu}{\kappa^2}\right\}\exp\left[i\kappa(x_1-x_2)\right]d^4\kappa.
$$

Since integration over any x_i gives a δ function, and since the computation of matrix elements always involves integration over all vertex coordinates x_i, one can formulate from the outset the rules for evaluation of Feynman's graphs in momentum space, as summarized in Table 24.1.

NOTES

Since the advent of the work by Wick [1], the treatment of the scattering matrix by perturbation expansion has become rather standardized and can be found in many texts with only small variations. Readers desiring to see this formalism applied to computation of cross sections are referred to Mandl [2], Källén [3], Bogoliubov and Shirkov [4], Jauch and Rohrlich [5], and Feynman [6].

TABLE 24.1

FEYNMAN'S RULES FOR EVALUATING DIRECTED GRAPHS

Sum over all indices appearing twice; integrate over the momenta of all internal lines along Feynman's contour; sum over polarizations in internal lines.

Graph	Particle	Direction	Factor in matrix element
	Electron	Entering	$[1/(2\pi)^{3/2}]A_+(k,r)$
	Electron	Leaving	$[1/(2\pi)^{3/2}]\bar{A}_+(k,r)$
	Positron	Entering	$[1/(2\pi)^{3/2}]\bar{A}_-(k,r)$
	Positron	Leaving	$[1/(2\pi)^{3/2}]A_-(k,r)$
	Photon	Entering	$[1/(2\pi)^{3/2}][\epsilon_\mu(S)/\sqrt{2\omega}]$ $\quad (S = 1, 2 \text{ only})$
	Photon	Leaving	$[1/(2\pi)^{3/2}][\epsilon_\mu^*(S)/\sqrt{2\omega}]$ $\quad (S = 1, 2 \text{ only})$
k_2 ... k_1	Vertex part		$ie\gamma_\mu(2\pi)^4\,\delta(k_2 - k_1 - \kappa) = \Gamma_\mu^0(k_1, k_1 + \kappa, -\kappa)$
k (1 ← 2)	Fermion propagating		$[1/i(2\pi)^4][\hat{k}+m]/[k^2 - m^2] = G^\circ(k)$
$\times$	Photon propagating		$[1/i(2\pi)^4\kappa^2]\left[\left(\delta_{\mu\nu} - \dfrac{\kappa_\mu\kappa_\nu}{\kappa^2}\right) + \dfrac{d_l\kappa_\mu\kappa_\nu}{\kappa^2}\right] = D_{\mu\nu}^0(\kappa)$

REFERENCES

[1] G. C. Wick, *Phys. Rev.* **80**, 268 (1950).

[2] F. Mandl, "Introduction to Quantum Field Theory." Wiley (Interscience), New York, 1959.

[3] G. Källén, Quantenelektrodynamik, *in* "Encyclopedia of Physics" (S. Flügge, ed.), Vol. 5, Pat. I. Springer, Berlin, 1958.

[4] N. N. Bogoliubov and D. V. Shirkov, "Introduction to the Theory of Quantized Fields." Wiley (Interscience), New York, 1959.

[5] J. M. Jauch and F. Rohrlich, "The Theory of Photons and Electrons." Addison-Wesley, Reading, Massachusetts, 1955.

[6] R. P. Feynman, "Quantum Electrodynamics." Benjamin, New York, 1961.

The Hierarchy of Propagators

The division of particles into "free" and "virtual" particles, corresponding to a separation of contributions to any Feynman graph into "external" and "internal" lines, is rather artificial and, from an operational point of view, in fact untenable. Since one cannot, in principle, "turn off" the interaction, it would appear more appropriate to view the world *sub specie aeternitatis* as an infinite concatenation of propagators, representable graphically as an infinite network of internal lines only.

Consider, for example, an experimental situation in which one can meaningfully introduce an initial and a final state containing just one fermion. In the language of Feynman graphs one would represent this by an external fermion line entering into and emerging from a region of internal lines, as drawn in Fig. 25.1. However, before one can label an external line by some quantum numbers, the corresponding state must be prepared somehow, i.e. a measurement of the quantum numbers must be performed. Now any such measurement requires interaction of some kind, and in a pure quantum electrodynamics one is forced to infer the existence of two vertices, involving emission or absorption of photons, which are the terminals of the supposedly "external" lines, so that one

Fig. 25.1. Hypothetical propagation of single fermion initial state into single fermion final state.

should draw Fig. 25.1 more realistically as in Fig. 25.2, with the usual convention that $t_2 > t_1$ describes the propagation of an electron and $t_2 < t_1$ the propagation of a positron from x_1 to x_2. Since any single fermion state can be considered as being raised from the vacuum state

FIG. 25.2. More realistic graph representing single fermion propagation.

by the application of the appropriate creation operator, allowing one to write any matrix element of S between an initial state "fermion at x_i" denoted $|x_i\rangle = \bar{\psi}(x_i)|0\rangle$ and a final state "fermion at x_j" denoted $\langle x_j| = \langle 0|\,\psi(x_j)$ as a vacuum expectation value

$$(25.1) \qquad \langle x_j|S|x_i\rangle = \langle 0|\,\psi(x_j)\,S\bar{\psi}(x_i)\,|0\rangle,$$

one is led to introduce the concept of the "*true fermion propagator*," being the amplitude for fermion propagation between vertices x_1 and x_2, denoted $G(x_2, x_1)$ and defined by

$$(25.2) \qquad G(x_2, x_1) = \langle 0|\,T[\psi(x_2)\,S\bar{\psi}(x_1)]\,|0\rangle$$

as a chronologically ordered amplitude of the scattering matrix. The ordinary fermion propagator (24.27), now denoted $G^0(x_2, x_1)$, is then obtained by omitting in the expansion of S all but the zero-order term, which is the identity, so that

$$(25.3) \qquad G^0(x_2, x_1) = \langle 0|\,T[\psi(x_2)\,\bar{\psi}(x_1)]\,|0\rangle.$$

In momentum space the true propagator $G(k)$ will be represented graphically as in Fig. 25.3, with a "region of ignorance," hatched by

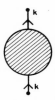

FIG. 25.3. Graph representing the true fermion propagator (region of ignorance hatched from SW to NE).

convention from SW to NE for fermion propagation, which symbolizes all possible networks of internal lines which will contribute to $G(x_2, x_1)$ and its momentum representation $G(k)$ in accordance with the expansion of S in (25.2). The term "region of ignorance" for the region of internal lines has been adopted, because no measurements beyond establishment of the value **k** at the terminal vertices are envisaged, and it should therefore be possible to find a measurement symbol, denoted $M(k;k)$ after the precedent of Section 5, so that

(25.4) $$G(k) = \text{trace } M(k;k)$$

and

(25.4′) $$G^0(k) = [1/i(2\pi)^4][(\hat{k}+m)/(k^2-m^2)].$$

Similarly, the "*true photon propagator*" can be introduced as the chronologically ordered amplitude of the scattering matrix between single photon states, which may be represented in analogy to (25.2) as a vacuum expectation value,

(25.5) $$D_{\mu\nu}(x_2, x_1) = \langle 0| T[A_\mu(x_2) SA_\nu(x_1)] |0\rangle$$

containing the ordinary propagator (24.42) as first term in the expansion of S,

(25.5′) $$D^0_{\mu\nu}(x_2, x_1) = \langle 0| T[A_\mu(x_2) A_\nu(x_1) |0\rangle.$$

In momentum space the true photon propagator may be rendered graphically as in Fig. 25.4, the region of ignorance in this case being

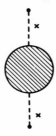

FIG. 25.4. Graph representing the true photon propagator (region of ignorance hatched from SE to NW).

hatched by convention from SE to NW. One anticipates existence of a measurement symbol $M_{\mu\nu}(\kappa;\kappa)$ associated with the momentum representation $D_{\mu\nu}(\kappa)$ of $D_{\mu\nu}(x_2, x_1)$ so that

(25.6) $$D_{\mu\nu}(\kappa) = \text{trace } M_{\mu\nu}(\kappa;\kappa)$$

and

(25.6′) $$D^0_{\mu\nu}(\kappa) = \frac{1}{i(2\pi)^4 \kappa^2}\left[\left(\delta_{\mu\nu} - \frac{\kappa_\mu \kappa_\nu}{\kappa^2}\right) + \frac{d^0_l \kappa_\mu \kappa_\nu}{\kappa^2}\right].$$

Of particular interest for later development is further the "*true vertex part*"; rendered graphically in Fig. 25.5, with conventional crosshatching

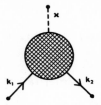

FIG. 25.5. Graph representing the true vertex part (region of ignorance crosshatched).

of the region of ignorance in this case. Denoting its momentum representation by $\Gamma_\mu(k_1, k_2, \kappa)$, there should exist a measurement symbol $M_\mu(k_1; k_2; \kappa)$ so that

$$(25.7) \qquad \Gamma_\mu(k_1, k_2, \kappa) = \text{trace } M_\mu(k_1; k_2; \kappa)$$

and in lowest order

$$(25.7') \qquad \Gamma_\mu^0(k_1, k_2, \kappa) = ie\gamma_\mu(2\pi)^4\,\delta(k_1 - k_2 - \kappa).$$

Finally, it is instructive to give special consideration to the true vacuum propagator, being the vacuum expectation value of the scattering operator, rendered graphically in Fig. 25.6, as a pure region of

FIG. 25.6. Graph representing the true vacuum propagator (region of ignorance hatched horizontally).

ignorance, hatched horizontally by convention. The corresponding measurement symbol will be denoted $M(0;0)$, and the vacuum propagator $C(0)$ should be obtainable from it by

$$(25.8) \qquad C(0) = \text{trace } M(0;0).$$

The actual construction of any measurement symbol can be carried out by noting that any matrix element of S can be written

$$(25.9) \qquad \langle\tau'|S|\tau\rangle = \text{trace }(S|\tau\rangle\langle\tau'|).$$

Thus the measurement symbol associated with the true vacuum propagator is simply

$$(25.10) \qquad M(0;0) = S|0\rangle\langle0|.$$

Perturbation theory can now be viewed as an attempt to illuminate the spheres of ignorance associated with each true propagator by a series of successive approximations. Thus, up to and including the third order in the expansion of S, the true propagators could be rendered graphically as follows:

(25.11)

(25.12)

(25.13)

(25.14)

Inspection reveals that each graph always appears in conjunction with the same series of vacuum graphs. Assuming that these expansions make sense, one usually disregards all terms containing any *disconnected* vacuum graphs on the grounds that doing so amounts simply to multi-plying all state vectors by an unobservable constant phase factor. Indeed, the eigenvalue $C(0)$ of S in the vacuum state,

$$(25.15) \qquad\qquad S|0\rangle = C(0)|0\rangle$$

must have the modulus unity because of the unitarity of S,

$$(25.16) \qquad |C|^2 = 1 \qquad \text{or} \qquad C = e^{i\alpha} \qquad (\alpha \text{ real})$$

and in a series expansion C is numerically equal to the sum of all con-tributions to S with no unpaired operators. But this factor is the same for all matrix elements of S, because disconnectedness of the vacuum graphs is analytically equal to writing, for example

$$(25.17)$$

so that omitting all vacuum graphs is equivalent to replacing S by

$$(25.18) \qquad\qquad S \rightarrow S' = C^{-1}S = e^{-i\alpha}S.$$

At this point in the development one cannot avoid any longer facing up to a most perplexing difficulty which infects all perturbation theory: *The second terms on the right in expansions* (25.11)–(25.14) *represent infinite contributions* to the true propagators. Evaluation of these contributions with the rules laid down in Table 24.1 leads to a quad-ratically divergent integral for the second order vacuum graph, and to logarithmically divergent integrals for the second order fermion and photon propagator, and the third order vertex part.

That these divergences appearing in the perturbation expansions of the scattering operator have not discredited this approach entirely and have not led to abolition of perturbation theory altogether is due to a number of facts among which the following looms most important. Employing the perturbation expansion and treating all infinite contri-butions as if they were finite (which formally can always be arranged by some invariant cutoff procedure), one can derive by an iteration pro-cedure a number of simultaneous equations satisfied by the thus evalu-ated true propagators, for which there is reasonable hope of being correct even though the expansion evaluation of the propagators turns out to be meaningless.

The existence of integral equations linking in hierarchical fashion the various spheres of ignorance in the presence of an interaction can be made plausible by the following elementary consideration. The basic interpretational postulate of quantum mechanics, as stated in Section 1, can be rewritten for the purpose of this section as follows:

If P_{ab} is the probability that if measurement of A gave the result a then measurement of B will give the result b, P_{bc} the probability for finding the value c of C if B is known to have the value b, and P_{ac} the probability for finding the value c of C if A is known to have the value a, then there exist complex numbers

$\langle a|b \rangle$, $\langle b|c \rangle$, $\langle a|c \rangle$ so that $P_{ab} = |\langle a|b \rangle|^2$, $P_{bc} = |\langle b|c \rangle|^2$,
$$P_{ac} = |\langle a|c \rangle|^2$$

and

(25.19) $$\langle a|c \rangle = \sum_b \langle a|b \rangle \langle b|c \rangle$$

if no attempt is made to measure B between measurements of A and C.

The sum in (25.19) goes over all possible "channels," that is all possible values of B through which the object may reach value c of C starting from value a of A.

In the language of measurement symbols one can write this famous "addition of probability amplitudes" according to (6.2) and (6.3)

(25.20)
$$\text{trace}\,[M(c,a)] = \sum_b \langle a|M(b)|c \rangle = \sum_b \text{trace}\,[M(c,a)\,M(b)]$$
$$= \sum_b \text{trace}\,[M(b,a)]\,\text{trace}\,[M(c,b)].$$

In the language of regions of ignorance this decomposition corresponds to a dissection of the original region of ignorance associated with $M(c,a)$ into a network of *other* regions of ignorance, namely the ones associated with $M(b,a)$ and $M(c,b)$, thus

(25.21)

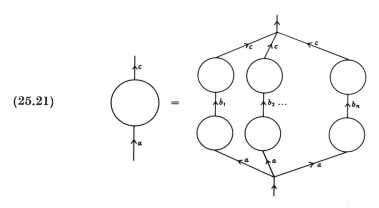

This equation is, of course, not sufficient to determine $M(c, a)$, because all regions of ignorance associated with $M(b_n, a)$ and $M(c, b_n)$ can be dissected after the fashion of Eq. (25.21), and so on and so forth, corresponding to decomposing in (25.19) each probability amplitude in turn after the same recipe, thus $\langle b|c \rangle = \sum_d \langle b|d \rangle \langle d|c \rangle$, etc.

Equation (25.19) *and its graphical representation* (25.21) *encompasses, by iteration, an in principle infinite hierarchy of spheres of ignorance.* By "hierarchy" is meant a simultaneous system of integral equations.

Now, a remarkable feature of quantum electrodynamics is the possibility of writing down hierarchical equations for the regions of ignorance associated with the true propagators for fermions and photons *in closed form*, involving only these propagators themselves and the true vertex part. This possibility can be traced to the peculiar form of the interaction Hamiltonian, which amounts to the possibility of dissecting any interaction process into elementary acts, represented graphically by vertices, in which single photons are either emitted or absorbed under observation of conservation of momentum and of lepton number.

For the true fermion propagator one arrives by iteration at the hierarchical equation

(25.22)

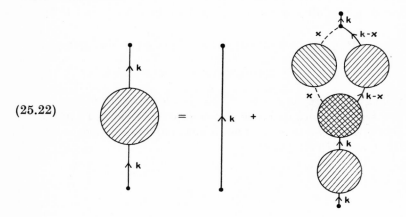

which may be rendered analytically in momentum space as

(25.22′)

$$G(k) = G^0(k) + i\, e (2\pi)^4\, G(k) \int \Gamma_\mu(k, k-\kappa, \kappa)\, G(k-\kappa)\, D_{\mu\nu}(\kappa)\, \gamma_\nu\, d^4\kappa\, G^0(k).$$

Similarly, for the true photon propagator one arrives at the hierarchical equation

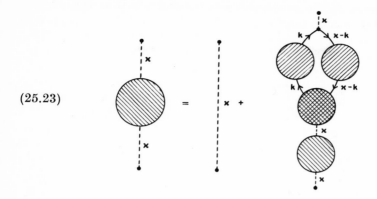

(25.23)

which may be rendered analytically in momentum space as

(25.23′)

$$D_{\mu\nu}(\kappa) = D^0_{\mu\nu}(\kappa) + i\,e(2\pi)^4\,D_{\mu\sigma}(\kappa)\,\text{trace}\left[\int \Gamma_\sigma(\kappa - k, k, \kappa)\,G(k)\,G(\kappa - k)\right.$$

$$\left.\times \gamma_\tau\,d^4\,k\right] D^0_{\tau\nu}(\kappa).$$

Before one can attempt a solution of this set of equations, one requires a similar expression for the true vertex part. *The sphere of ignorance associated with the true vertex part, however, has thus far eluded all efforts designed to write down a hierarchical equation in closed form.* One must therefore state in all honesty

(25.24)

The question mark in Eq. (25.24) constitutes one of the important mathematical challenges posed by the existence of the electromagnetic interaction.

Some authors, in particular Landau and his co-workers, have questioned the internal consistency of quantum electrodynamics on the basis of an approximation procedure, in which one substitutes for the question mark in (25.24) the expression

(25.25)

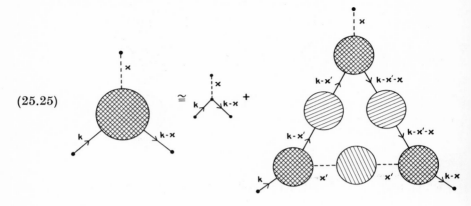

corresponding to adoption of an integral equation

(25.25′)

$$
\Gamma_\sigma(k, k-\kappa, \kappa) = \Gamma_\sigma^0(k, k-\kappa, \kappa) + \int \Gamma_\mu(k, k-\kappa', \kappa') D_{\mu\nu}(\kappa') G(k-\kappa')
$$
$$
\times \Gamma_\sigma(k-\kappa', k-\kappa'-\kappa, \kappa) G(k-\kappa'-\kappa)
$$
$$
\times \Gamma_\nu(k-\kappa'-\kappa, k-\kappa, \kappa') d^4\kappa'.
$$

The conclusions of these authors are open to doubt, because Eq. (25.25) cannot be a complete description of the true vertex part. It does not contain, for example, contributions to the true vertex part graph from decompositions of the type drawn in Fig. 25.7.

FIG. 25.7. Type of graphs omitted from the expansion (25.25).

In absence of any certain knowledge regarding the true propagators of quantum electrodynamics a concept known as "renormalization" has gained wide acceptance and been the object of a large body of learned literature. As to the meaning of renormalization, one finds in this literature two quite different schools of thought.

(i) An examination of the perturbation expansions (25.12)–(25.14) reveals that the infinities in the true propagators appear only as factors Z_1, Z_2, Z_3, and as an addition to the fermion mass, δm, so that we may write

$$(25.26) \qquad \begin{aligned} G &= Z_2\, G^{\text{reg}} \\ \Gamma_\mu &= Z_1^{-1}\, \Gamma_\mu^{\text{reg}} \\ D_{\mu\nu} &= Z_3\, D_{\mu\nu}^{\text{reg}} \end{aligned}$$

with the understanding that G^{reg}, Γ_μ^{reg}, and $D_{\mu\nu}^{\text{reg}}$ are regular functions, containing the fermion mass, if at all, in the combination

$$(25.27) \qquad m_{\text{exp}} = m + \delta m.$$

The fermion propagator, in particular, can be shown to have in perturbation theory the form

$$(25.28) \qquad G = Z_2/(\hat{k} - m_{\text{exp}})$$

which suggests the identification m_{exp} with the actually observed mass of the fermion. This procedure of making $m + \delta m$ finite by decree is called "mass renormalization" and amounts to replacing the "bare" mass m by the "dressed" mass m_{exp} which contains through δm all contributions to m_{exp} from the presence of interaction.

Furthermore, since any factor e^2 in the perturbation expansion will always appear multiplied with the constants Z_1, Z_2, and Z_3 in the combination $e^2 Z_3 Z_1^{-2} Z_2^2$, one identifies

$$(25.29) \qquad e_{\text{exp}} = e\, Z_3^{1/2}\, Z_1^{-1}\, Z_2$$

with the experimentally determined charge of the fermion, a procedure called "charge renormalization" in this approach.

Once the infinities of perturbation theory have been "taken care of" by this formal renormalization procedure, all contributions to the scattering matrix will be finite, and the calculation can be carried out, in principle, to any desired order in e_{exp}^2. Renormalization does not, of course, guarantee that the remaining perturbation expansion in powers of e_{exp}^2 converges.

Investigations into the convergence properties of the renormalized perturbation theory are impeded by the always remaining laboriousness of perturbation calculations beyond the second order which is already quite cumbersome. In the few cases where such calculations have been carried out to the third and fourth order, and where comparison with observation has been possible, the theoretical predictions have been confirmed with astonishing accuracy.

The empirical success of this recipe for performing perturbation calculations is remarkable because of the obvious meaninglessness of the divergent constants Z_1, Z_2, Z_3, and δm, which appear as power series in the unrenormalized charge e^2 with divergent coefficients. By adhering to perturbation theory one abandons, in fact, all hope of obtaining true propagators free from infinities, and the renormalization procedure, in this form, must be recognized as a recipe which works for as yet unknown reasons.

(ii) A completely different point of view regarding the meaning of "renormalization" has been advanced by a number of authors who suggest that the true propagators may not be uniquely determined by the hierarchical equations (25.22)–(25.24). The proponents of this approach *assume*, in fact, the existence of a set of true propagators G', Γ'_μ, and $D'_{\mu\nu}$ free from infinities and containing the coupling parameter e', and *propose* that transition to a second set of true propagators G'', Γ''_μ, and $D''_{\mu\nu}$ free from infinities and containing the coupling parameter e'', will not result in any observable consequences, provided the two sets are related through the equations

(25.30)
$$G'' = z_2 \, G'$$
$$\Gamma''_\mu = z_1^{-1} \, \Gamma'_\mu$$
$$D''_{\mu\nu} = z_3 \, D'_{\mu\nu}$$
$$e''^2 = e'^2 \, z_3^{-1} z_1^2 z_2^{-2}$$

where z_1, z_2, z_3 are arbitrary *finite* numbers. It must be stressed that the existence of such a "multiplicative renormalization group of transformations" cannot be proved by perturbation theory, because perturbation theory excludes, from the outset, the existence of true propagators free from infinities. The existence of relations (25.30) is at most *suggested* by perturbation theory, for if one replaces in (25.26) G by $z_2 G$, Γ_μ by $z_1^{-1}\Gamma_\mu$, and $D_{\mu\nu}$ by $z_3 D_{\mu\nu}$, then this leads to a change in the effective value of the charge, meaning replacement of $e_{\exp}^2$ by $e_{\exp}^2 z_3 z_1^{-2} z_2^2$. Therefore, if one simultaneously performs a *compensating* renormalization of charge $e^2 \rightarrow e^2 z_3^{-1} z_1^2 z_2^{-2}$, then as a result of all these operations *no change* in $e_{\exp}^2$ is obtained.

The very generality of this approach is its weakness. The invariance requirement expressed in Eqs. (25.30) is far too weak as a condition imposed on the unknown true propagators, and thus insufficient to lead to their unambiguous construction.

Before one can extract information regarding the analytic properties of the true propagators from (25.30), one has to make some assumptions about the dependence of the propagators on their arguments. It is, for example, customary to assume that the true photon propagator should be representable in the form

(25.31)
$$D_{\mu\nu}(\kappa^2) = \frac{d(\kappa^2)}{\kappa^2}\left(\delta_{\mu\nu} - \frac{\kappa_\mu \kappa_\nu}{\kappa^2}\right) + \frac{d_l(\kappa^2)}{\kappa^2}\frac{\kappa_\mu \kappa_\nu}{\kappa^2}$$

with unknown regular functions $d(\kappa^2)$ and $d_l(\kappa^2)$. Such a form is suggested by the free photon propagator $D^0_{\mu\nu}$, corresponding to $d(\kappa^2) = 1$. It should be kept in mind, however, that a specific form such as (25.31) contains drastic assumptions regarding the analytic properties of $D_{\mu\nu}(\kappa^2)$, as exemplified by the existence of a pole at $\kappa^2 = 0$. Similarly, the true fermion propagator is customarily assumed to have the form

(25.32)
$$G(k) = \frac{a(k^2)\,\hat{k} - b(k^2)\,m}{k^2 + m^2}$$

with unknown functions $a(k^2)$ and $b(k^2)$, showing once again the persuasiveness of perturbation theory, which contains the free fermion propagator as the lowest approximation in the form

$$G^0(k) = \frac{\hat{k} - m}{k^2 + m^2}.$$

Once specific forms such as (25.31) and (25.32) have been adopted, Eqs. (25.30) become functional equations for the unknown functions d, a, and b, giving opportunity for the employment of elegant mathematical techniques. This approach will not be pursued further here, because one may seriously doubt the existence of any analytical resemblance between true and free propagators. There just does not exist at present any detailed certain information about true propagators that does not have its roots in perturbation theory. Even the functional equations for d, a, and b have never been solved without relapse into perturbation-theoretical arguments, which are always invoked when particular values of the true propagators are needed to specify, for example, the limiting behavior of d for a small coupling parameter e^2.

In recent years ingenious attempts have been made to break out of this circle of frustration circumscribed by perturbation theory by imposing very general analytical features on true propagators, allowing one to infer general relationships between the real and imaginary parts of the true propagator, which have become known as "dispersion relations." For example, it has been conjectured that the true photon propagator may be decomposed into contributions from propagators for free particles with different masses. Thus $d(\kappa^2)$ is written

(25.33)
$$\frac{d(\kappa^2)}{\kappa^2} = Z\left[\frac{1}{\kappa^2} + \int\limits_{0}^{\infty} \frac{\rho(M^2)\,dM}{\kappa^2 + M^2}\right]$$

with unknown spectral function $\rho(M^2)$, allowing one to infer the dispersion relation

(25.34) $$\mathrm{Im}\,[d(\kappa^2)] = \frac{1}{\pi} \int\limits_{-\infty}^{+\infty} \mathrm{Re}\,\frac{d(\kappa'^2)\,d\kappa'^2}{\kappa'^2 - \kappa^2}.$$

The specific spectral representation (25.33) contains the tacit assumption that the true propagator will always have a pole at $\kappa^2 = 0$, corresponding to the pole exhibited by the free photon propagator, *independent of the value of the coupling parameter*. Some recent work by Schwinger (see end of Section 21) has thrown doubt, however, on the validity of this assumption, which seems so harmless and reasonable at first sight.

In a theory aimed at abandoning any reliance on perturbation theory, one cannot rely on starting with the assumed convergence of the iteration procedure, which in perturbation theory enables one to decompose any true propagators into three basic constituents, namely the fermion propagator G, the photon propagator $D_{\mu\nu}$, and the vertex part Γ_μ. Accordingly, the main effort in the theory of dispersion relations has been directed at guessing immediately analytic properties of more complicated true propagators, for example, the propagator associated with the four-fermion vertex as drawn in Fig. 25.8, representing all fermion-fermion

FIG. 25.8. Graph representing the four-fermion vertex.

scattering processes, with a region of ignorance encompassing all possible channels which may contribute to the amplitude of this particular process. Without trying to disparage these heroic efforts, one should admit that all so-called dispersion relations put forward to date have the character of ingenious guesswork regarding the analytic properties of true propagators.

The very existence of fermions and bosons with baffling numerical ratios of their masses would seem to indicate a much more intricate structure of even the simplest propagators than is commonly surmised. It should not surprise anyone if the enigma of the mass spectrum is in fact a self-consistency problem involving the *entire* hierarchy of propagators.

From this point of view any imminent breakthrough towards the solution of this problem seems improbable, and further arduous work should be anticipated.

NOTES

Integral equations connecting true propagators were first derived formally by Schwinger [1].

The concept of the true propagator is already contained in the work by Dyson [2] who initiated the current versions of renormalization theory. A comprehensive treatment is given in the book by Bogoliubov and Shirkov [3]. Examples of the point of view (i) regarding the meaning of renormalization are the papers by Källén [4] and Lehmann [5]. The point of view (ii) invoking the existence of a multiplicative renormalization group of transformations has been promulgated by Stueckelberg and Petermann [6], and Gell-Mann and Low [7].

The internal consistency of quantum electrodynamics has been questioned in a series of papers by Landau *et al.* [8]. See also Landau *et al.* [9] and Landau [10].

Readers interested in dispersion relations are referred to the article by Mandelstam [11].

REFERENCES

[1] J. Schwinger, *Proc. Natl. Acad. Sci. U.S.* **37**, 452 (1951).

[2] F. J. Dyson, *Phys. Rev.* **75**, 1736 (1949).

[3] N. N. Bogoliubov and D. V. Shirkov, "Introduction to the Theory of Quantized Fields." Wiley (Interscience), New York, 1959.

[4] G. Källén, *Helv. Phys. Acta* **25**, 416 (1952).

[5] H. Lehmann, *Nuovo Cimento* **11**, 342 (1954).

[6] E. C. G. Stueckelberg and A. Petermann, *Helv. Phys. Acta* **26**, 499 (1953).

[7] M. Gell-Mann and F. Low, *Phys. Rev.* **95**, 1300 (1954).

[8] L. D. Landau, A. A. Abrikosov, and I. M. Khalatnikov, *Doklady Akad. Nauk Uz.SSR* **95**, 733, 1177; **96**, 261 (1954).

[9] L. D. Landau, A. A. Abrikosov, and I. M. Khalatnikov, *Nuovo Cimento* **13**, *Suppl.* **3**, 80 (1956).

[10] L. D. Landau, On the quantum theory of fields, in "Niels Bohr and the Development of Physics" (W. Pauli, ed.), p. 52. Pergamon Press, New York, 1955.

[11] S. Mandelstam, *Rept. Progr. Phys.* **25**, 99 (1962).

On Selection Rules Due to Symmetry under Inversions and Rotations of Coordinates

Despite the perplexities besetting the internal consistency of quantum electrodynamics, it is, judged by the criterion of being able to make quantitative predictions, a satisfactory theory. The strong and weak interactions other than electromagnetism have not yet been grasped with a tool of similar usefulness. The absence of any profound dynamical insights contrasts sharply with the firm knowledge already gained about some symmetry properties of these otherwise ununderstood interactions. The purpose of this and the following three sections is to exhibit in a few examples, chosen for their simplicity, how elementary symmetry considerations can have the power of leading to detailed predictions. Since here, more than in any other section, any attempt at completeness would explode the resolve to keep this work within bounds, all effort will be directed towards the modest didactic aim of whetting the reader's appetite for more, thus providing him with a motivation for learning group theory, which is the language fitted to and almost indispensable for the systematic disentanglement of symmetry properties.

The consequences of symmetry can be conveniently categorized according to the different types of transformation operations associated with the various symmetries.

(I) Symmetries that can be formulated as transformation properties of state vectors under unitary operations in coordinate space. In this category belong conservation laws and selection rules generated by invariance under inversions and rotations of coordinates, which are the subject of this section.

(II) Symmetry under permutation of identical objects, which can be represented by a unitary transformation in occupation number space. This symmetry will be the subject of Section 27.

(III) Symmetries that must be formulated as transformation properties of state vectors under antiunitary operations. In this category belong

(a) Superselection rules, of which examples have already been mentioned in Sections 17 and 19.

(b) More subtle consequences of time-reversal invariance such as the principle of reciprocity and the principle of detailed balance.

(c) Consequences of invariance under particle-antiparticle conjugation. Details of iterms (b) and (c) are the subject of section 28.

(IV) Symmetries in state vector spaces representing abstract intrinsic properties such as isospin. These will be treated in Section 29.

A simple physical object which exhibits symmetries of categories (I) and (II) in a nontrivial fashion is the system of two photons. In particular, one of the simplest systems whose parity can be observed is the two-photon system in a state of total momentum zero.

As had been shown generally in Section 14, the operation of coordinate inversion can be represented by a unitary operator Π, and whenever the Hamiltonian of a physical system is invariant under Π, conservation of parity ensues. The conservation of parity in strong and electromagnetic interactions can often be used to monitor intrinsic properties of unstable objects by observation of the parity of their decay products.

A single photon state $|\varkappa, S\rangle$, defined as in (18.23), can never be a state of definite parity, because of the transformation properties

$$(26.1) \qquad \Pi|\varkappa, R\rangle = |-\varkappa, L\rangle; \qquad \Pi|\varkappa, L\rangle = |-\varkappa, R\rangle$$

in the combined momentum-polarization space. More generally, any single photon state (18.13) or (18.14) should transform according to

$$(26.2) \qquad \Pi|1_{\varkappa, R}\rangle = \eta|1_{-\varkappa, L}\rangle; \qquad \Pi|1_{\varkappa, L}\rangle = \eta|1_{-\varkappa, R}\rangle$$

where η is an as yet undetermined phase factor subject only to the condition

$$(26.3) \qquad \eta^2 = 1$$

imposed by the requirement that the operation Π^2 be equivalent to the identity. Consequently, the creation and annihilation operators should transform as

$$(26.4) \qquad \begin{aligned} \Pi b^+(\varkappa, R)\,\Pi^{-1} &= \eta b^+(-\varkappa, L); \qquad & \Pi b^+(\varkappa, L)\,\Pi^{-1} &= \eta b^+(-\varkappa, R) \\ \Pi b(\varkappa, R)\,\Pi^{-1} &= \eta b(-\varkappa, L); \qquad & \Pi b(\varkappa, L)\,\Pi^{-1} &= \eta b(-\varkappa, R) \end{aligned}$$

This phase factor η is sometimes given the unfortunate name the "intrinsic parity" of the single photon state. Its value depends on the transformation properties assigned to the operator (18.75) of the vector potential. With the usual convention of giving the magnetic field operator the transformation character of an axial vector, the vector potential must change sign under coordinate inversion,

$$(26.5) \qquad \Pi \mathbf{A}(\mathbf{q}, t)\,\Pi^{-1} = -\mathbf{A}(-\mathbf{q}, t).$$

The physical meaning of the factor η is obscure. It would be quite wrong to think of a photon "carrying" this "intrinsic parity" as it travels with momentum $\varkappa$, because the concept of a rest frame for a massless particle is meaningless, and only in such a frame can an object have definite parity. Fortunately, the transformation properties of any two photon state do not depend on η, since such a state is raised from vacuum by application of two creation operators, and therefore the operation Π will always generate only the factor $\eta^2 = 1$.

If the momentum $\varkappa$ of one photon is given, one can construct four different two-photon states of momentum zero, namely

(26.6)
$$|RR\rangle = b^+(\varkappa, R)\, b^+(-\varkappa, R)|0\rangle$$
$$|LL\rangle = b^+(\varkappa, L)\, b^+(-\varkappa, L)|0\rangle$$
$$|RL\rangle = b^+(\varkappa, R)\, b^+(-\varkappa, L)|0\rangle$$
$$|LR\rangle = b^+(\varkappa, L)\, b^+(-\varkappa, R)|0\rangle$$

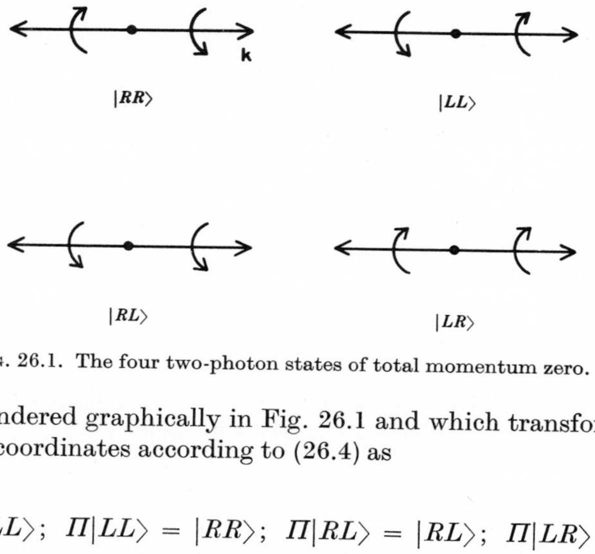

FIG. 26.1. The four two-photon states of total momentum zero.

which are rendered graphically in Fig. 26.1 and which transform under inversion of coordinates according to (26.4) as

(26.7)
$$\Pi|RR\rangle = |LL\rangle; \quad \Pi|LL\rangle = |RR\rangle; \quad \Pi|RL\rangle = |RL\rangle; \quad \Pi|LR\rangle = |LR\rangle.$$

The eigenstates of Π are therefore

(26.8a) $|RR\rangle - |LL\rangle$ with eigenvalue $P = -1$

and

(26.8b)
$|RR\rangle + |LL\rangle; \quad |RL\rangle; \quad |LR\rangle$ with eigenvalue $P = +1$.

The states $|RR\rangle - |LL\rangle$ and $|RR\rangle + |LL\rangle$ can be distinguished experimentally because the planes of polarization of the photons show distinctive correlations in each case. To see this, decompose in accord with the precedent given in (18.69), circular polarizations into linear polarizations by writing

(26.9)
$$b^+(\mathbf{\varkappa}, R) = (1/\sqrt{2})\,[b^+(\mathbf{\varkappa}, 1) + ib^+(\mathbf{\varkappa}, 2)]$$
$$b^+(\mathbf{\varkappa}, L) = (1/\sqrt{2})\,[b^+(\mathbf{\varkappa}, 1) - ib^+(\mathbf{\varkappa}, 2)]$$

yielding

(26.8a′)
$$|RR\rangle - |LL\rangle = [b^+(\mathbf{\varkappa}, R)\,b^+(-\mathbf{\varkappa}, R) - b^+(\mathbf{\varkappa}, L)\,b^+(-\mathbf{\varkappa}, L)]|0\rangle$$
$$= -i[b^+(\mathbf{\varkappa}, 2)\,b^+(-\mathbf{\varkappa}, 1) + b^+(\mathbf{\varkappa}, 1)\,b^+(-\mathbf{\varkappa}, 2)]|0\rangle$$

and

(26.8b′)
$$|RR\rangle + |LL\rangle = [b^+(\mathbf{\varkappa}, 1)\,b^+(-\mathbf{\varkappa}, 1) + b^+(\mathbf{\varkappa}, 2)\,b^+(-\mathbf{\varkappa}, 2)]|0\rangle.$$

This means the state $|RR\rangle - |LL\rangle$ is created whenever the emission of a photon polarized along one axis is accompanied by the emission of a photon polarized along the orthogonal axis, i.e. the planes of polarization in this state are always perpendicular to each other. State $|RR\rangle + |LL\rangle$, on the other hand, differs from the vacuum state by creation of two photons whose planes of polarization are always parallel. The states $|RL\rangle$ and $|LR\rangle$ show no such correlations, because one has, for example,

(26.8b″)
$$|RL\rangle = \tfrac{1}{2}[b^+(\mathbf{\varkappa}, 1)\,b^+(-\mathbf{\varkappa}, 1) + b^+(\mathbf{\varkappa}, 2)\,b^+(-\mathbf{\varkappa}, 2) + ib^+(\mathbf{\varkappa}, 1)\,b^+(-\mathbf{\varkappa}, 2)$$
$$- ib^+(\mathbf{\varkappa}, 2)\,b^+(-\mathbf{\varkappa}, 1)]|0\rangle$$

so that the probability for finding the planes of polarization parallel or perpendicular to each other is exactly $\tfrac{1}{2}$ in each case.

When an object of total angular momentum zero decays into two photons, then observation of the correlation between the planes of polarization will reveal the intrinsic parity of that object, provided the decay is mediated by an interaction invariant under inversion of coordinates so that parity is conserved. Examples of such objects are the neutral pion π^0 and positronium in the singlet ground state [see Eq. (27.22) below]. In both cases, observation has shown a correlation indicative of the state $|RR\rangle - |LL\rangle$, revealing both the π^0 and the positronium in the singlet ground state as pseudoscalar particles, i.e. having intrinsic parity $P = -1$.

The two-photon states of definite parity have interesting transformation properties under rotations. Consider first a single photon state $|\varkappa, S\rangle$ and perform a rotation, denoted $U_\perp$, of angle π around an axis perpendicular to $\varkappa$. It is intuitively clear that in the thus rotated coordinate frame the photon will appear to be travelling in direction $-\varkappa$, but with the same handedness. Thus single photon states transform under $U_\perp$ according to

$$(26.10) \qquad U_\perp|\varkappa, R\rangle = |-\varkappa, R\rangle; \qquad U_\perp|\varkappa, L\rangle = |-\varkappa, L\rangle.$$

Accordingly, the two-photon states (26.6) will transform as

$$(26.11) \qquad \begin{aligned} U_\perp|RR\rangle &= |RR\rangle; & U_\perp|LL\rangle &= |LL\rangle \\ U_\perp|RL\rangle &= |LR\rangle; & U_\perp|LR\rangle &= |RL\rangle. \end{aligned}$$

The simultaneous eigenstates of Π and $U_\perp$ are therefore

$(26.12a) \qquad |RR\rangle - |LL\rangle$

$(26.12b) \qquad |RR\rangle + |LL\rangle$ $\qquad$ with eigenvalue $\quad U_\perp = +1$

$(26.12c) \qquad |RL\rangle + |LR\rangle$

$(26.12d) \qquad |RL\rangle - |LR\rangle \qquad$ with eigenvalue $\quad U_\perp = -1$.

The linear combinations (26.12c) and (26.12d) are no longer eigenstates of the operator of rotation $U_\parallel$ around an axis parallel to $\varkappa$, which is not surprising because they are linear superpositions of states with different values of the component of angular momentum in direction $\varkappa$.

From these transformation properties one can deduce a *selection rule*: A particle of spin 1 and of definite parity cannot decay into two photons.

To prove this, one observes first of all that the spin of the particle can have, in the direction of $\varkappa$, only the values $m = +1, 0, -1$. Since the values of angular momentum in direction $\varkappa$ available in two-photon states (26.6) are $m = -2, 0, +2$, it follows, by conservation of angular momentum, that only the state $m = 0$ can decay into two photons. This leaves as possible final states $|RR\rangle - |LL\rangle$ and $|RR\rangle + |LL\rangle$. However, the state with $m = 0$ of the particle transforms as the spherical harmonic $Y_{1,0} = \cos\vartheta$ (ϑ is the angle with respect to $\varkappa$, see Appendix 1),

$$(26.13) \qquad U_\perp|m = 0\rangle = -|m = 0\rangle$$

exhibiting the eigenvalue $U_\perp = -1$, and thus removing the two remaining states (26.12a) and (26.12b) from the possible list of final states, because they belong to the eigenvalue $U_\perp = +1$.

Contrary to the case of massless particles, it should be possible to define the concept of intrinsic parity for single particles with mass unequal zero, because such particles can be found in states of definite parity provided they are at rest. Of particular interest are the fermions of spin $\frac{1}{2}$ whose ψ functions satisfy Dirac's equation. As had been shown in Section 19, the operation of coordinate inversion can be represented for these particles by

(26.14) $$\Pi = \eta_P \gamma_4 \Pi_D$$

where η_P is a phase factor which may have one of the four possible values ± 1 or $\pm i$ depending on whether the operation Π^2 of coordinate inversion applied twice has the value $+1$ or -1. In accordance with the discussion of Section 14 in connection with the double-valuedness of ψ functions for particles of spin $\frac{1}{2}$, the convention $\Pi^2 = +1$ is adopted here and only the case $\eta_P = \pm 1$ will be considered.

For a single electron or positron state at rest and in an eigenstate of Π_D with eigenvalue P_D, denoted $|e^\mp\rangle \equiv |\mathbf{k}=0, r, P_D, L = \pm 1\rangle$ where r is the spin label and L the lepton number, one can define the intrinsic parity P_i by

(26.15) $$\Pi|e^\mp\rangle = P_i P_D |e^\mp\rangle.$$

If the electron or positron is characterized by the orbital angular momentum quantum number l one will have, in general, $P_D = (-1)^l$ as had been shown in Section 14. Now from the representation of $|e^-\rangle$ in spin-chirality space by the ψ functions (see Table 19.2)

(26.16)
$$A_+(\mathbf{k}=0, r=1) = \text{const} \begin{pmatrix} 1 \\ 0 \\ 1 \\ 0 \end{pmatrix}; \qquad A_+(\mathbf{k}=0, r=2) = \text{const} \begin{pmatrix} 0 \\ 1 \\ 0 \\ 1 \end{pmatrix}$$

it follows that, with the representation (19.5) for γ_4,

(26.17) $$\Pi|e^-\rangle = \eta_P P_D |e^-\rangle.$$

Thus, the phase factor η_P can be identified as the intrinsic parity of the electron,

(26.18) $$P_i(e^-) = \eta_P.$$

On the other hand, a positron state $|e^+\rangle$ is represented in spin-chirality space by

(26.19)
$$A_-(\mathbf{k}=0, r=1) = \text{const} \begin{pmatrix} 0 \\ -1 \\ 0 \\ 1 \end{pmatrix}; \qquad A_-(\mathbf{k}=0, r=2) = \text{const} \begin{pmatrix} 1 \\ 0 \\ -1 \\ 0 \end{pmatrix}$$

so that

(26.20) $$\Pi|e^+\rangle = -\eta_P P_D|e^+\rangle$$

and one must conclude the intrinsic parity of the positron is

(26.21) $$P_i(e^+) = -\eta_P.$$

In other words: *Although the intrinsic parity of a fermion satisfying Dirac's equation is not uniquely determined, particle and antiparticle have always opposite intrinsic parity in this case.*

Consequently, positronium which can be described by a product state $|e^-\rangle \times |e^+\rangle$ has intrinsic parity $P_i = -1$. The total parity of positronium in a state of definite parity is therefore

(26.22) $$P(\text{positronium}) = -P_D.$$

One has thus the *selection rule*: Positronium in a state with even $l(P_D = +1)$ cannot decay into an n-photon state with even parity, and positronium in a state with odd $l(P_D = -1)$ cannot decay into an n-photon state with odd parity.

The optical transitions between different levels of positronium are not affected by the intrinsic parity, because only changes in P_D are observed in that case, and the selection rules following from conservation of parity and angular momentum are the same as in the corresponding hydrogen spectrum.

The possibility of establishing unambiguously the relative parity of electron and positron $P(e^-e^+)$ is intimately linked to the separation of electron and positron states by a superselection rule, which can in this case be associated with the conservation of lepton number, and which allows one [see Eqs. (15.52) and (15.53)] to speak meaningfully about the relative parity of the two states. It is not possible, on the other hand, to measure the relative parity of fermion states belonging to different particles in a given charge multiplet, such as neutron and proton states, because these are not separated by a superselection rule but rather by a conservation law such as the conservation of charge (which must be kept separate from conservation of lepton number or baryon number) following, presumably (see Section 20), from some kind of phase invariance, making the relative phases of the states belonging to a multiplet arbitrary. The relative parity $P(pn)$ must therefore be fixed by convention, and it is customary to assume that fermions belonging to the same multiplet have the same intrinsic parity, so that in particular *one assumes* $P(pn) = +1$. Similarly, the intrinsic parity of strange fermions, such as Λ, Σ, and Ξ, remains undetermined unless one fixes by convention the relative parity between nucleon N and one of them, $P(N\Lambda)$

say. The customary *assumption* is $P(N\Lambda) = +1$. Once these two assumptions have been made, the intrinsic parities of massive bosons, such as π and K, and of the baryons N, Λ, Σ, and Ξ can, in principle, be determined.

At the time of writing, the knowledge of intrinsic parities of so-called elementary particles is incomplete. It is interesting to note, however, that all massive bosons, including the "resonances" η, ω, ρ, etc., have odd intrinsic parity. All baryons, on the other hand, whose intrinsic parity is known at all, have even intrinsic parity. Whether this curious fact represents a general rule is unknown.

NOTES

Yang [1] first analyzed completely the symmetries of the two-photon state. The question of how one can observe in principle the correlation between the planes of polarization of the two photons emitted in the decay of the π^0 meson was also answered by Yang [2].

For a detailed discussion of how to obtain, in principle, information about the relative intrinsic parity of baryons and massive bosons see the article by Sakurai [3].

REFERENCES

[1] C. N. Yang, *Phys. Rev.* **77**, 242 (1950).
[2] C. N. Yang, *Phys. Rev.* **77**, 722 (1950).
[3] J. Sakurai, *in* "Brandeis Lectures in Theoretical Physics, 1961," Vol. 1, p. 231. Benjamin, New York, 1962.

Permutation Symmetry of Multiple Particle States

The two-photon states defined as in (26.6) have the flaw of not exhibiting the permutation symmetry characteristic of multiple particle states. Consider generally a two-particle boson state

(27.1) $$|1_{\tau_1}, 1_{\tau_2}\rangle = b^+(\tau_1)\,b^+(\tau_2)\,|0\rangle$$

where τ is a *complete* set of quantum numbers which can be used to characterize otherwise indistinguishable particles. "Indistinguishability" of particles means that it should not be possible to affix any other label to a particle beyond the set τ. Expression (27.1), however, does not yet take care of this requirement, because the order in which the operators b^+ have been written implies an additional labeling of the bosons, since the boson having quantum numbers τ_1 may, by a convention which counts the order of operators from left to right, be called the "first" boson and the one belonging to the set τ_2 the "second" boson. Physically, the state (27.1) should be indistinguishable from the state

(27.2) $$|1_{\tau_2}, 1_{\tau_1}\rangle = b^+(\tau_2)\,b^+(\tau_1)\,|0\rangle$$

in which, by the labeling convention employed above, the "first" boson has quantum numbers τ_2 and the "second" has quantum numbers τ_1. From this requirement of indistinguishability one can infer the existence of a unitary operator T_{12} of transposition connecting the states so that

(27.3) $$|1_{\tau_2}, 1_{\tau_1}\rangle = T_{12}|1_{\tau_1}, 1_{\tau_2}\rangle = T_{12}\,b^+(\tau_1)\,b^+(\tau_2)\,T_{12}^{-1}\,T_{12}|0\rangle$$

and thus

(27.4) $$T_{12}\,b^+(\tau_1)\,b^+(\tau_2)\,T_{12}^{-1} = b^+(\tau_2)\,b^+(\tau_1)$$

by the usual convention that the vacuum state is invariant, $T_{12}|0\rangle = |0\rangle$.

By an analogous argument, Eq. (27.3) will be found to exist for two fermion states as well, so that (27.4) will hold also if the operators b^+ are replaced by the corresponding fermion creation operators a^+.

Application of T_{12} twice must, up to a phase factor which is set equal to unity by convention, restore the initial state,

(27.5) $$T_{12}^2|1_{\tau_1}, 1_{\tau_2}\rangle = |1_{\tau_1}, 1_{\tau_2}\rangle.$$

241

The possible eigenvalues of T_{12} must therefore be $+1$ and -1. The corresponding eigenstates will be called "symmetric" and "antisymmetric" with respect to transposition and labeled $|1_{\tau_1}, 1_{\tau_2}\rangle_s$ and $|1_{\tau_1}, 1_{\tau_2}\rangle_a$, respectively, so that

(27.6) $$T_{12}|1_{\tau_1}, 1_{\tau_2}\rangle_s = +|1_{\tau_1}, 1_{\tau_2}\rangle_s$$

(27.7) $$T_{12}|1_{\tau_1}, 1_{\tau_2}\rangle_a = -|1_{\tau_1}, 1_{\tau_2}\rangle_a.$$

From the C.R.s of the boson creation operators b^+ and the anti-C.R.s of the fermion creation operators a^+ it follows immediately that two-particle boson states are symmetric and two-particle fermion states are antisymmetric under transposition. Indeed, for bosons Eq. (27.3) can be worked out to give with the help of the C.R.s (18.12)

(27.8)
$$T_{12}|1_{\tau_1}, 1_{\tau_2}\rangle_B = b^+(\tau_2)\, b^+(\tau_1)\,|0\rangle = +b^+(\tau_1)\, b^+(\tau_2)\,|0\rangle = +|1_{\tau_1}, 1_{\tau_2}\rangle_B$$

and for fermions the anti-C.R.s (17.17) yield

(27.9)
$$T_{12}|1_{\tau_1}, 1_{\tau_2}\rangle_F = a^+(\tau_2)\, a^+(\tau_1)\,|0\rangle = -a^+(\tau_1)\, a^+(\tau_2)\,|0\rangle = -|1_{\tau_1}, 1_{\tau_2}\rangle_F.$$

In terms of single particle states this transformation property can be made manifest by writing the symmetric two-boson state

(27.10) $$|1_{\tau_1}, 1_{\tau_2}\rangle_s = (1/\sqrt{2})[\,|1_{\tau_1}\rangle_1 |1_{\tau_2}\rangle_2 + |1_{\tau_2}\rangle_1 |1_{\tau_1}\rangle_2\,]$$

where now the transposition T_{12} operates on the subscript particle labels ($\alpha = 1, 2$) as in $|\ \rangle_\alpha$. Similarly, the two-fermion state can be written as an antisymmetrized combination of single particle states,

(27.11) $$|1_{\tau_1}, 1_{\tau_2}\rangle_a = (1/\sqrt{2})[\,|1_{\tau_1}\rangle_1 |1_{\tau_2}\rangle_2 - |1_{\tau_2}\rangle_1 |1_{\tau_1}\rangle_2\,].$$

Invariance of the Hamiltonian under transposition of particle labels means

(27.12) $$T_{12}HT_{12}^{-1} = H \quad \text{or} \quad T_{12}H - HT_{12} = 0.$$

It follows in this case that the eigenvalues of T_{12} must be constants of the system in time. If, thus, at any given instant the state vector of a system is either symmetric or antisymmetric under transposition of particle labels, it will remain so for all times, unless an interaction energy is introduced which is not invariant under this transposition, i.e. which introduces a feature allowing the particles to be distinguished. Such apparent "nonconservation of permutation symmetry" can, however, always be interpreted to mean that the set of quantum numbers considered is incomplete. Thus, if one transposes only the momentum label, but not the spin label, of two fermions described completely by momentum and spin, one may find the state to be either symmetric or antisymmetric under transposition of momentum labels alone, but transposition of *both* momentum and spin labels *must* in this case lead to antisymmetry of the state vector.

By straightforward generalization of the argument for two-particle states one finds any many-boson state $|n_{\tau_1}, n_{\tau_2}, \ldots\rangle$ to be symmetric under any permutation $\mathscr{P}$ of the particle labels,

$$(27.13) \qquad \mathscr{P}|n_{\tau_1}, n_{\tau_2}, \ldots\rangle = |n_{\tau_1}, n_{\tau_2}, \ldots\rangle$$

and any many-fermion state $|N_{\tau_1}, N_{\tau_2}, \ldots\rangle$ to be antisymmetric under that permutation,

$$(27.14) \qquad \mathscr{P}|N_{\tau_1}, N_{\tau_2}, \ldots\rangle = \delta_{\mathscr{P}}|N_{\tau_1}, N_{\tau_2}, \ldots\rangle$$

where

$$(27.15) \qquad \delta_{\mathscr{P}} = \begin{cases} +1 & \text{for even } \mathscr{P} \\ -1 & \text{for odd } \mathscr{P}. \end{cases}$$

Anyone familiar with the learned language of group theory will thus recognize boson and fermion states as belonging to the only two existing one-dimensional representations of the permutation group, namely the completely symmetrical and completely antisymmetrical representation, respectively. This is a direct consequence of the empirical facts embodied in the C.R.s and anti-C.R.s, namely that the occupation number of boson quantum states is unlimited and that the occupation number of fermion quantum states satisfies the exclusion principle.

Although there exist higher dimensional representations of the permutation group corresponding to different restrictions on the possible occupation numbers of quantum states, none of these appear to be realized in nature. The reasons for this peculiar empirical fact are not completely understood at present.

As an example illustrating in a nontrivial fashion the concepts developed in this section consider once again the two-photon state, but use instead of the quantum numbers $\varkappa$, S (momentum, polarization) the quantum numbers ω, j, m, P (energy, angular momentum, parity) (see Appendices 2 and 3) to characterize each single photon state contributing to the two-photon state. Any two-photon state is then a linear combination of products $|l, m_l\rangle |s, m_s\rangle$, where $|l, m_l\rangle$ represents a possible orbital state of two photons and $|s, m_s\rangle$ a possible spin state of the two photons. These linear combinations for given total angular momentum state $|j, m\rangle$ have to be chosen in accordance with the rules governing the addition of angular momenta, given in Appendix 2, where it is also shown that for $|s_1|, |s_2| = 1$ there are nine spin states $|s, m_s\rangle$ which transform under transposition of particle labels as

(27.16a) $T_{12}|s, m_s\rangle = +|s, m_s\rangle$ for $s = 0,2$

(27.16b) $T_{12}|s, m_s\rangle = -|s, m_s\rangle$ for $s = 1$.

The orbital state, on the other hand, transforms as

(27.17) $T_{12}|l, m_l\rangle = (-1)^l |l, m_l\rangle$.

This follows from the possibility of characterizing the orbital state completely by a ψ function $\psi_{l m_l}(\mathbf{x}_1 - \mathbf{x}_2) = \psi_{l m_l}(\mathbf{x})$ depending only on the relative momentum $\mathbf{x}_1 - \mathbf{x}_2 = \mathbf{x}$ of the two photons, because in momentum space the Hamiltonian for the two photons can be written

(27.18)

$$H = \pm(\mathbf{s}_1 \mathbf{x}_1 \pm \mathbf{s}_2 \mathbf{x}_2) = \pm[\tfrac{1}{2}(\mathbf{s}_1 \pm \mathbf{s}_2)(\mathbf{x}_1 + \mathbf{x}_2) + \tfrac{1}{2}(\mathbf{s}_1 \mp \mathbf{s}_2)(\mathbf{x}_1 - \mathbf{x}_2)]$$

which reduces to $\pm \tfrac{1}{2}(\mathbf{s}_1 \mp \mathbf{s}_2)\mathbf{x}$ in the center of mass frame. The operation of transposition of particle labels has thus the effect

(27.19) $T_{12}\psi_{l m_l}(\mathbf{x}) = \psi_{l m_l}(-\mathbf{x})$

and is therefore, as far as the orbital motion is concerned, equivalent to the effect of the inversion of coordinates,

(27.20) $T_{12}\psi_{l m_l}(\mathbf{x}) = \Pi \psi_{l m_l}(\mathbf{x})$.

Since the parity of any state of orbital angular momentum l is $(-1)^l$ (see Section 14), Eq. (27.17) follows, which may be written

(27.21) $T_{12}|l, m_l\rangle |s, m_s\rangle = (-1)^l |l, m_l\rangle T_{12}|s, m_s\rangle$.

Since the complete state is required to be symmetric,

(27.22) $T_{12}|j, m\rangle = +|j, m\rangle$

and the parity of the state is determined solely by its orbital part, one must conclude that *to a two-photon state of even parity* (i.e. l *even*) *only symmetric spin states* (i.e. $s = 0$ or $s = 2$) *can contribute, whereas to a state of odd parity* (i.e. l *odd*) *only antisymmetric spin states* (i.e. $s = 1$) *can contribute*.

There is an interesting alternative classification of the polarization states of two photons. In Section 18 the polarization space of the photon, spanned by the eigenstates of S_3, namely $|R\rangle$ and $|L\rangle$, had been shown to be isomorphic to the space spanned by the eigenstates $|\tfrac{1}{2}, +\tfrac{1}{2}\rangle$ and $|\tfrac{1}{2}, -\tfrac{1}{2}\rangle$ of the angular momentum operator s_3 belonging to spin $s = \tfrac{1}{2}$ [see Eqs. (18.57)–(18.60)]. Consequently, the rules governing addition of

angular momenta $j = \frac{1}{2}$ are applicable to the addition of photon polarizations. To bring out this analogy, the polarization states will be labeled
with the quantum numbers $\frac{1}{2}S_0$ and $\frac{1}{2}S_3$ for each particle, so that one
writes, in general,

(27.23) $$|S\rangle = |\tfrac{1}{2}S_0, \tfrac{1}{2}S_3\rangle$$

and, in particular, for single photons,

(27.24) $$|R\rangle = |\tfrac{1}{2}, +\tfrac{1}{2}\rangle; \qquad |L\rangle = |\tfrac{1}{2}, -\tfrac{1}{2}\rangle.$$

The polarization states for two photons follow now from the rules of
addition, laid down in Eq. (A2.9) of Appendix 2, giving rise to four
polarization states, consisting of an antisymmetric singlet and a symmetric triplet in polarization space, namely

(27.25) $\frac{1}{2}S_0 = 0$: $\qquad |0,0\rangle = (1/\sqrt{2})\,[\,|R\rangle_1|L\rangle_2 - |L\rangle_1|R\rangle_2\,]$

(27.26) $\frac{1}{2}S_0 = 1$:
$$\left\{ \begin{array}{l} |1,1\rangle = |R\rangle_1|R\rangle_2 \\[4pt] |1,0\rangle = (1/\sqrt{2})\,[\,|R\rangle_1|L\rangle_2 + |L\rangle_1|R\rangle_2\,] \\[4pt] |1,-1\rangle = |L\rangle_1|L\rangle_2 \end{array} \right.$$

Since the Hamiltonian for free photons commutes with the operator $\mathbf{S}$
of polarization, $\frac{1}{2}S_0$ and $\frac{1}{2}S_3$ are "good" quantum numbers to describe
any system of noninteracting photons. It is an intriguing and open
question to what extent one can use rotational symmetry in polarization
space, i.e. a "law of conservation of polarization," to restrict possible
choices of interaction Hamiltonians containing only photon creation and
annihilation operators designed to account phenomenologically for
possible photon-photon interactions. The question of whether emission
or absorption of photons by other objects "conserves polarization"
cannot be answered until one has defined the meaning of the polarization
operator for these other objects.

The operator of transposition T_{12}, which had been introduced sym-
bolically in (27.3), can actually be constructed in many cases if one
specifies the variable on whose particle label T_{12} should operate. One
can, for example, represent the transposition of spin labels for $j = \frac{1}{2}$ by
the operator

(27.27) $$T_{12} = \tfrac{1}{2}[I + \boldsymbol{\sigma}(1)\cdot\boldsymbol{\sigma}(2)]$$

where $\boldsymbol{\sigma}(1)$ and $\boldsymbol{\sigma}(2)$ are the Pauli spin matrices operating on the spin
state vectors $|\frac{1}{2}, \pm\frac{1}{2}\rangle_{1,2}$ of particle 1 and particle 2, respectively. Because
of the relations

(27.28) $$\sigma_i^2 = I; \qquad \sigma_i\sigma_j = i\sigma_k \qquad \text{(cyclically)},$$

the operator $\boldsymbol{\sigma}(1)\cdot\boldsymbol{\sigma}(2)$ satisfies the quadratic equation

(27.29) $$[\boldsymbol{\sigma}(1)\cdot\boldsymbol{\sigma}(2)]^2 + 2[\boldsymbol{\sigma}(1)\cdot\boldsymbol{\sigma}(2)] - 3 = 0$$

and therefore the eigenvalues of the operator $\boldsymbol{\sigma}(1)\cdot\boldsymbol{\sigma}(2)$, ξ (say), since they must also satisfy

(27.30) $$\xi^2 + 2\xi - 3 = 0,$$

are

(27.31) $$\xi^+ = +1 \qquad \text{and} \qquad \xi^- = -3.$$

The corresponding eigenvalues of T_{12}, η (say), are thus

(27.32)
$$\eta^+ = \tfrac{1}{2}(1+\xi^+) = +1$$
$$\eta^- = \tfrac{1}{2}(1+\xi^-) = -1$$

and T_{12} satisfies condition (27.5),

(27.33) $$T_{12}^2 = \tfrac{1}{4}\{I + 2[\boldsymbol{\sigma}(1)\cdot\boldsymbol{\sigma}(2)] + [\boldsymbol{\sigma}(1)\cdot\boldsymbol{\sigma}(2)]^2\} = I.$$

Relations (27.28) are also employed when one wishes to verify that

(27.34)
$$T_{12}\boldsymbol{\sigma}(1)T_{12}^{-1} = \boldsymbol{\sigma}(2)$$
$$T_{12}\boldsymbol{\sigma}(2)T_{12}^{-1} = \boldsymbol{\sigma}(1).$$

One finds, for example,

(27.35)
$$T_{12}\sigma_1(1) = \tfrac{1}{2}[\sigma_1(1)+\sigma_1(2)-i\sigma_3(1)\sigma_2(2)+i\sigma_2(1)\sigma_3(2)]$$
$$\sigma_1(2)T_{12} = \tfrac{1}{2}[\sigma_1(2)+\sigma_1(1)+i\sigma_2(1)\sigma_3(2)-i\sigma_3(1)\sigma_2(2)]$$

so that

(27.36) $$T_{12}\sigma_1(1) = \sigma_1(2)T_{12} \qquad \text{or} \qquad T_{12}\sigma_1(1)T_{12}^{-1} = \sigma_1(2)$$

which verifies the first of the six relations (27.34). The eigenstates of T_{12} are the antisymmetric singlet and the symmetric triplet (A2.9),

(27.37)
$$T_{12}|0,0\rangle = -|0,0\rangle$$
$$T_{12}|1,m_s\rangle = +|1,m_s\rangle; \qquad m_s = +1,0,-1;$$

as can again be seen by straightforward computation. For example, with the usual representations for $\boldsymbol{\sigma}$ and $|\tfrac{1}{2}, \pm\tfrac{1}{2}\rangle$, one finds

(27.38)
$$T_{12}|0,0\rangle = \frac{1}{2}\left[I + \begin{pmatrix}0&1\\1&0\end{pmatrix}_1\begin{pmatrix}0&1\\1&0\end{pmatrix}_2 + \begin{pmatrix}0&-i\\i&0\end{pmatrix}_1\begin{pmatrix}0&-i\\i&0\end{pmatrix}_2\right.$$
$$\left. + \begin{pmatrix}1&0\\0&-1\end{pmatrix}_1\begin{pmatrix}1&0\\0&-1\end{pmatrix}_2\right] \times \frac{1}{\sqrt{2}}\left[\begin{pmatrix}1\\0\end{pmatrix}_1\begin{pmatrix}0\\1\end{pmatrix}_2 - \begin{pmatrix}0\\1\end{pmatrix}_1\begin{pmatrix}1\\0\end{pmatrix}_2\right]$$

$$T_{12}|0,0\rangle = -\frac{1}{\sqrt{2}}\left[\begin{pmatrix}1\\0\end{pmatrix}_1\begin{pmatrix}0\\1\end{pmatrix}_2 - \begin{pmatrix}0\\1\end{pmatrix}_1\begin{pmatrix}1\\0\end{pmatrix}_2\right] = -|0,0\rangle.$$

Expressions analogous to (27.27) can be found for the operator of transposition of spin labels for values $j > \frac{1}{2}$. If one denotes the operators of angular momentum for particles 1 and 2 by $\mathbf{J}(1)$ and $\mathbf{J}(2)$, respectively, one requires, in general, for the representation of T_{12} a polynomial in $\mathbf{J}(1)\cdot\mathbf{J}(2)$ of order $2j$. This is so because the operator $\mathbf{J}(1)\cdot\mathbf{J}(2)$ satisfies, in general, an algebraic equation of order $2j + 1$. For $j = 1$, for example, one has the relations, readily obtained from representations (18.65)

(27.39)
$$J_i^3 = J_i; \quad J_iJ_jJ_i = 0; \quad J_1^2+J_2^2+J_3^2 = 2;$$
$$J_i^2J_j = iJ_iJ_k; \quad J_iJ_jJ_k = i(I-J_j^2); \quad J_iJ_j^2 = iJ_kJ_j \quad (ijk \text{ cycl.})$$

so that

(27.40) $$[\mathbf{J}(1)\cdot\mathbf{J}(2)]^3 + 2[\mathbf{J}(1)\cdot\mathbf{J}(2)]^2 - [\mathbf{J}(1)\cdot\mathbf{J}(2)] - 2 = 0.$$

The eigenvalues of $\mathbf{J}(1)\cdot\mathbf{J}(2)$ satisfy thus the cubic equation

(27.41) $$\xi^3 + 2\xi^2 - \xi - 2 = 0$$

which has the three solutions

(27.42) $$\xi_1 = +1; \quad \xi_2 = -1; \quad \xi_3 = -2.$$

The operator

(27.43) $$T_{12} = [\mathbf{J}(1)\cdot\mathbf{J}(2)]^2 + [\mathbf{J}(1)\cdot\mathbf{J}(2)] - 1$$

has then the eigenvalues

(27.44)
$$\eta^+ = +1 \quad \text{corresponding to } \xi_1 \text{ and } \xi_3$$
$$\eta^- = -1 \quad \text{corresponding to } \xi_2$$

and satisfies the condition

(27.45) $$T_{12}^2 = 1.$$

The eigenstates of T_{12} in this case are the multiplets (A2.10),

(27.46)
$$T_{12}|0,0\rangle = +|0,0\rangle$$
$$T_{12}|1,m_s\rangle = -|1,m_s\rangle \quad m_s = +1,0,-1$$
$$T_{12}|2,m_s\rangle = +|2,m_s\rangle \quad m_s = +2,+1,0,-1,-2.$$

It is interesting to note that the operator $\mathbf{J}(1)\cdot\mathbf{J}(2)$ removes the degeneracy with respect to T_{12} of the symmetric states $|0,0\rangle$ and $|2,m_s\rangle$, they belong to the eigenvalues ξ_3 and ξ_1, respectively, as follows from a simple computation:

(27.47) $\qquad \mathbf{J}(1) \cdot \mathbf{J}(2) |0, 0\rangle \; = \; -2|0, 0\rangle \; = \; \xi_3 |0, 0\rangle$

(27.48) $\qquad \mathbf{J}(1) \cdot \mathbf{J}(2) |2, m_s\rangle \; = \; +|2, m_s\rangle \; = \; \xi_1 |2, m_s\rangle.$

A representation can also be found for the operator of transposition of particle labels applied to a *complete* set of attributes. Such an operator would be required to have the property, for fermions,

(27.49) $\qquad T_{12}^+ \; = \; T_{12}^{-1}; \qquad T_{12}\, a(1)\, T_{12}^{-1} \; = \; a(2)$

so that also

$$T_{12}\, a^+(1)\, T_{12}^{-1} \; = \; a^+(2) \qquad \text{and} \qquad T_{12}\, a(2)\, T_{12}^{-1} \; = \; a(1) \qquad \text{etc.}$$

Here $a^+(1)$ refers to the creation operator of a fermion, $a^+(1) = a^+(\tau_1)$, which is characterized by a complete set of quantum numbers τ_1. The construction of the operator T_{12} in terms of creation and annihilation operators for fermions is relatively easy, because there is only a finite number of bilinear combinations containing the operators $a(1)$, $a^+(1)$, $a(2)$, and $a^+(2)$ from which one can construct the most general unitary operator involving $a(1)$, $a^+(1)$, $a(2)$, and $a^+(2)$ (see Appendix 5), of which T_{12}, if it exists at all, must be a special case. One finds

(27.50) $\quad T_{12} \; = \; -i\, e^{i(\pi/2)\, R_{12}}\, e^{i(\pi/2)\, S_{12}}; \qquad T_{12}^{-1} \; = \; +i\, e^{-i(\pi/2)\, S_{12}}\, e^{-i(\pi/2)\, R_{12}}$

with

(27.51) $\qquad\qquad R_{12} \; = \; [a^+(1)\, a(2) + a^+(2)\, a(1)]$

(27.52) $\qquad\qquad S_{12} \; = \; [1 - a^+(1)\, a(1) - a^+(2)\, a(2)]$

will do the job required by (27.49). Indeed, since

(27.53) $\qquad [S_{12}, a(1)] \; = \; a(1); \qquad [S_{12}, a(2)] \; = \; a(2)$

one has

$$e^{i(\pi/2)\, S_{12}}\, a(1)\, e^{-i(\pi/2)\, S_{12}} \; = \; a(1) + i(\pi/2)\, [S_{12}, a(1)] + (i(\pi/2))^2\, 1/2!$$

(27.54) $$\qquad\qquad\qquad \times\, [S_{12}, [S_{12}, a(1)]] + \cdots$$

$$\qquad\qquad = \; a(1)\, e^{i(\pi/2)} \; = \; ia(1),$$

and similarly

(27.55) $\qquad\qquad e^{i(\pi/2)\, S_{12}}\, a(2)\, e^{-i(\pi/2)\, S_{12}} \; = \; ia(2)$

so that, because

(27.56) $\qquad [R_{12}, a(1)] \; = \; -a(2); \qquad [R_{12}, a(2)] \; = \; -a(1)$

one has finally

(27.57)
$$T_{12}a(1)T_{12}^{-1} = i\,e^{i(\pi/2)\,R_{12}}\,a(1)\,e^{-i(\pi/2)\,R_{12}} = i\{a(1)+(i(\pi/2))\,[R_{12},a(1)]+\ldots\}$$
$$= i\{a(1)\cos(\pi/2)-ia(2)\sin(\pi/2)\} = a(2)$$

and similarly

(27.58) $T_{12}\,a(2)\,T_{12}^{-1} = i\{a(2)\cos(\pi/2)-ia(1)\sin(\pi/2)\} = a(1).$

The representation (27.50) for the transposition operator turns out to be equally valid for bosons, because the bilinear nature of the operators R_{12} and S_{12} leads to the same C.R.s (27.53) and (27.56) if the fermion operators are replaced everywhere by the corresponding boson operators, r_{12} and s_{12} (say). The decisive difference between the two cases is that T_{12} has with fermion operators only the eigenvalue -1 and with boson operators it has only the eigenvalue $+1$.

For fermion operators S_{12} and R_{12} satisfy the relations

(27.59)
$$S_{12}^2 + R_{12}^2 = 1; \qquad R_{12}S_{12} = 0; \qquad R_{12}^3 = R_{12}; \qquad S_{12}^3 = S_{12}$$

so that one can write

(27.60)
$$e^{i(\pi/2)\,S_{12}} = 1+iS_{12}\sin(\pi/2)+S_{12}^2[\cos(\pi/2)-1] = 1+iS_{12}-S_{12}^2$$
$$e^{i(\pi/2)\,R_{12}} = 1+iR_{12}-R_{12}^2$$

and thus, for fermions, the operator of transposition can be put in the somewhat more transparent form

(27.61)
$$T_{12} = S_{12}+R_{12} = 1-a^+(1)\,a(1)-a^+(2)\,a(2)+a^+(1)\,a(2)+a^+(2)\,a(1).$$

The factor $\pm i$ in (27.50) is now justified after the event, because

(27.62) $$T_{12}^2 = S_{12}^2 + R_{12}^2 = 1$$

requires this convention of phase.

No such simplified representation seems to exist for bosons.

NOTES

The importance of permutation symmetry in the quantum mechanical description of multiple particle states was apparently first appreciated by Heisenberg [1]. A comprehensive treatment of the application of permutation symmetry to the classification of multiple photon states is given in the book by Akhiezer and V. B. Berestetskii [2].

The explicit representation of the operator of transposition of spin labels for $j = \frac{1}{2}$ is due to Dirac [3].

REFERENCES

[1] W. Heisenberg, *Z. Physik* **38**, 411 (1926).
[2] A. I. Akhiezer and V. B. Berestetskii, "Quantum Electrodynamics." USAEC transl., 1st Russian ed., Oak Ridge, Tennessee, 1957; abbreviated English version, 2nd Russian ed., Oldbourne Press, London, 1963; complete English transl., 2nd Russian ed., Wiley, New York, 1964 (in preparation).
[3] P. A. M. Dirac, "Quantum Mechanics," 4th ed., §58. Oxford Univ. Press, London and New York, 1958.

Some Consequences of Symmetry under Particle Conjugation and Time Reversal

Whenever a particle differs from its antiparticle by an attribute, such as the fermion number F, a state describing such a single particle or antiparticle, i.e. a state with definite F, can never be an eigenstate of the operator of particle conjugation Γ, because by definition of the particle conjugation operation the operators Γ and F anticommute,

(28.1) $$\Gamma F + F\Gamma = 0$$

which is another way of saying that F changes sign under the operation Γ. For fermions of spin $\frac{1}{2}$ satisfying Dirac's equation there is an additional complication arising from the antiunitary nature of Γ, which had been established in Section 19. It is therefore not meaningful to introduce, in analogy to parity, the concept of "conjugality" meaning the eigenvalue of Γ, as long as only single particle states are considered.

However, in the special case of a many-fermion system whose total fermion number vanishes, i.e. in a system consisting of an equal number of fermions and antifermions, the state vector can become an eigenstate of Γ, [see the remark following Eq. (15.53)],

(28.2) $$\Gamma|F = 0\rangle = C|F = 0\rangle$$

and such states will be called "states of specific conjugality C." The value of C depends in a rather intriguing way on the orbital angular momentum quantum number l and the spin quantum number s of the state, through the intervention of permutation symmetry.

To bring this out consider as an example a positronium atom in the center of mass frame, so that its state vector can be described by

(28.3) $$|\text{positronium}\rangle = |\mathbf{k} = \mathbf{k}_1 - \mathbf{k}_2, s_1, s_2, L_1 = +1, L_2 = -1\rangle$$

where $\mathbf{k}_i$, s_i, L_i refer to momentum, spin quantum number, and lepton number of the respective particle. Now the operation of particle conjugation amounts in this case, by definition, to the transposition operation as far as the particle labels on L are concerned,

(28.4) $$\Gamma \equiv T_{12}(L).$$

Transposition of particle labels on s_i, on the other hand, has the eigen-value $+1$ for the triplet states $s_1 + s_2 = s = 1$ and the eigenvalue -1 for the singlet state $s_1 + s_2 = s = 0$ (see Appendix 2, Eqs. (A2.9)].

$$(28.5) \qquad\qquad T_{12}(s) = (-1)^{s+1}.$$

Finally, transposition of particle labels on $\mathbf{k}_i$ is equivalent to the operation of coordinate inversion Π_D and has in the center of mass frame the value

$$(28.6) \qquad\qquad T_{12}(\mathbf{k}) = (-1)^l.$$

If the attributes $\mathbf{k}$, s, and L form a complete set, then the operation of total transposition of particle labels T_{12} is the product

$$(28.7) \qquad T_{12} = T_{12}(\mathbf{k}) T_{12}(s) T_{12}(L) = (-1)^{l+s+1} \Gamma.$$

Since fermions of spin $\frac{1}{2}$ satisfy the exclusion principle, T_{12} must have the value -1, and one can conclude

$$(28.8) \qquad\qquad C(\text{positronium}) = (-1)^{l+s}.$$

This conclusion is not affected by introduction of a second lepton number L_μ, as had been done in Section 17, because electron and positron have the same lepton number L_μ and therefore $T_{12}\,(L_\mu) = +1$, so that (28.8) remains valid in this case.

If one assumes that the electromagnetic interaction (23.4) is invariant under the operation of particle conjugation, so that

$$(28.9) \qquad\qquad \Gamma A_\mu j_\mu \Gamma^{-1} = A_\mu j_\mu,$$

one arrives at a number of selection rules which can be used to test this assumption experimentally. Although the concept of particle conjuga-tion was originally only defined for objects with distinct particles and antiparticles by insisting, in accordance with (19.66), that the electric current density operator change sign,

$$(28.10) \qquad\qquad \Gamma j_\mu \Gamma^{-1} = -j_\mu,$$

the invariance requirement (28.9) imposes a definite transformation property on photon states under particle conjugation, because the operator A_μ must now also transform as

$$(28.11) \qquad\qquad \Gamma A_\mu \Gamma^{-1} = -A_\mu.$$

Now a single photon state $|1_\gamma\rangle$ is obtained, up to some factor, from the photon vacuum state $|0_\gamma\rangle$ by application of the operator A_μ, which by (18.90) contains photon creation and annihilation operators linearly,

$$(28.12) \qquad\qquad |1_\gamma\rangle = A_\mu |0_\gamma\rangle$$

so that

$$(28.13) \qquad \Gamma|1_\gamma\rangle = \Gamma A_\mu|0_\gamma\rangle = \Gamma A_\mu \Gamma^{-1} \Gamma|0_\gamma\rangle = -A_\mu \Gamma|0_\gamma\rangle.$$

With the convention, implicit already in the treatment of positronium above, that the vacuum state is a state of even conjugality, one concludes a single photon state has odd conjugality,

$$(28.13a) \qquad \Gamma|1_\gamma\rangle = C|1_\gamma\rangle \qquad \text{with} \quad C(1 \text{ photon}) = -1.$$

By induction one has then immediately the result for the n-photon state $|n_\gamma\rangle$,

$$(28.14) \qquad C(n \text{ photons}) = (-1)^n.$$

Upon comparison with (28.8) one arrives at the *selection rule*:

Positronium in a triplet state ($s = 1$) with odd l or in a singlet state ($s = 0$) with even l cannot decay into an odd number of photons, and positronium in a triplet state with even l or in a singlet state with odd l cannot decay into an even number of photons.

It is suprisingly difficult to devise a crucial experiment which would test the conservation of C alone in positronium decay, because if in the decay parity and angular momentum are conserved one has additional selection rules which follow from (27.22) and the classification of photon states. For decay into two photons, for example, one can read from Table 28.1, which has been constructed with the classification of two-photon

TABLE 28.1

SELECTION RULES FOR DECAY OF POSITRONIUM INTO TWO PHOTONS

Positronium state	J	$P = (-1)^{l+1}$	$C = (-1)^{l+s}$	Decay into two photons by conservation of:	
				C	P, J, J_3
$^1S_0$	0	-1	$+1$	Allowed	Allowed
$^3P_0$	0	$+1$	$+1$	Allowed	Allowed
$^1P_1$	1	$+1$	-1	Forbidden	Forbidden
$^3S_1, {}^3D_1$	1	-1	-1	Forbidden	Forbidden
$^3P_1$	1	$+1$	$+1$	Allowed	Forbidden
$^1D_2$	2	-1	$+1$	Allowed	Allowed
$^3P_2, {}^3F_2$	2	$+1$	$+1$	Allowed	Allowed
$^3D_2$	2	-1	-1	Forbidden	Allowed

states as given in Section 27 and Appendix 3, that the $^3D_2$ state of positronium is the lowest state from which one has, by observation of the absence or presence of two-photon decay, a crucial test of conservation or nonconservation of C in the interaction responsible for this decay.

Relations (28.8) and (28.14) lead also to a selection rule in optical transitions between various positronium states under emission or absorption of single photons. Denoting the spin and orbital quantum numbers of positronium in initial and final states by s, l and s', l', one must have

$$(28.15) \qquad (-1)^{l+s} = -(-1)^{l'+s'}$$

if C is to be conserved in the transition. This means, in particular, that quadrupole transitions $(l' - l = \pm 2, s' - s = 0)$ are forbidden in positronium, although they exist in the corresponding spectrum of hydrogen. Again, experimental verification of this prediction is difficult to obtain, because the electrical dipole transitions $(l' - l = \pm 1, s' - s = 0)$ between the corresponding energy levels are allowed. In this connection, it might be worth noting that these forbidden lines are not excluded if positronium is placed into a given fixed external electric field, because then the interaction ceases to be symmetric in the two particles of opposite charge, so that the expectation value of Γ is no longer conserved, or, to use a typical phrase, "C ceases to be a good quantum number."

The behavior of baryons and massive bosons under particle conjugation is complicated by the existence of yet another attribute, the isospin, which will be treated in Section 29, and consequences of symmetry under particle conjugation for these objects will be taken up there.

Turning now to the consequences of invariance under time reversal Θ, one might expect on first sight selection rules similar to the ones generated by invariance under particle conjugation on account of the antiunitary nature of Θ which had been established in Section 16. As in the case of particle conjugation symmetry, it is not meaningful to introduce, in analogy to parity, the concept of "reversality" meaning the eigenvalue of Θ, as long as single particle states are considered. The analog to the states of specific conjugality C (28.2), i.e. "states of specific reversality T" (say), can exist only if there were an additive quantum number, L_μ (say), which is odd under time reversal,

$$(28.16) \qquad \Theta L_\mu + L_\mu \Theta = 0$$

and if one could realize product states describing an equal number of particles with the attribute $L_\mu = +1$ and particles with the attribute $L_\mu = -1$, so that such states could have indeed the property

$$(28.17) \qquad \Theta |L_\mu = 0\rangle = T |L_\mu = 0\rangle.$$

Unless the speculations mentioned at the end of Section 17, where the second lepton number L_μ was assumed to have the property (28.16) making for example a state consisting of one ordinary neutrino ν and one muon neutrino ν_μ a candidate for having specific reversality according to (28.17), turn out to account faithfully for reality, no such situations seem to be realized in nature.

Despite the absence of selection rules analogous to the ones following from conservation of C and P, invariance under time reversal has some far-reaching and experimentally accessible consequences whenever transitions are observed which require a definite sequence of states in time.

Of immediate interest are situations, treated in Section 23, in which the amplitude for transition between a final state $|\tau'\rangle$ from an initial state $|\tau\rangle$ is given by the matrix element

$$(28.18) \qquad \langle \tau' | S | \tau \rangle$$

of the scattering operator S defined as the limit

$$(28.19) \qquad S = U(+\infty, -\infty)$$

of a unitary operator $U(t_2, t_1)$ which satisfies

(28.20)

$$i \frac{\partial U(t_2, t_1)}{\partial t_2} = H_{\text{int}}(t_2)\, U(t_2, t_1) \qquad \text{with} \qquad U(t_2, t_1) = 1 \qquad \text{for} \quad t_2 = t_1$$

where $H_{\text{int}}(t_2)$ is the interaction Hamiltonian in the interaction picture. The solution of (28.20) can always be put in the symbolical form [see Eq. (23.16)]

$$(28.21) \qquad U(t_2, t_1) = P\left[\exp\left(-i \int_{t_1}^{t_2} H_{\text{int}}(t)\, dt \right) \right]$$

where P is the time-ordering operator.

A general consequence of the invariance of Eq. (28.20) under time reversal is the *principle of reciprocity*:

The amplitude of transition from an initial state $|\tau\rangle$ to a final state $|\tau'\rangle$ is, up to a phase factor ± 1, equal to the amplitude for the "reverse" process, in which the sequence of initial and final states is reversed *and* each state $|\tau\rangle$ is replaced by its time reversed state $|\tau_T\rangle$,

$$(28.22) \qquad \langle \tau' | S | \tau \rangle = \pm \langle \tau_T | S | \tau'_T \rangle.$$

One must not confuse the "reverse" process $|\tau'_T\rangle \to |\tau_T\rangle$ with the "inverse" process $|\tau'\rangle \to |\tau\rangle$ in which only the sequence of states is interchanged.

For example, if τ stands for the set of quantum numbers $\mathbf{k}_A$, $\mathbf{k}_B$, ..., $\mathbf{s}_A$, $\mathbf{s}_B$, ... of momenta and spins of particles A, B, ... in the initial state, and τ' stands for the set of quantum numbers $\mathbf{k}_C$, $\mathbf{k}_D$, ..., $\mathbf{s}_C$, $\mathbf{s}_D$, ... of particles C, D, ... (which are in general different from particles A, B, ...) in the final state, one has, since both $\mathbf{k}$ and $\mathbf{s}$ change sign under time reversal,

the process $\qquad |\mathbf{k}_A, \mathbf{k}_B, \ldots, \mathbf{s}_A, \mathbf{s}_B, \ldots\rangle \to |\mathbf{k}_C, \mathbf{k}_D, \ldots, \mathbf{s}_C, \mathbf{s}_D, \ldots\rangle,$

the inverse process $\qquad |\mathbf{k}_C, \mathbf{k}_D, \ldots, \mathbf{s}_C, \mathbf{s}_D, \ldots\rangle \to$
$$|\mathbf{k}_A, \mathbf{k}_B, \ldots, \mathbf{s}_A, \mathbf{s}_B, \ldots\rangle, \qquad \text{and}$$

the reverse process $\qquad |-\mathbf{k}_C, -\mathbf{k}_D, \ldots, -\mathbf{s}_C, -\mathbf{s}_D, \ldots\rangle \to$
$$|-\mathbf{k}_A, -\mathbf{k}_B, \ldots, -\mathbf{s}_A, -\mathbf{s}_B, \ldots\rangle.$$

The proof of the principle of reciprocity (28.22) is expedited if one exploits the possibility of representing the antiunitary property of time reversal Θ by the operation of transposition in occupation number space. Thus, if with a convention of phase the vacuum state is assumed to transform as

$$(28.23) \qquad \Theta|0\rangle = \langle 0|$$

then any occupation state, which is raised from vacuum by application of the appropriate number of creation operators

$$(28.24) \qquad |\tau\rangle = a^+(\tau_1) \ldots a^+(\tau_n)|0\rangle$$

will transform as

$$(28.25) \qquad \Theta|\tau\rangle = \epsilon_n \langle 0| a[(\tau_n)_T] \ldots a[(\tau_1)_T] = \epsilon_n \langle \tau_T|$$

where ϵ_n is a phase factor ± 1 which can be chosen arbitrarily for given number n of particles without violating the time reversal invariance of the C.R.s or anti-C.R.s valid between the operators $a(\tau)$ and $a^+(\tau)$. [The operators a, a^+ in Eqs. (28.25) and (28.26) may be taken to represent fermions and/or bosons.] The effect of Θ on all dynamical variables is taken care of by the substitution of the time reversed quantum numbers τ_T for the original quantum numbers τ. Similarly one has

$$(28.26) \quad \langle \tau'|\Theta^{-1} = \langle 0| a(\tau'_m) \ldots a(\tau'_1)\Theta^{-1} = \epsilon_m a^+[(\tau'_1)_T] \ldots a^+[(\tau'_m)_T]|0\rangle$$
$$= \epsilon_m |\tau'_T\rangle.$$

With this understanding about the representation of Θ one can rewrite the transition amplitude (28.18)

$$(28.27) \qquad \langle \tau'|S|\tau\rangle = \langle \tau'|\Theta^{-1}\Theta S\Theta^{-1}\Theta|\tau\rangle = \epsilon_m \epsilon_n \langle \tau_T|S_T|\tau'_T\rangle$$

where S_T is the time reversed scattering operator

(28.28) $$S_T = \Theta S \Theta^{-1}.$$

To prove reciprocity one has therefore to prove that

(28.29) $$S_T = S$$

follows from the invariance of Eq. (28.20) under time reversal.

To this end remember the unitary nature of $U(t_2, t_1) = U^+(t_1, t_2)$, write down the adjoint of Eq. (28.20) and interchange the labels t_1 and t_2:

(28.20$^+$) $$-i \frac{\partial U(t_2, t_1)}{\partial t_1} = U(t_2, t_1) H_{int}(t_1).$$

Addition of Eqs. (28.20) and (28.20$^+$) gives, for the special values $t_2 = t$ and $t_1 = -t$

(28.30) $$2i \frac{\partial U(t, -t)}{\partial t} = H_{int}(t) U(t, -t) + U(t, -t) H_{int}(-t).$$

Time reversal invariance requires that if (28.30) is true, then

(28.30$_T$) $$2i \frac{\partial U_T(t, -t)}{\partial t} = U_T(t, -t) [H_{int}(t)]_T + [H_{int}(-t)]_T U_T(t, -t)$$

must also be true, where $U_T = \Theta U \Theta^{-1}$, etc. The antiunitary nature of time reversal has been used consistently, requiring, for example, that upon application of Θ to the right hand side of (28.30) the order of the factors U and H, both of which contain creation and annihilation operators, be reversed. By the definition of time reversal

(28.31) $$[H_{int}(t)]_T = H_{int}(-t)$$

and (28.30$_T$) may be written as an equation for $U_T(t, -t)$

(28.32) $$2i \frac{\partial U_T(t, -t)}{\partial t} = H_{int}(t) U_T(t, -t) + U_T(t, -t) H_{int}(-t)$$

which is seen to be identical with Eq. (28.30) for $U(t, -t)$. Therefore,

(28.33) $$U_T(t, -t) = U(t, -t)$$

and the observation that S is by (28.19) a special case of $U(t, -t)$ completes the proof of Eq. (28.29) and thus of the principle of reciprocity.

Under special circumstances the principle of reciprocity implies a simple relationship between the process and its inverse process, known as *the principle of detailed balance*:

(28.34) $$|\langle \tau' | S | \tau \rangle|^2 = |\langle \tau | S | \tau' \rangle|^2.$$

The validity of such a principle aids greatly the statistical analysis of states in thermal equilibrium and the approach to equilibrium, because it means transitions from a state τ to a state τ' can be balanced directly by the inverse transitions from the state τ' to state τ without having to invoke any intermediate states through which such a balancing might be effected. *It must be stressed*, however, *that the principle of detailed balance is not generally true.* Even in classical mechanics it is not true. For example, if the transition refers to a collision between two particles interacting by a force which manifests itself as a nonspherical "shape" of the particles, detailed balancing is not valid, as was already pointed out by Boltzmann.

Some of the special circumstances under which detailed balance *does* hold in quantum mechanics will now be recorded in a number of theorems.

Theorem I. If the interaction mediating transitions is weak and if a perturbation expansion of S after the recipe (24.1) is meaningful, so that all observable effects can be accounted for by the first approximation

$$(28.35) \qquad \langle\tau'|S|\tau\rangle = \langle\tau'|S_1|\tau\rangle$$

then the principle of detailed balance is valid.

This follows immediately from the representation of S_1 as the time integral over H_{int} and from the hermitean property of H_{int},

$$(28.36) \qquad \langle\tau'|H_{\text{int}}|\tau\rangle^* = \langle\tau|H_{\text{int}}|\tau'\rangle$$

so that

$$(28.37) \qquad |\langle\tau'|S_1|\tau\rangle|^2 = |\langle\tau|S_1|\tau'\rangle|^2.$$

Theorem II. If the process is invariant under inversion of coordinates Π, and if one measures only quantum numbers which change sign under *both* time reversal Θ and coordinate inversion Π, then the principle of detailed balance is applicable.

For example, consider a process between states characterized completely by momenta $\mathbf{k}$ and spins $\mathbf{s}$. The principle of reciprocity requires then

$$(28.38)$$
$$\langle\mathbf{k}'_C, \mathbf{s}'_C, \ldots |S|\mathbf{k}_A, \mathbf{s}_A, \ldots\rangle = \pm\langle-\mathbf{k}_A, -\mathbf{s}_A, \ldots |S|-\mathbf{k}'_C, -\mathbf{s}'_C, \ldots\rangle.$$

Invariance under inversion of coordinates means

$$(28.39)$$
$$\langle-\mathbf{k}_A, -\mathbf{s}_A, \ldots |S|-\mathbf{k}'_C, -\mathbf{s}'_C, \ldots\rangle = \langle\mathbf{k}_A, -\mathbf{s}_A, \ldots |S|\mathbf{k}'_C, -\mathbf{s}'_C, \ldots\rangle$$

so that

(28.40) $\quad \langle \mathbf{k}'_C, \mathbf{s}'_C, \ldots | S | \mathbf{k}_A, \mathbf{s}_A, \ldots \rangle = \pm \langle \mathbf{k}_A, -\mathbf{s}_A, \ldots | S | \mathbf{k}'_C, -\mathbf{s}'_C, \ldots \rangle.$

If the spins are not measured in either transition, one has

(28.41)
$$\sum_{\text{spins}} |\langle \mathbf{k}'_C, \mathbf{s}'_C, \ldots | S | \mathbf{k}_A, \mathbf{s}_A, \ldots \rangle|^2 = \sum_{\text{spins}} |\langle \mathbf{k}_A, \mathbf{s}_A, \ldots | S | \mathbf{k}'_C, \mathbf{s}'_C, \ldots \rangle|^2$$

which is the principle of detailed balance applied to initial and final states in which only the momenta are known.

A famous application of Theorem II is the determination of the spin of the π^+ meson, which is obtained by comparing the cross sections for the reaction $p + p \to d + \pi^+$ and its inverse $\pi^+ + d \to p + p$ which are measured for given momenta without observation of spins. The ratio of the two cross sections depends then essentially only on factors in phase space, which contain as only unknown the spin degeneracy $(2s_\pi + 1)$ of pion states.

Another pair of reactions for which the principle of detailed balance, as required by time reversal invariance and Theorem II, has been checked is the photodisintegration of He^3, $\gamma + \text{He}^3 \to \text{H}^2 + p$ and the radiative capture of protons by deuterium, $p + \text{H}^2 \to \text{He}^3 + \gamma$.

Theorem III. If in a reaction involving two particles in the initial state and in the final state the spins of the particles lie in the reaction plane, then detailed balance holds, provided the interaction is invariant under rotations in space. This is true even if parity is not conserved in the reaction.

The validity of this theorem is most easily established by the following geometrical reasoning. Draw a graph of the process, as indicated in Fig. 28.1, with the understanding that none of the spins $\mathbf{s}_A, \mathbf{s}_B, \mathbf{s}_C, \mathbf{s}_D$ have any components perpendicular to the plane formed by the momentum vectors $\mathbf{k}_A$ and $\mathbf{k}_B$. By conservation of momentum this plane must be identical with the plane formed by $\mathbf{k}_C$ and $\mathbf{k}_D$. Now draw the reverse process, as indicated in Fig. 28.2. If the process is invariant under rotations in space,

FIG. 28.1. The process.

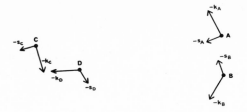

FIG. 28.2. The reverse process.

FIG. 28.3. The inverse process.

then the amplitude for the reverse process is equal to the amplitude for the process which is obtained by rotating all vectors in the reverse process around an axis perpendicular to the plane of reaction by an angle π, as indicated in Fig. 28.3, and comparison with Fig. 28.1 shows that the result is the inverse process. This means the amplitudes for the process and its inverse are equal, and therefore the principle of detailed balance holds in this case. Since the invariance under inversions of coordinates has not been invoked, Theorem III will be true even if parity is not conserved in the reaction.

Theorem IV. If in a reaction initial and final state are characterized only by the *total* angular momentum quantum numbers j, m and sets of other *scalar* quantum numbers τ_A, τ_B, ... and τ_C, τ_D, ... which do not change sign under time reversal, then the principle of detailed balance holds provided the reaction is invariant under rotations in space.

This follows from the observation that under the stated conditions only the quantum number m changes sign under time reversal, so that one has, on account of the principle of reciprocity,

$$(28.42) \quad \langle \tau_C, \tau_D, \ldots, j, m | S | \tau_A, \tau_B, \ldots, j, m \rangle$$
$$= \pm \langle \tau_A, \tau_B, \ldots, j, -m | S | \tau_C, \tau_D, \ldots, j, -m \rangle$$

and, therefore, if S is invariant under rotations, the matrix element on the right-hand side is equal to the one obtained by a rotation of co-ordinates which transforms m into $-m$ but leaves the scalars τ, j invariant,

$$(28.43) \quad \langle \tau_C, \tau_D, \ldots, j, m | S | \tau_A, \tau_B, \ldots, j, m \rangle$$
$$= \pm \langle \tau_A, \tau_B, \ldots, j, m | S | \tau_C, \tau_D, \ldots, j, m \rangle.$$

Since Eq. (28.43) connects the amplitude of the process with the amplitude of the inverse process, the principle of detailed balance holds in this case.

It is rather unsettling to realize how stringent the conditions are under which detailed balance is a valid principle, in view of the great burden this principle has to carry in all proofs for the existence of stationary states in statistical physics. It seems almost impossible to realize experimentally systems devoid of any magnetic interaction properties, and since any magnetic interaction will in general remove the degeneracy of energy levels due to spins, the labeling of energy levels by some spin quantum number s becomes practically unavoidable. But under such circumstances the principle of detailed balance *cannot* be valid, and therefore all proofs which invoke detailed balance for existence of stationary states are suspect because their applicability to reality is in doubt.

Nevertheless, in the absence of detailed balance, some kind of *overall balance* for transitions seems to be required if stationary states are to exist at all. The question of under which conditions overall balance generally follows from the principle of reciprocity for physically realizable systems poses one of the unsolved problems of statistical physics.

NOTES

Early treatments of the selection rules for positronium following from symmetry under particle conjugation are contained in papers by Wolfenstein and Ravenhall [1], and Michel [2]. The conjugality property of the photon is already implicit in the work of Furry [3]. Explicitly it seems to have been stated first by Gell-Mann and Pais [4].

A lucid discussion of the principle of reciprocity and of conditions under which detailed balance holds is contained in the book by Williams [5]. Implications of time-reversal symmetry for strong interactions have been studied in detail by Henley and Jacobsohn [6].

The proof for the principle of reciprocity given here follows closely that of Mandl [7].

REFERENCES

[1] L. Wolfenstein and D. G. Ravenhall, *Phys. Rev.* **88**, 279 (1952).

[2] L. Michel, *Nuovo Cimento* **10**, 319 (1953).

[3] W. Furry, *Phys. Rev.* **51**, 125 (1937).

[4] M. Gell-Mann and A. Pais, *Phys. Rev.* **97**, 1387 (1955).

[5] W. S. C. Williams, "An Introduction to Elementary Particles." Academic Press, New York, 1961.

[6] E. M. Henley and B. A. Jacobsohn, *Phys. Rev.* **113**, 225, 234 (1959).

[7] F. Mandl, "Symmetry Properties of Particles and Fields." CERN Lectures, Geneva, 1960.

Attributes Characteristic of Objects Engaging in Strong Interactions

All baryons (N, Λ, Σ, Ξ, ...) and massive mesons (π, K, ...) partaking in the strong interactions possess a number of intrinsic attributes not shared, for example, by the leptons which engage only in weak and electromagnetic interactions. Most prominent among these attributes ranks the "isospin" $\mathbf{T}$ which derives its name and its status as a full-fledged dynamical variable from a curious symmetry property of strongly interacting objects in an abstract "isospace," resembling formally the symmetry under rotations in three-dimensional coordinate space, and leading through this isomorphism with ordinary rotations to conservation of two new quantum numbers T and T_3 in processes mediated by strong interactions, in analogy to the conservation of two angular momentum quantum numbers s and s_3 (or j and m) in processes mediated by interactions invariant under rotations in ordinary space.

The "isospace" was originally conceived by Heisenberg as a convenient device to distinguish the neutron state and the proton state of the nucleon by a dichotomic attribute τ_3 (say) attached to the nucleon, which is given the value $\tau_3 = +1$ if it is found to be a proton and the value $\tau_3 = -1$ if the nucleon is found to be a neutron. Once this abstraction of considering proton and neutron as but two, and the only two possible, states of a generic entity called "nucleon" is accepted, the mathematical machinery set up in Sections 1 and 2 for the purpose of grasping the dichotomic attribute of fermion spin can be adapted to this new situation by changing nothing but the interpretation of symbols.

Thus the state of a nucleon can be written in isospace as

$$(29.1) \qquad |\chi\rangle = a|p\rangle + b|n\rangle$$

where $|p\rangle$ and $|n\rangle$ are pure proton and neutron states defined as the eigenstates

$$(29.2) \qquad |p\rangle = \begin{pmatrix}1\\0\end{pmatrix}; \qquad |n\rangle = \begin{pmatrix}0\\1\end{pmatrix}$$

of an "isospin" operator

(29.3)
$$\tau_3 = \begin{pmatrix} 1 & 0 \\ 0 & -1 \end{pmatrix}$$

with eigenvalues $+1$ and -1, respectively, so that a and b in (29.1) can be looked upon as ψ functions in the isospace spanned by the vectors (29.2) with the interpretation that

$|a|^2$ is the probability for finding the nucleon as a proton and

$|b|^2$ is the probability for finding the nucleon as a neutron.

In this space the electric charge Q (in units of the electronic charge $-e$) carried by the nucleon can be represented by the operator

(29.4)
$$Q = \tfrac{1}{2}(\tau_3 + I); \qquad I = \begin{pmatrix} 1 & 0 \\ 0 & 1 \end{pmatrix}$$

which has $|p\rangle$ and $|n\rangle$ as eigenstates with eigenvalues $+1$ and 0, respectively. The analogs of the operators σ_+ and σ_- (see Appendix 1) constructed from the spin matrices σ_1 and σ_2 may therefore be called the "charge creation and annihilation" operators,

(29.5)
$$\tau_+ = \tfrac{1}{2}(\tau_1 + i\tau_2) = \begin{pmatrix} 0 & 1 \\ 0 & 0 \end{pmatrix}; \qquad \tau_- = \tfrac{1}{2}(\tau_1 - i\tau_2) = \begin{pmatrix} 0 & 0 \\ 1 & 0 \end{pmatrix}.$$

They are constructed from the isospin matrices

(29.6)
$$\tau_1 = \begin{pmatrix} 0 & 1 \\ 1 & 0 \end{pmatrix}; \qquad \tau_2 = \begin{pmatrix} 0 & -i \\ i & 0 \end{pmatrix}$$

and have the properties

(29.7)
$$\tau_+|n\rangle = |p\rangle; \qquad \tau_-|n\rangle = 0$$
$$\tau_+|p\rangle = 0; \qquad \tau_-|p\rangle = |n\rangle.$$

Also useful are the "projection" operators

(29.8)
$$\tau_p = \tfrac{1}{2}(I + \tau_3) = \begin{pmatrix} 1 & 0 \\ 0 & 0 \end{pmatrix}; \qquad \tau_n = \tfrac{1}{2}(I - \tau_3) = \begin{pmatrix} 0 & 0 \\ 0 & 1 \end{pmatrix}$$

which project out of the general state $|\chi\rangle$ the proton and neutron components respectively,

(29.9)
$$\tau_p|\chi\rangle = a|p\rangle; \qquad \tau_n|\chi\rangle = b|n\rangle.$$

In the terminology developed here, the state vector needed for the complete description of a single nucleon will then consist of a direct product of two vectors $|\varphi\rangle$ and $|\chi\rangle$,

$$(29.10) \qquad |\text{nucleon}\rangle = |\varphi\rangle \times |\chi\rangle$$

where $|\varphi\rangle$ accounts for all dynamical attributes of the nucleon such as momentum and spin, and $|\chi\rangle$ is specified by whatever quantum numbers are needed to completely label the nucleon in isospace. In view of the isomorphism of the isospin operators, τ_1, τ_2, and τ_3 with the Pauli spin operators σ_1, σ_2, and σ_3 one can hardly avoid trying at this stage to label the nucleon state in isospace by two quantum numbers $T = \frac{1}{2}$ and $T_3 = \pm \frac{1}{2}$ in analogy to the two quantum numbers $s = \frac{1}{2}$ and $s_3 = \pm \frac{1}{2}$ needed to completely specify an ordinary fermion spin. Instead of saying "there are only two nucleons" it will be said from now on "the nucleon has isospin $T = \frac{1}{2}$, with two possible orientations $T_3 = +\frac{1}{2}$ and $T_3 = -\frac{1}{2}$ in isospace." Thus the states (29.2) will be labeled

$$(29.11)$$

$$|p\rangle = |T = \tfrac{1}{2}; T_3 = +\tfrac{1}{2}\rangle; \qquad |n\rangle = |T = \tfrac{1}{2}; T_3 = -\tfrac{1}{2}\rangle,$$

with the implication that addition of isospin quantum numbers for several nucleons shall be effected by applying the rules governing the addition of ordinary angular momenta, as given in Appendix 2.

In particular, for an atomic nucleus containing A nucleons the total isospin will be represented by the operator $\hat{\mathbf{T}} = (\hat{T}_1, \hat{T}_2, \hat{T}_3)$ with

$$(29.12) \qquad \hat{T}_i = \tfrac{1}{2} \sum_{N=1}^{A} \tau_i(N)$$

so that

$$(29.13) \qquad \hat{\mathbf{T}}^2 = \hat{T}_1^2 + \hat{T}_2^2 + \hat{T}_3^2$$

and the total charge of the nucleus will be the eigenvalue of the operator

$$(29.14)$$

$$Z = \sum_{N=1}^{A} Q(N) = \tfrac{1}{2}\left[\sum_{N=1}^{A} \tau_3(N) + \sum_{N=1}^{A} I\right] = \hat{T}_3 + (A/2)\,I.$$

Since $\hat{\mathbf{T}}$ satisfies the C.R.s governing the components of angular momentum, the conclusions of Appendix 1 are applicable, and one concludes there must exist in isospace a state vector $|T, T_3\rangle$ satisfying

$$(29.15)$$

$$\hat{T}^2|T, T_3\rangle = T(T+1)|T, T_3\rangle; \qquad \hat{T}_3|T, T_3\rangle = T_3|T, T_3\rangle$$

so that any many-nucleon system can be characterized by two isospin quantum numbers T, T_3 if the single nucleon can be so classified.

However, as soon as one, in this fashion, admits isospin as a legitimate attribute of nucleons, without which any set of attributes would be incomplete, most profound consequences ensue for the classification of many-nucleon states because of the intervention of the exclusion principle.

To illustrate this consider a two-nucleon state which will, in general, be the direct product

$$(29.16) \qquad |\tau(1)\,\tau'(2)\rangle = |s(1)\,s'(2)\rangle \times |\mathbf{T}(1)\,\mathbf{T}'(2)\rangle$$

of a state vector characterized completely by the spatial quantum numbers s and s' (such as momentum, spin, etc.) of nucleon 1 and nucleon 2, and a state vector in isospace characterized by the isospin quantum numbers $\mathbf{T}$ and $\mathbf{T}'$ of the two nucleons. The exclusion principle requires that the operation of transposition of particle labels T_{12} applied to the particle labels 1 and 2 of the *complete* set of quantum numbers τ and τ' result in a change of sign of the state vector,

$$(29.17) \qquad T_{12}|\tau(1)\,\tau'(2)\rangle = -|\tau(1)\,\tau'(2)\rangle$$

and therefore a given isospin state $|\mathbf{T}(1)\,\mathbf{T}'(2)\rangle$ by its transformation property under T_{12} in isospace will impose severe restrictions on the possible quantum numbers s and s' of the two nucleons if (29.17) should be valid.

More specifically, a two nucleon system may, in accordance with the rules governing the addition of two spins $\tfrac{1}{2}$ [see Appendix 2, Eqs. (A2.9)], belong in isospace either to the antisymmetric singlet of total isospin $T = 0$,

$$(29.18) \qquad |T = 0, T_3 = 0\rangle = (1/\sqrt{2})\,(|p\rangle|n\rangle - |n\rangle|p\rangle)$$

or to the symmetric triplet of total isospin $T = 1$,

$$|T = 1, T_3 = +1\rangle = |p\rangle|p\rangle$$

$$(29.19) \qquad |T = 1, T_3 = 0\rangle = (1/\sqrt{2})\,(|p\rangle|n\rangle + |n\rangle|p\rangle)$$

$$|T = 1, T_3 = -1\rangle = |n\rangle|n\rangle$$

By working in the center of mass frame, denoting the coordinate state of two nucleons symmetric under transposition of particle labels as $|\text{even}\rangle$ and the antisymmetric coordinate state as $|\text{odd}\rangle$, and remembering that the possible ordinary spin states of a system of two nucleons are either the antisymmetric singlet denoted $|1\rangle$ or the symmetric triplet denoted $|3\rangle$, one can write down immediately the possible totally antisymmetric states for a system of two nucleons, namely

(29.20a) $$|\text{even}\rangle \times |1\rangle \times |T = 1\rangle$$

(29.20b) $$|\text{odd}\rangle \times |3\rangle \times |T = 1\rangle$$

(29.20c) $$|\text{even}\rangle \times |3\rangle \times |T = 0\rangle$$

(29.20d) $$|\text{odd}\rangle \times |1\rangle \times |T = 0\rangle$$

because the symmetric isospin triplet $|T = 1\rangle$ *must* be combined with a state antisymmetric in the combined coordinate-spin space, and the antisymmetric isospin singlet $|T = 0\rangle$ *must* be combined with a state symmetric in coordinate-spin space to insure the validity of (29.17).

The terminology of calling symmetrical coordinate states $|\text{even}\rangle$ and the antisymmetric ones $|\text{odd}\rangle$ stems from the observation, used already in preceding sections, that in the case of two particles the operation T_{12} of transposition of particle labels is identical with the operation of coordinate inversion Π in the center of mass frame. If the relative intrinsic parity of the particles is taken to be even, then the coordinate state has the parity $(-1)^l$, l being the orbital angular momentum of the relative motion. It follows that the antisymmetric states must be states of odd l and the symmetric states must be states of even l.

As a consequence of the particular combinations (29.20) of coordinate-spin-isospin quantum numbers demanded by the exclusion principle, both the diproton and the dineutron, which by (29.19) necessarily belong to the isospin triplet, cannot exist in $|\text{even}\rangle|3\rangle$ or $|\text{odd}\rangle|1\rangle$ states with respect to spatial and spin variables. The deuteron, on the other hand, can exist in all four combinations of spatial and spin states.

Experimentally, dineutron and diproton, which belong to $T = 1$, are not found to exist in stable states, whereas the deuteron is found to be stable, but only if it is in an $|\text{even}\rangle|3\rangle$ state, namely the ^3S-state which belongs to $T = 0$. The lowest $|\text{even}\rangle|1\rangle$ state of the deuteron, namely the ^1S-state belonging to $T = 1$, is not stable. These facts indicate that the specific nuclear interactions, giving rise to stable states among nuclei, depend only on the absolute value of the isospin and do not distinguish between the $2T + 1$ multiplets. Moreover, a classification of the lowest energy levels of nuclei (be they stable states or resonant states of positive energy) suggests that the lowest possible isospin T results in the strongest bond, as indicated in Table 29.1.

Accordingly, in all theories of nuclear interactions which start from

TABLE 29.1

The Lowest Energy Levels of Nuclei up to $A = 6$ Classified According to Total Isospin T

the concept of an interaction Hamiltonian, the invariance of H_{int} under rotations in isospace is assumed, being equivalent to assuming that in nuclear interactions isospin is conserved. Since changing the orientation in isospace means changing T_3, which in turn means changing the total electric charge value of the nucleus, assuming the invariance under rotations in isospace amounts to assuming the charge independence of nuclear forces. It is now generally believed that the actual splitting observed between the energy levels belonging to a given isospin multiplet can, in principle, be accounted for by the remaining electromagnetic interaction between nucleons, which is by definition charge dependent, so that the electromagnetic interaction Hamiltonian depends on T_3 and is thus not invariant under rotations in isospace. This is the classic example of a symmetry, valid for a strong interaction, which is "broken" by a weaker interaction. In other words, conservation of isospin is not an absolute conservation law, it is valid only in reactions mediated by strong interactions, but is "violated" if electromagnetic interactions are taken into account.

In recent years it has become experimentally possible to substitute in nuclei a Λ particle for a neutron, and the observation of thus obtained "hypernuclei" has turned up another piece of evidence supporting the view that strong interactions, to which belongs the $\Lambda - N$ interaction, conserve isospin, allowing the resulting energy levels to depend only on the total isospin T, which is a scalar in isospace. Since the Λ particle has isospin $T = 0$ (it comes only as an electrically neutral particle), its substitution for a neutron will result in the formation of a nucleus with isospin T lowered by $\frac{1}{2}$. Since the isotopic number A of the nucleus is not changed by the substitution, one should expect a multiplet structure of energy levels in hypernuclei similar to that of the nuclei, but shifted one up in the A scale, as indicated in Table 29.2.

This table has been confirmed by observation of the hypernuclei $_\Lambda\text{H}^3$, $_\Lambda\text{He}^4$, $_\Lambda\text{He}^5$, and $_\Lambda\text{Li}^7$ in multiplets as predicted. One should not expect numerical agreement with the binding energies of the corresponding nuclear levels, because the total number of particles A partaking in the interaction is different, and besides the Λ particle differs from the neutron by another intrinsic attribute, the "strangeness" to be taken up later in this section, which may have dynamical significance. Nevertheless, it is interesting to note that the observed binding energy of about 2.6 Mev for the hypernucleus $_\Lambda\text{H}^3$ is very close to the binding energy of about 2.2 Mev for the deuteron H^2, which correspond to each other in the two level schemes of Tables 29.1 and 29.2.

It is perhaps instructive to set down here the most general interaction Hamiltonian for two nucleons which satisfies the requirement of charge

TABLE 29.2

THE LOWEST ENERGY LEVELS OF HYPERNUCLEI UP TO $A = 7$ CLASSIFIED ACCORDING TO TOTAL ISOSPIN T

independence of nuclear forces, i.e. invariance under rotations in iso-space. Prior to imposing this symmetry, one would have as most general interaction Hamiltonian an expression of the form

$$(29.21) \qquad H_{\text{int}} = H_{1,1}\,\mathscr{O}_{1,1} + H_{1,0}\,\mathscr{O}_{1,0} + H_{1,-1}\,\mathscr{O}_{1,-1} + H_{0,0}\,\mathscr{O}_{0,0},$$

where H_{T,T_3} are the interaction operators acting on the spatial and spin variables of the two nucleons if they are in the isospin state $|T,T_3\rangle$, and $\mathscr{O}_{T,T_3}$ stands for the projection operators which project out of the general isospin state of two nucleons,

$$(29.22) \qquad |\chi\rangle = a|1,1\rangle + b|1,0\rangle + c|1,-1\rangle + d|0,0\rangle$$

the components belonging to quantum numbers T, T_3, respectively. These operators are explicitly

$$\mathscr{O}_{1,1} = [\tfrac{1}{2}+\hat{T}_3(1)][\tfrac{1}{2}+\hat{T}_3(2)]$$
$$\text{so that} \qquad \mathscr{O}_{1,1}|\chi\rangle = a|1,1\rangle$$
$$\mathscr{O}_{1,0} = \tfrac{1}{4}+\hat{T}_1(1)\hat{T}_1(2)+\hat{T}_2(1)\hat{T}_2(2)-\hat{T}_3(1)\hat{T}_3(2)$$
$$(29.23) \qquad \text{so that} \qquad \mathscr{O}_{1,0}|\chi\rangle = b|1,0\rangle$$
$$\mathscr{O}_{1,-1} = [\tfrac{1}{2}-\hat{T}_3(1)][\tfrac{1}{2}-\hat{T}_3(2)]$$
$$\text{so that} \qquad \mathscr{O}_{1,-1}|\chi\rangle = c|1,-1\rangle$$
$$\mathscr{O}_{0,0} = \tfrac{1}{4}-\hat{T}_1(1)\hat{T}_1(2)-\hat{T}_2(1)\hat{T}_2(2)-\hat{T}_3(1)\hat{T}_3(2)$$
$$\text{so that} \qquad \mathscr{O}_{0,0}|\chi\rangle = d|0,0\rangle.$$

The hypothesis of rotational symmetry in isospace means to put

$$(29.24) \qquad H_{1,1} = H_{1,0} = H_{1,-1} = H_a \qquad \text{(say)}$$

and

$$(29.25) \qquad H_{0,0} = H_s \qquad \text{(say)}$$

where H_a and H_s are, in general, different. In fact, one knows from experiment that the H_a having as lowest eigenstate the virtual $|\text{even}\rangle|1\rangle$ state of the deuteron must give rise to a potential energy which is about $\tfrac{1}{2}$ the potential energy due to H_s which has as its ground state the stable $|\text{even}\rangle|3\rangle$ state of the deuteron. By substitution of (29.23)–(29.25) into (29.21) one finds as general isospin-conserving inter-action Hamiltonian for two nucleons

$$(29.26) \qquad H_{\text{int}} = H_a\{\tfrac{3}{4}+[\hat{\mathbf{T}}(1)\,\hat{\mathbf{T}}(2)]\} + H_s\{\tfrac{1}{4}-[\hat{\mathbf{T}}(1)\,\hat{\mathbf{T}}(2)]\}$$
$$= [(3H_a+H_s)/4] + (H_a-H_s)\,[\hat{\mathbf{T}}(1)\,\hat{\mathbf{T}}(2)]$$

which is obviously a scalar in isospace.

It is interesting to see in detail how the "law of conservation of isospin" is "violated" if the electromagnetic interactions are added to the interaction Hamiltonian. Taking, as an example, the Coulomb interaction which acts only between two protons, i.e. between two nucleons which are by (29.19) necessarily in the $|T = 1, T_3 = 1\rangle$ state, one has to add to (29.26) the term

$$(29.27) \quad H_{\text{Coul}} = (e^2/r)\, \mathcal{O}_{1,1} = (e^2/r) \{ \tfrac{1}{4} + \tfrac{1}{2}[\hat{T}_3(1) + \hat{T}_3(2)] + \hat{T}_3(1)\,\hat{T}_3(2) \}$$

which is no longer invariant under rotations in isospace. Indeed,

$$(29.28) \quad H_{\text{Coul}}\hat{T}_1 - \hat{T}_1 H_{\text{Coul}} \neq 0 \qquad \text{and} \qquad H_{\text{Coul}}\hat{T}_2 - \hat{T}_2 H_{\text{Coul}} \neq 0$$

so that T is no longer conserved. However, T_3 is still conserved, because

$$(29.29) \quad\quad\quad\quad H_{\text{Coul}}\hat{T}_3 - \hat{T}_3 H_{\text{Coul}} = 0.$$

This relation guarantees conservation of electric charge by virtue of the general definition (29.14), which reads here

$$(29.30) \quad\quad\quad\quad \hat{T}_3 = Z - 1$$

where Z is the charge number of the two-nucleon system.

In the 1940's there were lingering doubts in the minds of many physicists regarding the necessity of introducing the isospin as a genuine attribute. After all, one may conceivably arrive at a theory of nuclear forces by considering from the very beginning neutron and proton as different particles (namely differing in electric charge) and try to obtain the structure of the Hamiltonian (29.26) by assuming a peculiar coordinate and spin dependence of the specifically charge-independent nuclear interaction. This is logically possible if one *postulates* that the interaction between two nucleons in $|\text{even}\rangle|1\rangle$ and $|\text{odd}\rangle|3\rangle$ states is different from the interaction in $|\text{even}\rangle|3\rangle$ and $|\text{odd}\rangle|1\rangle$ states, but is otherwise charge independent.

These doubts were largely dispelled, however, when in the early 1950's extension of the isospin formalism to pions and their interactions began to account elegantly for many features of the pion-nucleon interaction which would have been hard to explain in any other way. More influential than any other single piece of evidence in turning the tide in favor of isospin as an acceptable legitimate attribute probably was Brueckner's explanation of a general feature governing the scatterings of pions by protons in 1952. Experimentally there had been established at that time a ratio $\sigma(\pi^+):\sigma(\pi^-) = 3:1$ for the respective total scattering cross section of positive and negative pions on protons up to 300 Mev incident pion energy in the laboratory frame of reference. Brueckner showed how this empirical relation can be understood simply as a consequence of the rules governing addition of isospins, if one assumes conservation of isospin to hold in pion-nucleon interactions. The line of reasoning runs as follows.

Since there are three pions, π^+, π^0, and π^-, one requires in isospace a vector with three components to describe the charge state of a pion, being a linear combination of the three basis vectors

$$(29.31) \qquad |\pi^+\rangle = \begin{pmatrix} 1 \\ 0 \\ 0 \end{pmatrix}; \qquad |\pi^0\rangle = \begin{pmatrix} 0 \\ 1 \\ 0 \end{pmatrix}; \qquad |\pi^-\rangle = \begin{pmatrix} 0 \\ 0 \\ -1 \end{pmatrix}$$

denoted alternatively

$$|T = 1, T_3 = +1\rangle; \qquad |T = 1, T_3 = 0\rangle; \qquad |T = 1, T_3 = -1\rangle;$$

and chosen with a convention of phase, which are the eigenstates of the operator

$$(29.32) \qquad \rho_3 = \begin{pmatrix} 1 & 0 & 0 \\ 0 & 0 & 0 \\ 0 & 0 & -1 \end{pmatrix}$$

with eigenvalues $+1$, 0, and -1, respectively. Thus ρ_3 may be taken as the operator representing the charge Q of the pion. By analogy with the operators of angular momentum belonging to $j = 1$, the pion may be said to have isospin $\rho = 1$, the remaining two components of the isospin operator $\boldsymbol{\rho}$ given by

$$(29.33) \qquad \rho_1 = \frac{1}{\sqrt{2}} \begin{pmatrix} 0 & 1 & 0 \\ 1 & 0 & 1 \\ 0 & 1 & 0 \end{pmatrix}; \qquad \rho_2 = \frac{1}{\sqrt{2}} \begin{pmatrix} 0 & -i & 0 \\ i & 0 & -i \\ 0 & i & 0 \end{pmatrix}.$$

There are also, in analogy to (29.5), the charge creation and annihilation operators

$$(29.34)$$

$$\rho_+ = \tfrac{1}{2}(\rho_1 + i\rho_2) = \frac{1}{\sqrt{2}} \begin{pmatrix} 0 & 1 & 0 \\ 0 & 0 & 1 \\ 0 & 0 & 0 \end{pmatrix}; \qquad \rho_- = \tfrac{1}{2}(\rho_1 - i\rho_2) = \frac{1}{\sqrt{2}} \begin{pmatrix} 0 & 0 & 0 \\ 1 & 0 & 0 \\ 0 & 1 & 0 \end{pmatrix}$$

having the properties

$$(29.35)$$

$$\rho_+|\pi^+\rangle = 0; \qquad \rho_+|\pi^0\rangle = (1/\sqrt{2})\,|\pi^+\rangle; \qquad \rho_+|\pi^-\rangle = (-1/\sqrt{2})\,|\pi^0\rangle$$
$$\rho_-|\pi^+\rangle = (1/\sqrt{2})\,|\pi^0\rangle; \qquad \rho_-|\pi^0\rangle = (-1/\sqrt{2})\,|\pi^-\rangle; \qquad \rho_-|\pi^-\rangle = 0.$$

In this notation the total isospin operator $\hat{\mathbf{T}}$ of a system of A nucleons and B pions will be represented by

$$(29.36) \qquad \hat{\mathbf{T}} = \tfrac{1}{2} \sum_{i=1}^{A} \boldsymbol{\tau}(i) + \sum_{k=1}^{B} \boldsymbol{\rho}(k)$$

and the operator of the electric charge of this system is

$$(29.37) \qquad Q = \hat{T}_3 + (A/2) = \tfrac{1}{2} \sum_{i=1}^{A} \tau_3(i) + \sum_{k=1}^{B} \rho_3(k) + (A/2)$$

It must then be possible, as in the case of a pure nucleon system, to characterize a system of A nucleons and B pions by two quantum numbers T, T_3 which fix the isospin state as in (29.15). Demanding "conservation of isospin" in strong interactions is then equivalent to requiring that the dynamical properties of a system consisting of pions and nucleons are dependent on the value of T only and do not depend on T_3.

Consider as a simple example a system consisting of one pion and one nucleon. By the rules of addition of angular momenta (see Table A2.1) one has as possible states in isospace a quartet belonging to total isospin $T = \tfrac{3}{2}$ and a doublet belonging to $T = \tfrac{1}{2}$,

(29.38)

$$
\begin{cases}
|T = \tfrac{3}{2}, T_3 = \tfrac{3}{2}\rangle = |\tfrac{1}{2}, \tfrac{1}{2}\rangle |1, 1\rangle = |p\rangle |\pi^+\rangle \\[4pt]
|\tfrac{3}{2}, \tfrac{1}{2}\rangle = (\sqrt{\tfrac{2}{3}}) |\tfrac{1}{2}, \tfrac{1}{2}\rangle |1, 0\rangle + (\sqrt{\tfrac{1}{3}}) |\tfrac{1}{2}, -\tfrac{1}{2}\rangle |1, 1\rangle \\[4pt]
\qquad = (\sqrt{\tfrac{2}{3}}) |p\rangle |\pi^0\rangle + (\sqrt{\tfrac{1}{3}}) |n\rangle |\pi^+\rangle \\[4pt]
|\tfrac{3}{2}, -\tfrac{1}{2}\rangle = (\sqrt{\tfrac{2}{3}}) |\tfrac{1}{2}, -\tfrac{1}{2}\rangle |1, 0\rangle + (\sqrt{\tfrac{1}{3}}) |\tfrac{1}{2}, \tfrac{1}{2}\rangle |1, -1\rangle \\[4pt]
\qquad = (\sqrt{\tfrac{2}{3}}) |n\rangle |\pi^0\rangle + (\sqrt{\tfrac{1}{3}}) |p\rangle |\pi^-\rangle \\[4pt]
|\tfrac{3}{2}, -\tfrac{3}{2}\rangle = |\tfrac{1}{2}, -\tfrac{1}{2}\rangle |1, -1\rangle = |n\rangle |\pi^-\rangle
\end{cases}
$$

(29.39)

$$
\begin{cases}
|\tfrac{1}{2}, \tfrac{1}{2}\rangle = (\sqrt{\tfrac{1}{3}}) |\tfrac{1}{2}, \tfrac{1}{2}\rangle |1, 0\rangle - (\sqrt{\tfrac{2}{3}}) |\tfrac{1}{2}, -\tfrac{1}{2}\rangle |1, 1\rangle \\[4pt]
\qquad = (\sqrt{\tfrac{1}{3}}) |p\rangle |\pi^0\rangle - (\sqrt{\tfrac{2}{3}}) |n\rangle |\pi^+\rangle \\[4pt]
|\tfrac{1}{2}, -\tfrac{1}{2}\rangle = (\sqrt{\tfrac{2}{3}}) |\tfrac{1}{2}, \tfrac{1}{2}\rangle |1, -1\rangle - (\sqrt{\tfrac{1}{3}}) |\tfrac{1}{2}, -\tfrac{1}{2}\rangle |1, 0\rangle \\[4pt]
\qquad = (\sqrt{\tfrac{2}{3}}) |p\rangle |\pi^-\rangle - (\sqrt{\tfrac{1}{3}}) |n\rangle |\pi^0\rangle
\end{cases}
$$

For comparison with the experimental evidence cited above the isospin properties of proton-pion states will now be examined. Inspection of the combinations (29.38) and (29.39) reveals that a $|p\rangle |\pi^+\rangle$ state is always a pure isospin state belonging to $T = \tfrac{3}{2}$, whereas the states $|p\rangle |\pi^0\rangle$ and $|p\rangle |\pi^-\rangle$ are always isospin mixtures,

$$
\begin{aligned}
|p\rangle |\pi^+\rangle &= |\tfrac{3}{2}, \tfrac{3}{2}\rangle \\[4pt]
(29.40) \qquad |p\rangle |\pi^0\rangle &= (\sqrt{\tfrac{1}{3}}) |\tfrac{1}{2}, \tfrac{1}{2}\rangle + (\sqrt{\tfrac{2}{3}}) |\tfrac{3}{2}, \tfrac{1}{2}\rangle \\[4pt]
|p\rangle |\pi^-\rangle &= (\sqrt{\tfrac{2}{3}}) |\tfrac{1}{2}, -\tfrac{1}{2}\rangle + (\sqrt{\tfrac{1}{3}}) |\tfrac{3}{2}, -\tfrac{1}{2}\rangle
\end{aligned}
$$

If one now supposes that isospin is conserved in the pion-nucleon interaction, then there must exist, for given total energy and angular momentum, two probability amplitudes $a(\frac{1}{2})$ and $a(\frac{3}{2})$, depending on T only, which govern, respectively, the scattering of an initial pion-nucleon state $|T = \frac{1}{2}\rangle$ into a final pion-nucleon state $|T = \frac{1}{2}\rangle$ and the scattering from an initial state $|T = \frac{3}{2}\rangle$ into a final state $|T = \frac{3}{2}\rangle$. In particular, a $|p\rangle|\pi^+\rangle$ initial state, which by (29.40) is necessarily a $|T = \frac{3}{2}\rangle$ state, can be scattered only into a final state $|T = \frac{3}{2}\rangle$ with amplitude $a(\frac{3}{2})$,

$$(29.41) \qquad |p\rangle|\pi^+\rangle = |\tfrac{3}{2}, \tfrac{3}{2}\rangle \to a(\tfrac{3}{2})|\tfrac{3}{2}, \tfrac{3}{2}\rangle = a(\tfrac{3}{2})|p\rangle|\pi^+\rangle$$

whereas a $|p\rangle|\pi^-\rangle$ initial state, which is a mixture of $|T = \frac{1}{2}\rangle$ and $|T = \frac{3}{2}\rangle$ states, will be scattered into a final state to which contribute the original $|T = \frac{1}{2}\rangle$ component with amplitude $a(\frac{1}{2})$ and the $|T = \frac{3}{2}\rangle$ component with amplitude $a(\frac{3}{2})$,

(29.42)

$$
\begin{aligned}
|p\rangle|\pi^-\rangle &= (\sqrt{\tfrac{2}{3}})|\tfrac{1}{2}, -\tfrac{1}{2}\rangle + (\sqrt{\tfrac{1}{3}})|\tfrac{3}{2}, -\tfrac{1}{2}\rangle \\
&\to (\sqrt{\tfrac{2}{3}})\,a(\tfrac{1}{2})|\tfrac{1}{2}, -\tfrac{1}{2}\rangle + (\sqrt{\tfrac{1}{3}})\,a(\tfrac{3}{2})|\tfrac{3}{2}, -\tfrac{1}{2}\rangle \\
&= (\tfrac{1}{3})\sqrt{2}[a(\tfrac{3}{2}) - a(\tfrac{1}{2})]|n\rangle|\pi^0\rangle + (\tfrac{1}{3})[a(\tfrac{3}{2}) + 2a(\tfrac{1}{2})]|p\rangle|\pi^-\rangle.
\end{aligned}
$$

The content of Eq. (29.42) is summarized in Table 29.3.

TABLE 29.3

AMPLITUDES FOR PION-PROTON SCATTERING PROCESSES

Process	Amplitude
$|p\rangle|\pi^+\rangle \to |p\rangle|\pi^+\rangle$	$a(\frac{3}{2})$
$|p\rangle|\pi^-\rangle \to |p\rangle|\pi^-\rangle$	$(\frac{1}{3})a(\frac{3}{2}) + (\frac{2}{3})a(\frac{1}{2})$
$|p\rangle|\pi^-\rangle \to |n\rangle|\pi^0\rangle$	$(\frac{1}{3})\sqrt{2}a(\frac{3}{2}) - (\frac{1}{3})\sqrt{2}a(\frac{1}{2})$

Suppose now the pion-nucleon interaction favors, for as yet unknown reasons, the state $T = \frac{3}{2}$ in the region below 300 Mev incident energy, so that $a(\frac{3}{2}) \gg a(\frac{1}{2})$. Then, since the cross section for a process is proportional to the square of the corresponding amplitude, one has the relationship

(29.43)

$$
\sigma(p\pi^+ \to p\pi^+) : \sigma(p\pi^- \to p\pi^-) : \sigma(p\pi^- \to n\pi^0) = |a(\tfrac{3}{2})|^2 : (\tfrac{1}{9})|a(\tfrac{3}{2})|^2 :
$$
$$
(\tfrac{2}{9})|a(\tfrac{3}{2})|^2
$$
$$
= 9 : 1 : 2
$$

which, if one lumps the processes $(p\pi^- \to p\pi^-)$ and $(p\pi^- \to n\pi^0)$ into the total π^- scattering cross section $\sigma(\pi^-)$, implies the ratio

$$(29.44) \qquad \sigma(\pi^+):\sigma(\pi^-) = 3:1$$

in accordance with observation. If, on the other hand, the pion-nucleon interaction had the property $a(\frac{1}{2}) = a(\frac{3}{2})$, then this would have been exhibited as a ratio $1:1:0$ instead of the observed $9:1:2$ and $a(\frac{1}{2}) \gg a(\frac{3}{2})$ would have resulted in a ratio $0:2:1$, which is again excluded by experiment.

The experimental evidence favors, thus, the notion that the pion-nucleon interaction conserves isospin and in the energy region below 300 Mev takes place predominantly in the state $|T = \frac{3}{2}\rangle$. Clearly, the analysis does not explain *why* the $|T = \frac{3}{2}\rangle$ state is favored in this way. From the angular distribution of the scattered pions one can also infer a spin dependence of the interaction, favoring the $P_{3/2}$ state of the pion-nucleon system, again for as yet unknown reasons. The cross sections also show a pronounced maximum near 200 Mev incident energy. One lumps these curious experimental facts into the phrase "there is a $(T = \frac{3}{2}, P_{3/2})$ resonance in the pion-nucleon system at about 200 Mev." Other evidence suggests that if $T = \frac{3}{2}$ then in the P state the interaction is attractive, whereas in the S state the interaction is repulsive (this last fact is summarized in the phrase "the interaction has a hard core"), for reasons which also remain obscure. In short, one badly needs a theory which explains the features of the strong interaction between pions and nucleons, which, one would hope, encompasses the nucleon-nucleon interaction through the pions as the "glue" mediating that interaction. Despite ingenious patch- and guesswork by many workers, no such theory seems to be within sight at the time of writing.

In absence of a dynamical theory, exploitation of the symmetries in isospace exhibited by the strong interaction remains the most reliable tool for making predictions and giving "explanations" for observed "branching ratios." Further examples of such ratios are obtained by comparing the reactions $|n\rangle|p\rangle \to |\pi^0\rangle|d\rangle$ and $|p\rangle|p\rangle \to |\pi^+\rangle|d\rangle$ which require $T = 1$ by isospin conservation, so that

$$(29.45) \qquad \sigma(np \to \pi^0 d):\sigma(pp \to \pi^+ d) = 1:2$$

and the reactions $|p\rangle|d\rangle \to |\pi^+\rangle|H^3\rangle$ and $|p\rangle|d\rangle \to |\pi^0\rangle|He^3\rangle$ resulting in

$$(29.46) \qquad \sigma(pd \to \pi^+ H^3):\sigma(pd \to \pi^0 He^3) = 2:1.$$

Conservation of isospin may also result in strict selection rules. For example, the reaction

$$(29.47) \qquad |d\rangle|d\rangle \to |\pi^0\rangle|He^4\rangle$$

is strictly forbidden by conservation of **T**. If one ever were to observe this reaction one would have to conclude it is mediated by an interaction which does not conserve isospin, such as the electromagnetic interaction. For further details the reader is referred to the extensive literature on this subject.

The relationship (29.37), among the operators of electric charge Q, the third component of isospin $\hat{T}_3$, and the nucleon number A summarizes the curious fact that there is no negatively charged nucleon, making $Q = \frac{1}{2}$ the center of charge, so to speak, for the nucleon, whereas the pion can exist in three charge states around $Q = 0$ as center of charge. This relation can be extended to include antinucleons, if one introduces the baryon number B which is given the value $+1$ for the nucleon, -1 for the antinucleon, and 0 for the boson, because existence of two and only two antinucleons, the antineutron $\bar{n}(Q = 0)$ and the antiproton $\bar{p}(Q = -1)$, can be summarized by writing

$$(29.48) \qquad Q - \hat{T}_3 - (B/2) = 0.$$

This equation implies an assignment of isospin quantum numbers T, T_3 to antinucleons, following the conventions employed for nucleons and pions, by associating a declining sequence of numbers T_3 with the declining sequence of charge values Q. Thus the isospin states of antineutron and antiproton are labeled $|\bar{n}\rangle = |T = \frac{1}{2}, T_3 = +\frac{1}{2}\rangle$ and $|\bar{p}\rangle = |T = \frac{1}{2}, T_3 = -\frac{1}{2}\rangle$.

In 1953 Gell-Mann and Nishijima showed how one can accommodate the curious displacements of the center of charge found empirically among the so-called "strange" hyperons and kaons by introduction of yet another attribute, the "strangeness" S, defined as

$$(29.49) \qquad S = 2[Q - \hat{T}_3 - (B/2)]$$

resulting in the now famous classification scheme of baryons and massive bosons laid out in Table 29.4. In this scheme the assignment of baryon number $+1$ to baryons, -1 to antibaryons, and 0 to bosons is assumed.

An alternative, but equivalent, scheme which has gained some popularity in recent years is obtained by introducing instead of the strangeness S the "hypercharge" Y defined as

$$(29.50) \qquad Y = S + B$$

resulting in a possible classification of particles according to isospin **T**, hypercharge Y, and baryon number B. For the discussion of conservation laws it is, however, most convenient to adhere to the strangeness concept as defined in Eq. (29.49).

Since there is quite overwhelming evidence for the separate conservation of electric charge Q and baryon number B in *all* interactions, conservation of isospin in strong interactions will result in conservation of strangeness. *For reasons as yet completely unknown*, experiments reveal conservation of strangeness (and thus of T_3) in all processes mediated by strong interactions, and nonconservation of strangeness in weak interactions. Further, there is some experimental evidence indicating, again for unknown reasons, charge independence (meaning conservation of *both* T and T_3) in all strong interactions involving baryons and massive

TABLE 29.4.

THE SCHEME OF GELL-MANN AND NISHIJIMA

T	T_3	-2	-1	0	$+1$	$+2$
0	0		Λ^0		$\overline{\Lambda}^0$	
$\tfrac{1}{2}$	$\tfrac{1}{2}$	Ξ^0	$\overline{K}^0$	p | $\bar{n}$	K^+	$\overline{\Xi}^+$
	$-\tfrac{1}{2}$	Ξ^-	K^-	n | $\bar{p}$	K^0	$\overline{\Xi}^0$
	1		Σ^+	π^+	$\overline{\Sigma}^+$	
1	0		Σ^0	π^0	$\overline{\Sigma}^0$	
	-1		Σ^-	π^-	$\overline{\Sigma}^-$	

bosons. In addition, there is some evidence for existence of two selection rules, namely $|\Delta T_3| = 1$ and $|\Delta T| = \tfrac{1}{2}$, governing the isospin-nonconserving weak interactions, whose origin is also obscure.

Even supposing one can understand conservation of charge Q as a consequence of a gauge-invariance principle, as indicated in Sections 20 and 21, there remains for discussion the empirically very well-established conservation of baryon number B, and raises once again the question of whether this conservation law is not in fact a superselection rule, generated by invariance under an antiunitary symmetry operation, such as combined inversion Σ or time reversal Θ, which have already been recognized in Section 17 as possible origins for the analogous conservation of lepton numbers.

To settle this question, an examination of the transformation properties of isospin $\mathbf{T}$ and baryon number B under the operations of coordinate inversion Π, time reversal Θ and, in particular, under particle conjugation

$\varGamma$ is required, because once isospin and baryon number have been admitted as legitimate quantum numbers, the complete representation of each operator will consist of a direct product of the representations in the various subspaces needed to accommodate the complete set of quantum numbers. For example, the operator of particle conjugation for a baryon state will consist of the direct product

$$(29.51) \qquad \varGamma_{\text{baryons}} = \varGamma_D \times \varGamma_{SC} \times \varGamma_{IS} \times \varGamma_B$$

where $\varGamma_D$ and $\varGamma_{SC}$ refer, respectively, to the representations in momentum space and in spin-chirality space already given in Section 19, and $\varGamma_{IS}$ and $\varGamma_B$ refer to the representations in isospace and baryon number space which will now be established.

To carry as far as possible the analogy with the corresponding treatment for leptons given in Section 28, both nucleon and antinucleon states will be represented as four-component vectors in the product space spanned by the simultaneous eigenstates of T_3 and B. Thus, if one introduces with a convention of phase as basis the four states

$$(29.52)$$

$$|p\rangle = \begin{pmatrix} 1 \\ 0 \\ 0 \\ 0 \end{pmatrix}; \qquad |n\rangle = \begin{pmatrix} 0 \\ 1 \\ 0 \\ 0 \end{pmatrix}; \qquad |\bar{n}\rangle = \begin{pmatrix} 0 \\ 0 \\ 1 \\ 0 \end{pmatrix}; \qquad |\bar{p}\rangle = \begin{pmatrix} 0 \\ 0 \\ 0 \\ -1 \end{pmatrix}$$

one has for isospin, baryon number, and electric charge in this combined isospin–baryon-number space the representations

$$(29.53)$$

$$\hat{T}_1 = \tfrac{1}{2}\tau_1 \times I = \frac{1}{2}\begin{pmatrix} 0 & 1 & 0 & 0 \\ 1 & 0 & 0 & 0 \\ 0 & 0 & 0 & 1 \\ 0 & 0 & 1 & 0 \end{pmatrix}; \qquad \hat{T}_2 = \tfrac{1}{2}\tau_2 \times I;$$

$$\hat{T}_3 = \tfrac{1}{2}\tau_3 \times I; \qquad \hat{\mathbf{T}}^2 = \tfrac{3}{4}I \times I$$

$$(29.54) \qquad B = I \times \begin{pmatrix} 1 & 0 \\ 0 & -1 \end{pmatrix} = \begin{pmatrix} 1 & 0 & 0 & 0 \\ 0 & 1 & 0 & 0 \\ 0 & 0 & -1 & 0 \\ 0 & 0 & 0 & -1 \end{pmatrix}$$

$$(29.55) \qquad Q = \hat{T}_3 + (B/2) = \begin{pmatrix} 1 & 0 & 0 & 0 \\ 0 & 0 & 0 & 0 \\ 0 & 0 & 0 & 0 \\ 0 & 0 & 0 & -1 \end{pmatrix}$$

It is readily seen that $\mathbf{T}$, B, and Q all commute, as required if $\hat{T}_3$, $\hat{T}^2$, B, and Q are to be diagonal simultaneously.

Now the operator of particle conjugation Γ should have, by definition, the properties

(29.56)
$$\Gamma|p\rangle = |\bar{p}\rangle; \qquad \Gamma|n\rangle = |\bar{n}\rangle; \qquad \Gamma|\bar{n}\rangle = |n\rangle; \qquad \Gamma|\bar{p}\rangle = |p\rangle$$

implying the space spanned by (29.52) the representation

(29.57)
$$\Gamma = \begin{pmatrix} 0 & 0 & 0 & -1 \\ 0 & 0 & 1 & 0 \\ 0 & 1 & 0 & 0 \\ -1 & 0 & 0 & 0 \end{pmatrix} = \Gamma_{IS} \times \Gamma_{B}.$$

As in the corresponding case of a system composed of leptons and antileptons, it is not possible for a system composed of nucleons and antinucleons with eigenvalues $B \neq 0$ and/or $Q \neq 0$ to be in a state of specific conjugality, because Γ anticommutes with both B and Q,

(29.58)
$$\Gamma B + B\Gamma = 0; \qquad \Gamma Q + Q\Gamma = 0.$$

If, on the other hand, $B = 0$ and $Q = 0$, for example in an object composed of a proton and antiproton, the system *may* be in an eigenstate of Γ. This is true even though Γ may not be representable as a unitary operator, for the general reasons stated in Section 15 following Eq. (15.53). However, the conjugality properties of baryons are set apart from the conjugality properties of leptons by a complication arising from the existence of isospin as an attribute needed to characterize baryons.

At this point in the development it becomes essential to distinguish between the possibilities of representing Γ either as a unitary or as an antiunitary operator.

On the assumption that Γ can be represented by a unitary operator, Lee and Yang in 1956 first analyzed the particle-antiparticle symmetry of objects possessing isospin by a line of reasoning which will now be retraced here. The conclusions reached by these authors should, however, be approached with caution, until the consequences of the alternative representation of Γ by an antiunitary operator have been scrutinized, as will be done later in this section. If the unitary representation of Γ in isospin–baryon-number space is denoted Γ_U, then the C.R.s follow from the representations (29.57) and (29.53) (assuming Γ_U *not* to contain an operation of complex conjugation as would be necessary if Γ were to be antiunitary):

(29.59)
$$\Gamma_U \hat{T}_1 + \hat{T}_1 \Gamma_U = 0; \qquad \Gamma_U \hat{T}_2 - \hat{T}_2 \Gamma_U = 0; \qquad \Gamma_U \hat{T}_3 + \hat{T}_3 \Gamma_U = 0.$$

Therefore, a system composed of baryons and antibaryons can never be in a simultaneous eigenstate of Γ_U and T, T_3. For example, the eigenstates of T, T_3 for a nucleon-antinucleon pair are (note the phase convention implied by (29.52))

$$(29.60) \qquad T = 1 : \begin{cases} |1,1\rangle = |p\rangle|\bar{n}\rangle \\ |1,0\rangle = (1/\sqrt{2})\,(|p\rangle|\bar{p}\rangle - |n\rangle|\bar{n}\rangle) \\ |1,-1\rangle = |n\rangle|\bar{p}\rangle \end{cases}$$

$$(29.61) \qquad T = 0 : \quad |0,0\rangle = (1/\sqrt{2})\,(|p\rangle|\bar{p}\rangle + |n\rangle|\bar{n}\rangle).$$

The two possible eigenstates of Γ_U, namely $|p\rangle|\bar{p}\rangle$ and $|n\rangle|\bar{n}\rangle$, are therefore necessarily mixtures of states belonging to $T = 0$ and $T = 1$,

(29.62)

$$|p\rangle|\bar{p}\rangle = (1/\sqrt{2})\,(|0,0\rangle + |1,0\rangle); \qquad |n\rangle|\bar{n}\rangle = (1/\sqrt{2})\,(|0,0\rangle - |1,0\rangle).$$

Despite this impossibility of using eigenvalues of Γ_U to characterize the conjugality properties of a state with given quantum numbers T, T_3, Lee and Yang pointed out that there is an operator involving Γ which can be used for this purpose, namely

(29.63)

$$G = \Gamma \exp(i\pi \hat{T}_2) = \Gamma_D \times \Gamma_{SC} \times \Gamma_U \exp(i\pi \hat{T}_2) = \Gamma_D \times \Gamma_{SC} \times G_U$$

consisting of a rotation by angle π around the 2-axis in isospace combined with the operation of particle conjugation. G has the commutation properties

$$(29.64) \qquad\qquad G\hat{\mathbf{T}} - \hat{\mathbf{T}}G = 0,$$

$$(29.65) \qquad\qquad GB + BG = 0,$$

can therefore be diagonalized simultaneously with T, T_3, and a state with $B = 0$ can be an eigenstate of G, called a state of specific G conjugality.

To obtain an explicit representation in the space (29.52), one utilizes the relations, valid for $T = \frac{1}{2}$,

$$(29.66) \qquad \hat{T}_2^{2n+1} = \tfrac{1}{2}{}^n \hat{T}_2 \qquad \text{and} \qquad \hat{T}_2^{2n} = \tfrac{1}{2}{}^n I$$

which follow from (29.53), yielding

(29.67)

$$G_U = \Gamma_U \,[\cos(\pi/2)\,I + 2i\,\sin(\pi/2)\,\hat{T}_2]$$

$$= \begin{pmatrix} 0 & 0 & 0 & -1 \\ 0 & 0 & 1 & 0 \\ 0 & 1 & 0 & 0 \\ -1 & 0 & 0 & 0 \end{pmatrix} \begin{pmatrix} 0 & 1 & 0 & 0 \\ -1 & 0 & 0 & 0 \\ 0 & 0 & 0 & 1 \\ 0 & 0 & -1 & 0 \end{pmatrix}$$

$$= \begin{pmatrix} 0 & 0 & 1 & 0 \\ 0 & 0 & 0 & 1 \\ -1 & 0 & 0 & 0 \\ 0 & -1 & 0 & 0 \end{pmatrix}.$$

Single particle states transform thus under G as

(29.68)

$$G_U|p\rangle = -|\bar{n}\rangle; \qquad G_U|n\rangle = |\bar{p}\rangle; \qquad G_U|\bar{n}\rangle = |p\rangle; \qquad G_U|\bar{p}\rangle = -|n\rangle.$$

G_U is by definition unitary and satisfies

(29.69) $$G_U^2 = -I \qquad \text{for} \qquad B = \pm 1.$$

Since in a system containing many particles and antiparticles **T** and B are additive, whereas G is multiplicative, one has generally

(29.70)

$$G\hat{\mathbf{T}} - \hat{\mathbf{T}}G = 0; \qquad GB + BG = 0; \qquad G^2 = C^2 G_U^2 = (-1)^B$$

provided $C^2 = +1$, where C denotes the eigenvalue of $\Gamma_D \times \Gamma_{SC}$, and is identical with the ordinary conjugality introduced in Section 28.

Consequently, states with $B = 0$ can be assigned specific G conjugality. Components of the same isospin multiplet T have always the same conjugality G_U. For example, one reads immediately from (29.68) the G_U conjugality of the nucleon-antinucleon states (29.60) and (29.61), as summarized in Table 29.5. Since for the ordinary conjugality C the same analysis applies as in the case of positronium (see Table 28.1), the total G conjugality of a nucleon-antinucleon pair is given by

(29.71) $$G = CG_U = (-1)^{l+s}(-1)^T.$$

From the invariance of strong interactions under particle conjugation and under rotations in isospace, the conservation of **G** *in all processes mediated by strong interactions* now follows. As the concept of G conjugality can be extended to pions and strange particles, a number of interesting selection rules are engendered by this conservation law. To obtain some of these consider next the conjugality properties of pions.

Since the π^0 decays into two photons, and since the interaction responsible for this decay is assumed to leave the conjugality C invariant, it follows from (28.14) that the state $|\pi^0\rangle$ must be a state of even conjugality,

(29.72) $$\Gamma|\pi^0\rangle = +|\pi^0\rangle.$$

There is no need to distinguish between Γ and Γ_U here, because pions have spin 0 and one can represent Γ in coordinate space by the identity

TABLE 29.5

THE G_U CONJUGALITY OF NUCLEON-ANTINUCLEON PAIRS

T	T_3	State	G_U
1	+1	$\lvert p\rangle\lvert\bar{n}\rangle$	-1
	0	$(1/\sqrt{2})(\lvert p\rangle\lvert\bar{p}\rangle-\lvert n\rangle\lvert\bar{n}\rangle)$	-1
	-1	$\lvert n\rangle\lvert\bar{p}\rangle$	-1
0	0	$(1/\sqrt{2})(\lvert p\rangle\lvert\bar{p}\rangle+\lvert n\rangle\lvert\bar{n}\rangle)$	$+1$

operation. The charged pions, on the other hand, may be considered as each other's particle conjugate, so that Γ satisfies

$$(29.73) \qquad \Gamma\lvert\pi^+\rangle = \lvert\pi^-\rangle \qquad \text{and} \qquad \Gamma\lvert\pi^-\rangle = \lvert\pi^+\rangle.$$

This implies, in the isospace spanned by the vectors (29.31), the representation

$$(29.74) \qquad \Gamma = \begin{pmatrix} 0 & 0 & -1 \\ 0 & 1 & 0 \\ -1 & 0 & 0 \end{pmatrix}$$

and the C.R.s of Γ with the operators of isospin (29.32) and (29.33) are, as in the case of the nucleon-antinucleon system,

(29.75)

$$\Gamma\hat{T}_1+\hat{T}_1\Gamma = 0$$

$$\Gamma\hat{T}_2-\hat{T}_2\Gamma = 0$$

$$\Gamma\hat{T}_3+\hat{T}_3\Gamma = 0 \qquad \text{i.e.} \qquad \Gamma Q+Q\Gamma = 0 \quad \text{for pions,}$$

giving rise once again to the construction of an operator G defined as in (29.63), which will still satisfy (29.70), and which can be represented, because of the relations (valid for $T = 1$)

(29.76)

$$\hat{T}_2^{2n+1} = \hat{T}_2 \quad \text{for} \quad n \geqslant 0; \qquad \hat{T}_2^{2n} = \hat{T}_2^2 \quad \text{for} \quad n \geqslant 1$$

in the isospace of the pions (29.31) as

(29.77)
$$G = \Gamma \exp(i\pi \hat{T}_2) = \Gamma\{I + i\sin(\pi)\hat{T}_2 + [\cos(\pi) - 1]\hat{T}_2^2\}$$

$$= \Gamma\left\{\begin{pmatrix} 1 & 0 & 0 \\ 0 & 1 & 0 \\ 0 & 0 & 1 \end{pmatrix} - \begin{pmatrix} 1 & 0 & -1 \\ 0 & 2 & 0 \\ -1 & 0 & 1 \end{pmatrix}\right\}$$

$$= \begin{pmatrix} 0 & 0 & -1 \\ 0 & 1 & 0 \\ -1 & 0 & 0 \end{pmatrix}\begin{pmatrix} 0 & 0 & 1 \\ 0 & -1 & 0 \\ 1 & 0 & 0 \end{pmatrix} = \begin{pmatrix} -1 & 0 & 0 \\ 0 & -1 & 0 \\ 0 & 0 & -1 \end{pmatrix}.$$

This means all pion states are eigenstates of G; *the pion has odd G conjugality.*

If now G is conserved, a number of selection rules can be inferred immediately by consulting Eq. (29.71), for example:

(I) The system $|\bar{p}\rangle|n\rangle$ which belongs necessarily to isospin $T = 1$ cannot decay through strong interactions into an ($_{\text{odd}}^{\text{even}}$) number of pions from a state with $l+s$ ($_{\text{odd}}^{\text{even}}$) such as the

$$\begin{pmatrix} ^1S_0, & ^3P_0, & \ldots \\ ^3S_1, & ^1P_1, & \ldots \end{pmatrix}$$

state.

(II) An even number of pions cannot by strong interactions go into an odd number of pions, and *vice versa.*

Since Eq. (29.49) is invariant under particle conjugation, strangeness anticommutes with the operator Γ_U, and extension of the foregoing consideration to strange particles requires incorporation of the additional C.R.

(29.78) $$GS + SG = 0$$

leading to

(29.79) $$G^2 = (-1)^{B+S}$$

and the conclusion that only systems with *both* $B = 0$ and $S = 0$ can be states of specific G conjugality. Further details can be found in the comprehensive work of Goldhaber, Lee, and Yang, quoted in the references at the end of this section.

The treatment of the conjugality properties of particles possessing isospin, given above, has the defect of not taking into account the antiunitary nature, already recognized in Section 19, of the particle-conjugation operator Γ applied to fermion states governed dynamically by the Dirac equation. The consequences of this complication have never been fully analyzed, because the transformation properties of isospin **T** under particle conjugation are not known with certainty, and one can infer from the invariance of (29.49) only that T_3 must change sign under Γ. Suppose, for example, *all* isospin components anticommute with the correct antiunitary operator of particle conjugation, denoted Γ_A,

$$(29.80) \qquad \Gamma_A \hat{\mathbf{T}} + \hat{\mathbf{T}} \Gamma_A = 0.$$

By analogy with the corresponding property (15.14) of the antiunitary operator of time reversal Θ—with respect to ordinary spin **J**—Γ_A can be represented in isospace by the operator

$$(29.81) \qquad (\Gamma_A)_{IS} = [\exp(i\pi \hat{T}_2)] K.$$

Whether invariance under Γ_A engenders a superselection rule or not depends on whether $\Gamma_A^2 = -I$ or $\Gamma_A^2 = +I$. Now the operator Γ_A for baryons is the direct product

$$(29.82) \qquad \Gamma_A = \Gamma_D \times \Gamma_{SC} \times \Gamma_{IS}$$

of the representations in coordinate space, spin-chirality space, and isospace. Since in the combined coordinate–spin–chirality space, according to Eq. (19.69),

$$(29.83) \qquad (\Gamma_D \times \Gamma_{SC})^2 = +I$$

and since it follows from (29.81), as in the analogous case governing the representation of time reversal, that

$$(29.84) \qquad [(\Gamma_A)_{IS}]^2 = \begin{cases} -I & \text{if } T = \frac{1}{2}, \frac{3}{2}, \ldots \\ +I & \text{if } T = 0, 1, \ldots \end{cases}$$

one concludes

$$(29.85) \qquad (\Gamma_A)^2 = \begin{cases} -I & \text{for nucleons and } \Xi \text{ hyperons} \\ +I & \text{for } \Lambda \text{ and } \Sigma \text{ hyperons.} \end{cases}$$

This means invariance under particle conjugation engenders a superselection rule only for nucleons and Ξ hyperons. Moreover, since invariance under Γ does not hold in weak interactions, the attribute associated with this superselection rule will be conserved only in strong interactions. An attribute which fits this description is the hypercharge Y defined in (29.50), because Y is unequal zero only for nucleons and Ξ hyperons and is conserved in strong interactions only. It is thus not unreasonable to conjecture that conservation of hypercharge is, in fact, due to a superselection rule generated by invariance under particle conjugation.

The invariance under combined inversion $\Sigma_A = \Pi \Gamma_A$, on the other hand, since by Eq. (19.71) for fermions

$$(29.86) \qquad (\Sigma_D \times \Sigma_{SC})^2 = -I,$$

will yield a superselection rule only for Λ and Σ hyperons, because $\Sigma_A = \Sigma_D \times \Sigma_{SC} \times \Sigma_{IS}$ satisfies (assuming $\Pi^2 = +I$ in isospace)

$$(29.87) \qquad (\Sigma_A)^2 = \begin{cases} +I & \text{for} \quad T = \tfrac{1}{2}, \tfrac{3}{2}, \ldots \\ -I & \text{for} \quad T = 0, 1, \ldots \quad \text{i.e. for } \Lambda \text{ and } \Sigma \text{ hyperons.} \end{cases}$$

One is thus led to conclude that Λ and Σ hyperons possess an attribute not shared by the other baryons, which is conserved through a super-selection rule in all interactions invariant under combined inversion, thus including, presumably, the weak interactions. The nature of this attribute, if it exists, is obscure.*

In any case, *the conservation of baryon number* (an attribute shared by *all* baryons) valid in *all* interactions, *cannot be understood as a super-selection rule engendered by particle-conjugation symmetry.*

This leaves as the only symmetry which may be made responsible for the conservation of baryon number through a superselection rule the invariance under time reversal Θ as had been suggested at the end of Section 15.

Fortunately, the conclusions reached earlier regarding selection rules following from conservation of G conjugality in strongly interacting systems with $B = 0$ and $S = 0$ are not invalidated, if the two-dimensional baryon-number space is identified with the two-dimensional reversality space spanned by the baryon states $|B\rangle$ and their time reversed analog $\Theta|B\rangle$. In particular, the representation (29.57) for the operator of particle conjugation can be looked upon as the direct product

$$(29.88) \qquad \Gamma = \begin{pmatrix} 0 & -1 \\ 1 & 0 \end{pmatrix}_{IS} \times \begin{pmatrix} 0 & 1 \\ -1 & 0 \end{pmatrix}_B$$

made up out of the representation in isospace and in reversality space in analogy to the representation (15.44).

* There is one more symmetry operation (valid only for strong interactions), the "weak reflection," $\Pi\Theta$, which is antiunitary and whose square is $\gamma_4\gamma_1\gamma_3 K\gamma_4\gamma_1\gamma_3 K = -1$ for all fermions. The nature of the corresponding attribute, which ought to be conserved in strong interactions by a superselection rule, is equally obscure.

The conclusions reached in this work regarding possible superselection rules engendered by antiunitary symmetry operations and their possible interpretation are summarized in Table 29.6.

TABLE 29.6

LIST OF POSSIBLE ATTRIBUTES CONSERVED THROUGH A SUPERSELECTION RULE

Symmetry operation generating superselection rule	Applicable to:	Conserved attribute	Validity
Time reversal Θ	Leptons	Muon number L_μ	All interactions
	Baryons	Baryon number B	All interactions
Combined inversion $\Sigma = \Pi\Gamma$	Leptons	Lepton number L	All interactions
	Λ and Σ hyperons	?	All interactions
Particle conjugation Γ	Nucleons and Ξ hyperons	Hypercharge $Y = B + S$	Strong interactions
Weak reflection $\Pi\Theta$	Fermions	?	Strong interactions

NOTES

Brueckner [1] turned the tide in favor of isospin as a legitimate attribute.

Fermi's article [2] contains a complete treatment of the isospin formalism and its application to problems involving nucleons and pions, as well as references to earlier work on this subject. An analysis of the energy levels of hypernuclei in terms of isospin labels was given by Morrison [3].

For details on branching ratios and selection rules following from isospin conservation see the lecture notes by Sakurai [4]. The concept of strangeness as a quantum number is due to Gell-Mann [5]. See also Nishijima [6]. The significance of hypercharge as a quantum number was pointed out by d'Espagnat and Prentki [7]. The concept of G conjugality, although already used by Michel [8], was introduced generally by Lee and Yang [9]. A comprehensive treatment of selection rules following from particle conjugation symmetry for systems of vanishing baryon number B is contained in the paper by Goldhaber *et al.* [10].

Robertson [11] has analyzed some consequences of the assumption that the anti-unitary operator of particle conjugation anticommutes with the isospin operator, and that isospin is invariant under reversal of motion.

REFERENCES

[1] K. Brueckner, *Phys. Rev.* **86**, 106 (1952).
[2] E. Fermi, *Nuovo Cimento, Suppl.* **2**, 18 (1955).
[3] P. Morrison, Lecture Notes Canadian Association of Physicists Seminar. Edmonton, Alberta, 1957.
[4] J. Sakurai, *in* "Brandeis Lectures in Theoretical Physics, 1961," Vol. 1, p. 231. Benjamin, New York, 1962.
[5] M. Gell-Mann, *Phys. Rev.* **92**, 833 (1953).
[6] K. Nishijima, *Progr. Theoret. Phys. (Kyoto)* **12**, 107 (1954).
[7] B. d'Espagnat and J. Prentki, *Nucl. Phys.* **1**, 33 (1956).
[8] L. Michel, *Nuovo Cimento* **10**, 319 (1953).
[9] T. D. Lee and C. N. Yang, *Nuovo Cimento* **3**, 749 (1956).
[10] M. Goldhaber, T. D. Lee, and C. N. Yang, *Phys. Rev.* **111**, 1796 (1958).
[11] D. A. Robertson, M.Sc. Thesis (unpublished), Univ. British Columbia, Vancouver, 1963.

The Quasi Particle Concept

Some of the most startling manifestations of the quantum mechanical principles which apparently govern the actual physical world are *macroscopic* phenomena encountered in liquids and solids at low temperatures, known as "superfluidity" and "superconductivity." They can be understood by application of the principle of superposition of probability amplitudes in conjunction with the permutation symmetry characteristic of many-particle states. Although a theory of supertransfer phenomena could have been developed immediately after the inception of quantum mechanics, and although Einstein provided, through discovery of the "Einstein condensation" of bosons, an important clue to the understanding of superfluidity even before the meaning of the superposition principle had been fully appreciated, it took more than twenty years before the now generally accepted ideas needed to explain these phenomena assumed shape. This is rather surprising, after the event, and may perhaps partly stem from a tendency of physicists, raised on a diet of classical notions regarding the particle concept, to relegate applicability of quantum mechanics to a strictly microscopic domain.

A substantial intrusion of quantum mechanical concepts into the macroscopic domain should be expected, for objects in a state of low temperature containing many particles, as a result of quite elementary considerations. Take, as an example, a liquid made up out of n bosons enclosed in a volume V. To account for the low compressibility of fluids, in general, assume for simplicity's sake as the only interaction between any two bosons a "hard core" potential as indicated in Fig. 30.1. In the liquid state each boson will be hemmed in by its neighbors so that it occupies a volume of order $(V/n) \sim r_0^3$. Consequently, each boson will have in the state of lowest energy a rather high "zero-point energy" of order

$$(30.1) \qquad \epsilon_0 \sim (k_0^2/2m) \sim (1/2mr_0^2) \cong (1/2m)(n/V)^{2/3}$$

and the spacing between the ground state and the first excited state of a single boson will be of the same order of magnitude. Therefore, if the temperature T of the liquid is less than a finite critical temperature, identical with Einstein's "condensation temperature" (natural units with Boltzmann's constant $k = 1$ are used)

$$(30.2) \qquad\qquad T < T_0 \sim (1/2m)\,(n/V)^{2/3}$$

practically no single-particle excited states will exist in the liquid. By putting the known density and atomic mass for liquid helium into (30.2) one finds for T_0 a value of about $3°$K.

On first sight one might conclude from this that there is no mechanism by which the fluid can take up energy below T_0, resulting in a practically vanishing specific heat at temperatures below T_0. Upon second thoughts

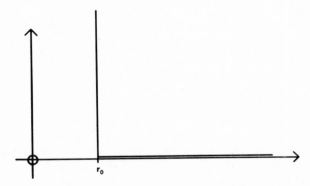

FIG. 30.1. The "hard core" potential.

it will be realized, however, that there is a mechanism for imparting energy to the fluid by excitation of *collective modes*, such as sound waves, which may be described quantum mechanically by *admitting single particle states which are linear superpositions of ground and excited states*, the amplitude of the ground state being dominant and the excited states represented by small admixtures only. The spacing of energy levels characteristic of these collective modes will then be determined by the size of the *entire* volume V, and one anticipates an almost continuous spectrum of "phonons" in analogy with the practically continuous spectrum of photons representing black body radiation enclosed in a volume V. By this analogy the specific heat of the liquid should be expected to depend on temperature T and the velocity of sound c_0 according to

$$(30.3) \qquad\qquad (dE/dT) \sim (T^3/c_0^3).$$

This result is borne out by observation in liquid helium below $0.6°$K, with an accuracy astonishing in view of the simplifying assumption made above about the intermolecular forces.

The critical temperature T_0 (30.2) has another significance which leads one to expect drastic consequences of the restrictions imposed by permutation symmetry on state vectors describing n indistinguishable particles at temperatures $T < T_0$. *The measurement of the number of particles* n, *in the classical sense of counting spatially separate objects, and the measurement of the temperature* T, *are incompatible at temperatures* $T < T_0$, for the following reason. If one wishes to "count," in the classical sense of spatial separation from its neighbors, a particle in a volume V occupied by n particles, one must confine it to a region of volume less than (V/n). This corresponds to a localization of coordinate

$$(30.4) \qquad \Delta q \lesssim (V/n)^{1/3},$$

and an uncertainty in the knowledge of the particle momentum

$$(30.5) \qquad \Delta p \gtrsim (1/\Delta q) \gtrsim (n/V)^{1/3}$$

is engendered by the uncertainty relations (10.5). Thus, there will be an uncertainty in the knowledge of the energy of each particle

$$(30.6) \qquad \Delta \epsilon \sim [(\Delta p)^2/2m] \gtrsim (1/2m)(n/V)^{2/3}$$

corresponding to an uncertainty in the knowledge of the temperature

$$(30.7) \qquad \Delta T \sim \Delta \epsilon \gtrsim T_0.$$

One may look upon this intrusion of specifically quantum mechanical features into the behavior of physical objects at temperatures $T < T_0$ from yet another point of view which might be instructive. Suppose one is in possession of that supremely intelligent agent known as "Maxwell's demon," and instructs it to separate, by judicious opening and closing of a door of diameter d between two compartments of V, the fast and the slow particles in a gas consisting of n particles (see Fig. 30.2). The

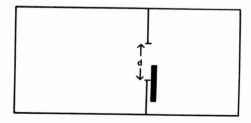

Fig. 30.2. The door operated by Maxwell's demon.

particles are assumed to have the usual velocity distribution of average velocity $\bar{v}$ and width $\sim (T/m)^{1/2}$. If the little fellow is instructed to separate particles with $v > \bar{v}$ from particles with $v < \bar{v}$, then his uncertainty in the knowledge of the velocity of each particle must be less than the natural width of the velocity distribution,

$$(30.8) \qquad (T/m)^{1/2} \gtrsim \Delta v.$$

The uncertainty relations impose a lower limit on Δv, however. During time τ the door of area d^2 will be bombarded by approximately $(n/V)d^2\bar{v}\tau$ particles. If the demon wants to let a particle pass, he must not let the door be open longer than the average time $\Delta\tau$ between successive arrivals of particles,

$$(30.9) \qquad \Delta\tau \sim (V/nd^2\bar{v})$$

engendering thus an uncertainty in the knowledge of the kinetic energy of the particle of order

$$(30.10) \qquad \Delta\epsilon \gtrsim (1/\Delta\tau) \sim (nd^2\bar{v}/V)$$

corresponding to an uncertainty in the knowledge of the particle velocity

$$(30.11) \qquad \Delta v \sim (\Delta\epsilon/m\bar{v}) \gtrsim (nd^2/mV).$$

The door itself produces an uncertainty because of the localization of the particle to diameter d during passage,

$$(30.12) \qquad (\Delta v)_d \gtrsim (1/md).$$

In order that $(\Delta v)_d$ will not be larger than (30.11), the opening d should have at least the size

$$(30.13) \qquad d > (1/m\Delta v).$$

By substitution into (30.11) one obtains the inequality

$$(30.14) \qquad \Delta v \gtrsim (1/m)(n/V)^{1/3}$$

and therefore from (30.8)

$$(30.15) \qquad T \gtrsim (1/m)(n/V)^{2/3} \sim T_0.$$

This means: Maxwell's demon can begin to do his job only if the temperature of the gas is at least equal to the characteristic temperature T_0. Below T_0 his intelligence will be completely frustrated by the uncertainty relations.

The foregoing considerations will not be qualitatively affected if one replaces the crude hard core potential by a more realistic interaction including an attractive potential well, as indicated in Fig. 30.3. In one

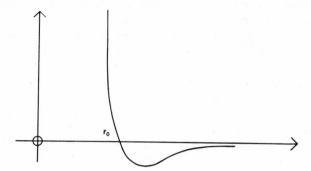

FIG. 30.3. Realistic interaction potential between two bosons.

respect, however, an attraction in addition to the repulsive core between two bosons should lead to an important new feature. The state representing a physical situation in which all single particles are in their ground state need no longer be the state of lowest energy for the entire system. Presence of a few phonons can minimize the average potential energy between particles such that the increase in average kinetic energy

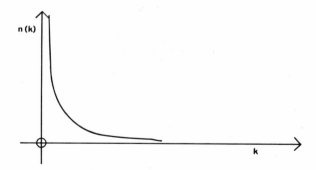

FIG. 30.4. Distribution of bosons in momentum space at absolute zero of temperature in the presence of an interaction of the type of that shown in Fig. 30.3.

occasioned by these collective modes is more than compensated. Thus even at the absolute zero of temperature the "Einstein condensation" into the ground state, represented by a δ function in momentum space in case of the ideal gas, will be modified by the actual interaction into a distribution spread out somewhat as indicated in Fig. 30.4. The expectation that in the absence of attractive interaction the ground state of the entire system would be the state in which all particles are "condensed" into the state $\mathbf{k} = 0$ (apart from a zero-point energy which will from now on be taken as the origin of the energy scale), whereas "switching on" the weak attractive interaction results in a ground state for the entire system in which the single particle states acquire admixtures of excited states with $\mathbf{k} \neq 0$, is borne out by the following more rigorous argument due to Bogoliubov.

A system of weakly interacting bosons, subject to the usual conservation laws and the condition that the total number of bosons is a given constant and that all interactions can be accounted for entirely by specific two-body central forces, can be described by the Hamiltonian

$$(30.16) \quad H = \sum_{\mathbf{k}} \omega(k) \, b^+(\mathbf{k}) \, b(\mathbf{k})$$
$$+ \sum_{\mathbf{k}'''} \sum_{\mathbf{k}''} \sum_{\mathbf{k}'} \sum_{\mathbf{k}} F(\mathbf{k}''' - \mathbf{k}') \, \delta_{\mathbf{k}'''+\mathbf{k}'', \, \mathbf{k}'+\mathbf{k}} \, b^+(\mathbf{k}''') \, b^+(\mathbf{k}'') \, b(\mathbf{k}') \, b(\mathbf{k}).$$

The bosons are assumed to have no intrinsic properties and to be characterized completely by the momentum quantum numbers $\mathbf{k}$.

It is perhaps worth elaborating a bit, parenthetically, on how the various conditions imposed phenomenologically on the actual system enforce the specific form (30.16). Considering first a set of bosons without interactions, one could begin in a more systematic vein by considering the most general bilinear expression

$$(30.17) \qquad\qquad H_0 = \sum_{\mathbf{k}'} \sum_{\mathbf{k}} w(\mathbf{k}', \mathbf{k}) \, b^+(\mathbf{k}') \, b(\mathbf{k})$$

which will guarantee conservation of particle number because it contains for each annihilation operator exactly one creation operator, and has no matrix elements between states of a different total number of particles. In the language of Feynman graphs, each term of H_0 can be represented as indicated in Fig. 30.5. Imposing now one by one the

FIG. 30.5. Graph for (30.17).

various invariance requirements on H_0, one arrives at successive restrictions imposed on the function $w(\mathbf{k}', \mathbf{k})$, namely

(i) Invariance under reversal of motion requires $w(\mathbf{k}', \mathbf{k}) = w^*(-\mathbf{k}', -\mathbf{k})$
(ii) Invariance under inversion of coordinates requires $w(\mathbf{k}', \mathbf{k}) = w(-\mathbf{k}', -\mathbf{k})$
(iii) Invariance under displacement in time requires w does not contain t explicitly
(iv) Invariance under displacement in space requires $w(\mathbf{k}', \mathbf{k}) = \omega(\mathbf{k}', \mathbf{k}) \, \delta_{\mathbf{k}', \mathbf{k}}$
(v) Invariance under rotations in space requires $\omega(\mathbf{k}', \mathbf{k}) = \omega[(\mathbf{k}' \cdot \mathbf{k})]$

and one obtains the usual form for the Hamiltonian of a system of noninteracting bosons

(30.18) $$H_0 = \sum_{\mathbf{k'}} \sum_{\mathbf{k}} \omega[(\mathbf{k'} \cdot \mathbf{k})] \delta_{\mathbf{k'},\mathbf{k}} b^+(\mathbf{k'}) b(\mathbf{k}) = \sum_{\mathbf{k}} \omega(k) b^+(\mathbf{k}) b(\mathbf{k}); \qquad \omega \text{ real.}$$

The analogous expression for noninteracting fermions is similarly obtained.

If all Hamiltonians which are *not* of the form (30.18) are called "interaction Hamiltonians," then all possible interaction Hamiltonians can be divided into

(A) particle conserving Hamiltonians, containing in each term as many creation operators as there are annihilation operators, and

(B) particle nonconserving Hamiltonians which do not satisfy the condition stated under (A).

A simple interaction Hamiltonian of type (B), conserving the number of fermions but not the number of bosons and reminiscent of the expression (23.6) employed in quantum electrodynamics, is

(30.19) $$H'_{\text{int}} = \sum_{\mathbf{k''}} \sum_{\mathbf{k'}} \sum_{\mathbf{k}} G(\mathbf{k''}, \mathbf{k'}, \mathbf{k})[a^+(\mathbf{k''}) a(\mathbf{k'}) b^+(\mathbf{k}) + \text{hermitean conjugate}]$$

FIG. 30.6 (left and FIG. 30.7 right). Graphs representing (30.19).

corresponding to the graph given in Fig. 30.6 and the reversed graph given in Fig. 30.7. Invariance requirements again impose restrictions on G, demanding that it be of the form

(30.20) $$G(\mathbf{k''}, \mathbf{k'}, \mathbf{k}) = g[(\mathbf{k'} - \mathbf{k})^2; \mathbf{k}^2] \delta_{\mathbf{k''}+\mathbf{k},\mathbf{k'}}.$$

One of the basic ideas employed in the theory of interactions is to replace all direct particle interactions by intermediate boson interactions, so that all graphs are viewed as made up out of graphs containing only three-particle vertices such as the ones rendered graphically in Figs. 30.6 and 30.7. The perplexities caused by the infinite hierarchy of basic interaction vertices have been exhibited in Section 25. For the purpose of the present section, phenomena will be accounted for by introduction of a boson conserving interaction Hamiltonian

(30.21) $$H_{\text{int}} = \sum_{\mathbf{k'''}} \sum_{\mathbf{k''}} \sum_{\mathbf{k'}} \sum_{\mathbf{k}} W(\mathbf{k'''}, \mathbf{k''}, \mathbf{k'}, \mathbf{k}) b^+(\mathbf{k'''}) b^+(\mathbf{k''}) b(\mathbf{k'}) b(\mathbf{k})$$

corresponding to the graph given in Fig. 30.8. No attempt will be made to justify this expression by deriving it through perturbation theory or some other formal device from

FIG. 30.8. Graph representing (30.21).

some other supposedly more basic interaction such as (30.19). In fact, the conditions under which an expression such as (30.21) will follow from a basic interaction such as (30.19) are, in general, unknown, and it is even an open question whether such conditions exist at all.

A possible dissection of the graph in Fig. 30.8 in terms of the graphs in Figs. 30.6 and 30.7 is given in Fig. 30.9. It is transparent that such a decomposition of, say, the empirical

FIG. 30.9. Possible dissection of the graph Fig. 30.8 in terms of the graphs rendered in Fig. 30.6 and 30.7.

interaction between two helium atoms is quite unrealistic, because if this interaction is essentially electromagnetic in origin, one has to solve, in principle, a six-body problem on account of the two nuclei and the four electrons involved in this interatomic interaction.

Casting aside then all doubts regarding a possible deeper origin of the interaction (30.21), one concludes from the invariance requirements imposed by symmetry under translations and rotations that W be of the form

$$(30.22) \qquad W(\mathbf{k}''', \mathbf{k}'', \mathbf{k}', \mathbf{k}) = F[(\mathbf{k} \cdot \mathbf{k}'); (\mathbf{k} \cdot \mathbf{k}''); (\mathbf{k}' \cdot \mathbf{k}'')] \, \delta_{\mathbf{k}'''+\mathbf{k}'', \mathbf{k}'+\mathbf{k}}.$$

Instead of the three scalars $(\mathbf{k} \cdot \mathbf{k}')$, $(\mathbf{k} \cdot \mathbf{k}'')$, $(\mathbf{k}' \cdot \mathbf{k}'')$, one can use equivalently as scalar parameters characterizing the function F the barycentric energy $(\mathbf{k} + \mathbf{k}')^2$ and the two momentum transfers $(\mathbf{k} - \mathbf{k}'')^2$ and $(\mathbf{k}' - \mathbf{k}'')^2$. If one demands further that F represent a central force between any two bosons, expression (30.16) for the total Hamiltonian results.

Suppose now, in accordance with the qualitative considerations stated earlier, there is an *average* number $\bar{n}_0$ of particles in their ground states $\mathbf{k} = 0$, so that the actual number of particles belonging to $\mathbf{k} = 0$ is $n_0 = \bar{n}_0 + n'$, where n' may have positive or negative values so that $\bar{n}' = 0$ but in any case $|n'| \ll \bar{n}_0$. The number of excited particles with $\mathbf{k} \neq 0$ is then $n_1 = n - \bar{n}_0 - n'$. One expects the lowest states $|\tau(n)\rangle$ of the system of n interacting bosons to be a linear superposition of states $|n_0, 0_1\rangle$

$$(30.23) \quad |\tau(n)\rangle = \sum_{n'} c_\tau(n_1) |n_0, 0_1\rangle = \sum_{n'} c_\tau(n - \bar{n}_0 - n') |\bar{n}_0 + n', 0_1\rangle$$

where $c_\tau(n_1)$ is a properly normalized amplitude involving operators $b^+(\mathbf{k})$ which create n_1 particles from the state $|\ldots, 0_1\rangle$ and where

$$(30.24) \qquad\qquad b^+(0) \, b(0) |n_0, 0_1\rangle = n_0 |n_0, 0_1\rangle.$$

The sum $\sum_{n'}$ in (30.23) goes over positive and negative values of n'. The label τ corresponds to some order of the lowest states, $\tau = 0$ denoting the ground state (say) of the entire system, $\tau = 1$ the first excited state, etc. One has further, from the representations for the boson creation and annihilation operators (18.4),

$$(30.25) \qquad b(0)|\tau(n)\rangle = \sum_{n'} \sqrt{n_0}\, c_\tau(n_1)|n_0 - 1, 0_1\rangle$$

$$(30.26) \qquad b^+(0)|\tau(n)\rangle = \sum_{n'} \sqrt{n_0 + 1}\, c_\tau(n_1)|n_0 + 1, 0_1\rangle$$

$$(30.27) \qquad b^+(0)\, b(0)|\tau(n)\rangle = \sum_{n'} n_0\, c_\tau(n_1)|n_0, 0_1\rangle$$

$$= \sum_{n'} (\bar{n}_0 + n')\, c_\tau(n_1)|\bar{n}_0 + n', 0_1\rangle.$$

Thus far no approximations have been made. Since one will have $|n'| \ll \bar{n}_0$ for all appreciable amplitudes $c_\tau(n_1) = c_\tau(n - \bar{n}_0 - n')$ one may write instead of (30.27) the *approximation*

$$(30.27') \qquad b^+(0)\, b(0)|\tau(n)\rangle = \bar{n}_0|\tau(n)\rangle.$$

By the same reasoning, and assuming $c_\tau(n_1)$ does not depend sensitively on the precise value of the total number of particles n, one may write further the approximations

$$(30.25')$$

$$b(0)|\tau(n)\rangle = \sqrt{\bar{n}_0} \sum_{n'} c_\tau(n - \bar{n}_0 - n')|\bar{n}_0 + n' - 1, 0_1\rangle$$

$$= \sqrt{\bar{n}_0} \sum_{n'} c_\tau[(n-1) - (\bar{n}_0 + n' - 1)]\,|\bar{n}_0 + n' - 1, 0_1\rangle$$

$$= \sqrt{\bar{n}_0}|\tau(n-1)\rangle$$

and

$$(30.26') \qquad b^+(0)|\tau(n)\rangle = \sqrt{\bar{n}_0}|\tau(n+1)\rangle.$$

Since one wants to apply the Hamiltonian (30.16) to such lowest states (or superpositions of them if n is not fixed), it is useful to separate in H_{int} the terms with $\mathbf{k} = 0$ from those with $\mathbf{k} \neq 0$ so that the approximations (30.25′), (30.26′), and (30.27′) can be applied. One finds, first of all, the exact expression ($\sum_{\mathbf{k}}'$ means sum over all $\mathbf{k} \neq 0$)

(30.28)

$$H_{\text{int}} = b^+(0)\,b^+(0)\,b(0)\,b(0)\,F(0)$$

$$+ 2b^+(0)\,b(0)\left[F(0)\,{\sum_{\mathbf{k}}}'\,b^+(\mathbf{k})\,b(\mathbf{k}) + {\sum_{\mathbf{k}}}'\,F(\mathbf{k})\,b^+(\mathbf{k})\,b(\mathbf{k})\right]$$

$$+ b(0)\,b(0)\,{\sum_{\mathbf{k}'}}'\,F(\mathbf{k})\,b^+(\mathbf{k})\,b^+(-\mathbf{k}) + b^+(0)\,b^+(0)\,{\sum_{\mathbf{k}}}'\,F(\mathbf{k})\,b(\mathbf{k})\,b(-\mathbf{k})$$

$$+ b(0)\,{\sum_{\substack{\mathbf{k}' \quad \mathbf{k} \\ \mathbf{k}' \neq \mathbf{k}}}}'\,[F(\mathbf{k}) + F(\mathbf{k}' - \mathbf{k})]\,b^+(\mathbf{k})\,b^+(\mathbf{k}' - \mathbf{k})\,b(\mathbf{k}')$$

$$+ b^+(0)\,{\sum_{\substack{\mathbf{k}' \quad \mathbf{k} \\ \mathbf{k}' \neq \mathbf{k}}}}'\,[F(\mathbf{k}) + F(\mathbf{k}')]\,b^+(\mathbf{k}' + \mathbf{k})\,b(\mathbf{k}')\,b(\mathbf{k})$$

$$+ {\sum_{\mathbf{k}'''}}'\,{\sum_{\substack{\mathbf{k}'' \\ \mathbf{k}'' \neq \mathbf{k}' + \mathbf{k}}}}'\,{\sum_{\mathbf{k}'}}'\,{\sum_{\mathbf{k}}}'\,F(\mathbf{k}''' - \mathbf{k})\,\delta_{\mathbf{k}'''+\mathbf{k}'',\,\mathbf{k}'+\mathbf{k}}\,b^+(\mathbf{k}''')\,b^+(\mathbf{k}'')\,b(\mathbf{k}')\,b(\mathbf{k}).$$

Since this interaction Hamiltonian will be applied to low-lying states $|\tau(n)\rangle$ only, the entire Hamiltonian will now be approximated in accordance with Eqs. (30.25′), (30.26′), and (30.27′) by putting

$$b^+(0)\,b(0) = b^+(0)\,b^+(0) = b(0)\,b(0) = \bar{n}_0$$

and neglecting terms of lower order in $b(0)$ and $b^+(0)$, terms linear in $b(0)$ and $b^+(0)$ being of order $\sqrt{\bar{n}_0}$, and terms containing no operators $b(0)$ and $b^+(0)$ being of order 1 compared to the terms with $\bar{n}_0$ in them. One obtains, thus,

(30.29)

$$\tilde{H} = \bar{n}_0\,\omega(0) + {\sum_{\mathbf{k}}}'\,\omega(k)\,b^+(\mathbf{k})\,b(\mathbf{k}) + \bar{n}_0^2 F(0) + 2\bar{n}_0 F(0)\,{\sum_{\mathbf{k}}}'\,b^+(\mathbf{k})\,b(\mathbf{k})$$

$$+ 2\bar{n}_0\,{\sum_{\mathbf{k}}}'\,F(\mathbf{k})\,b^+(\mathbf{k})\,b(\mathbf{k}) + \bar{n}_0\,{\sum_{\mathbf{k}}}'\,F(\mathbf{k})\,[b^+(\mathbf{k})\,b^+(-\mathbf{k}) + b(\mathbf{k})\,b(-\mathbf{k})]$$

This Hamiltonian will be applicable even if the total number n of particles is not fixed, so that any low-lying state $|\tau\rangle$ can be written as a linear superposition of states with various n, $|\tau\rangle = \sum_n d(n)|\tau(n)\rangle$. However, if only state vectors describing a definite number n of particles are considered, one may replace $\bar{n}_0$ by using

(30.30) $\qquad n = b^+(0)\,b(0) + {\sum_{\mathbf{k}}}'\,b^+(\mathbf{k})\,b(\mathbf{k}) = \bar{n}_0 + {\sum_{\mathbf{k}}}'\,b^+(\mathbf{k})\,b(\mathbf{k})$

and obtain, in the same approximation,

(30.31) $\quad \tilde{H} = n\omega(0) + n^2 F(0) + {\sum_{\mathbf{k}}}'\,[\omega(k) - \omega(0) + 2nF(\mathbf{k})]\,b^+(\mathbf{k})\,b(\mathbf{k})$

$$+ n\,{\sum_{\mathbf{k}}}'\,F(\mathbf{k})\,[b^+(\mathbf{k})\,b^+(-\mathbf{k}) + b(\mathbf{k})\,b(-\mathbf{k})].$$

This Hamiltonian does not conserve the number of *excited* particles, which is not surprising because the entire approximation is just aimed at considering the single particle state with $\mathbf{k} = 0$ as an effectively unlimited reservoir from which particles may be lifted by the interaction into excited states with $\mathbf{k} \neq 0$.

To diagonalize this Hamiltonian, Bogoliubov introduced new operators

(30.32)
$$\tilde{b}(\mathbf{k}) = u(\mathbf{k})\, b(\mathbf{k}) - v(\mathbf{k})\, b^+(-\mathbf{k})$$
$$\tilde{b}^+(\mathbf{k}) = u(\mathbf{k})\, b^+(\mathbf{k}) - v(\mathbf{k})\, b(-\mathbf{k})$$

where $u(\mathbf{k})$ and $v(\mathbf{k})$ are *real* numbers satisfying the relations

(30.33) $\quad u(-\mathbf{k}) = u(\mathbf{k}); \qquad v(-\mathbf{k}) = v(\mathbf{k}); \qquad u^2(\mathbf{k}) - v^2(\mathbf{k}) = 1.$

The minus sign in front of v in (30.32) is purely conventional, v need not be positive. With these conditions the transformation (30.32) is canonical i.e. $\tilde{b}, \tilde{b}^+$ satisfy the C.R.s as do the b, b^+. Indeed,

(30.34)
$$[\tilde{b}(\mathbf{k})\, \tilde{b}^+(\mathbf{k}')] = u(\mathbf{k})\, u(\mathbf{k}')\, [b(\mathbf{k})\, b^+(\mathbf{k}')] + v(\mathbf{k})\, v(\mathbf{k}')\, [b^+(-\mathbf{k})\, b(-\mathbf{k}')]$$
$$- u(\mathbf{k})\, v(\mathbf{k}')\, [b(\mathbf{k})\, b(-\mathbf{k}')] - u(\mathbf{k}')\, v(\mathbf{k})\, [b^+(-\mathbf{k})\, b^+(\mathbf{k}')]$$
$$= \{u(\mathbf{k})\, u(\mathbf{k}') - v(\mathbf{k})\, v(\mathbf{k}')\}\delta_{\mathbf{k}, \mathbf{k}'} = \begin{cases} u^2(\mathbf{k}) - v^2(\mathbf{k}) & \text{for } \mathbf{k} = \mathbf{k}' \\ 0 & \text{for } \mathbf{k} \neq \mathbf{k}' \end{cases}$$

$$[\tilde{b}(k)\, \tilde{b}(k')] = -u(\mathbf{k})\, v(\mathbf{k}')\, [b(\mathbf{k})\, b^+(-\mathbf{k}')] - u(\mathbf{k}')\, v(\mathbf{k})\, [b^+(-\mathbf{k})\, b(\mathbf{k}')]$$
$$= \{-u(\mathbf{k})\, v(\mathbf{k}') + u(\mathbf{k}')\, v(\mathbf{k})\}\, \delta_{\mathbf{k}, -\mathbf{k}'}$$
$$= \begin{cases} -u(\mathbf{k})\, v(-\mathbf{k}) + u(-\mathbf{k})\, v(\mathbf{k}) & \text{for } \mathbf{k} = -\mathbf{k}' \\ 0 & \text{for } \mathbf{k} \neq -\mathbf{k}' \end{cases}$$

$$[\tilde{b}^+(k)\, \tilde{b}^+(k')] = \{-u(\mathbf{k}')\, v(\mathbf{k}) + u(\mathbf{k})\, v(\mathbf{k}')\}\, \delta_{\mathbf{k}, -\mathbf{k}'}$$
$$= \begin{cases} -u(-\mathbf{k})\, v(\mathbf{k}) + u(\mathbf{k})\, v(-\mathbf{k}) & \text{for } \mathbf{k} = -\mathbf{k}' \\ 0 & \text{for } \mathbf{k} \neq \mathbf{k}' \end{cases}$$

which gives the usual C.R.s for boson operators if (30.33) is satisfied.

The operators $\tilde{b}(\mathbf{k})$ and $\tilde{b}^+(\mathbf{k})$ will now be referred to as the annihilation and creation operators of quasi particles of momentum $\mathbf{k}$. The Hamiltonian H can be expressed entirely in terms of quasi-particle operators, because (30.32) can be solved for the b, b^+ to give

(30.35)
$$b(\mathbf{k}) = u(\mathbf{k})\, \tilde{b}(\mathbf{k}) + v(\mathbf{k})\, \tilde{b}^+(-\mathbf{k})$$
$$b^+(\mathbf{k}) = u(\mathbf{k})\, \tilde{b}^+(\mathbf{k}) + v(\mathbf{k})\, \tilde{b}(-\mathbf{k})$$

so that

$$(30.36) \quad b^+(\mathbf{k})\, b(\mathbf{k}) = v^2(\mathbf{k}) + u^2(\mathbf{k})\, \tilde{b}^+(\mathbf{k})\, \tilde{b}(\mathbf{k}) + v^2(\mathbf{k})\, \tilde{b}^+(-\mathbf{k})\, \tilde{b}(-\mathbf{k})$$

$$+ u(\mathbf{k})\, v(\mathbf{k})\, [\tilde{b}^+(\mathbf{k})\, \tilde{b}^+(-\mathbf{k}) + \tilde{b}(\mathbf{k})\, \tilde{b}(-\mathbf{k})]$$

$$b^+(\mathbf{k})\, b^+(-\mathbf{k}) = u(\mathbf{k})\, v(\mathbf{k}) + u(\mathbf{k})\, v(\mathbf{k})\, [\tilde{b}^+(\mathbf{k})\, \tilde{b}(\mathbf{k}) + \tilde{b}^+(-\mathbf{k})\, \tilde{b}(-\mathbf{k})]$$

$$+ u^2(\mathbf{k})\, \tilde{b}^+(\mathbf{k})\, \tilde{b}^+(-\mathbf{k}) + v^2(\mathbf{k})\, \tilde{b}(\mathbf{k})\, \tilde{b}(-\mathbf{k})$$

$$b(\mathbf{k})\, b(-\mathbf{k}) = u(\mathbf{k})\, v(\mathbf{k}) + u(\mathbf{k})\, v(\mathbf{k})\, [\tilde{b}^+(\mathbf{k})\, \tilde{b}(\mathbf{k}) + \tilde{b}^+(-\mathbf{k})\, \tilde{b}(-\mathbf{k})]$$

$$+ u^2(\mathbf{k})\, \tilde{b}(\mathbf{k})\, \tilde{b}(-\mathbf{k}) + v^2(\mathbf{k})\, \tilde{b}^+(\mathbf{k})\, \tilde{b}^+(-\mathbf{k}).$$

One obtains then, using the symmetry properties $\omega(-\mathbf{k}) = \omega(\mathbf{k})$ and $F(-\mathbf{k}) = F(\mathbf{k})$

$$(30.37) \qquad\qquad \tilde{H} = \tilde{H}_0 + \tilde{H}_1 + \tilde{H}_2$$

with

$$(30.37a) \qquad \tilde{H}_0 = n\omega(0) + n^2 F(0) + \sideset{}{'}\sum_{\mathbf{k}} \{[\omega(\mathbf{k}) - \omega(0)$$

$$+ 2nF(\mathbf{k})]\, v^2(\mathbf{k}) + 2nF(\mathbf{k})\, u(\mathbf{k})\, v(\mathbf{k})\}$$

$$(30.37b) \qquad \tilde{H}_1 = \sideset{}{'}\sum_{\mathbf{k}} \{[\omega(\mathbf{k}) - \omega(0) + 2nF(\mathbf{k})]\, [u^2(\mathbf{k}) + v^2(\mathbf{k})]$$

$$+ 4nF(\mathbf{k})\, u(\mathbf{k})\, v(\mathbf{k})\}\, \tilde{b}^+(\mathbf{k})\, \tilde{b}(\mathbf{k})$$

$$(30.37c) \qquad \tilde{H}_2 = \sideset{}{'}\sum_{\mathbf{k}} \{[\omega(\mathbf{k}) - \omega(0) + 2nF(\mathbf{k})]\, u(\mathbf{k})\, v(\mathbf{k}) + nF(\mathbf{k})\, [u^2(\mathbf{k})$$

$$+ v^2(\mathbf{k})]\}\, [\tilde{b}^+(\mathbf{k})\, \tilde{b}^+(-\mathbf{k}) + \tilde{b}(\mathbf{k})\, \tilde{b}(-\mathbf{k})].$$

This Hamiltonian is obviously diagonal provided $u(\mathbf{k})$ and $v(\mathbf{k})$ are arranged such that $\tilde{H}_2 = 0$. If this is done, then

$$(30.38) \qquad\qquad \tilde{H} = \tilde{H}_0 + \sideset{}{'}\sum_{\mathbf{k}} \omega(\mathbf{k})\, \tilde{b}^+(\mathbf{k})\, \tilde{b}(\mathbf{k})$$

where now

$$(30.39) \quad \tilde{\omega}(k) = [\omega(\mathbf{k}) - \omega(0) + 2nF(\mathbf{k})]\, [u^2(\mathbf{k}) + v^2(\mathbf{k})] + 4nF(\mathbf{k})\, u(\mathbf{k})\, v(\mathbf{k})$$

can be looked upon as the energy of a free quasi particle of momentum $\mathbf{k}$.

The condition for the vanishing of H_2 is

$$(30.40) \quad [\omega(\mathbf{k}) - \omega(0) + 2nF(\mathbf{k})]\, u(\mathbf{k})\, v(\mathbf{k}) + nF(\mathbf{k})\, [u^2(\mathbf{k}) + v^2(\mathbf{k})] = 0$$

which has to be solved in conjunction with (30.33). The solution can be obtained as follows. To satisfy (30.33) identically let

$$(30.41) \qquad\qquad u(\mathbf{k}) = \cosh(x); \qquad v(\mathbf{k}) = \sinh(x)$$

so that [using $\sinh(x)\cosh(x) = \frac{1}{2}\sinh(2x)$; $\sinh^2(x) + \cosh^2(x) = \cosh(2x)$]

$$(30.42) \quad u(\mathbf{k})\,v(\mathbf{k}) = \tfrac{1}{2}\sinh(2x); \qquad u^2(\mathbf{k}) + v^2(\mathbf{k}) = \cosh(2x).$$

Equation (30.40) then reads

$$(30.43) \qquad \tanh(2x) = -\frac{2nF(\mathbf{k})}{[\omega(\mathbf{k}) - \omega(0) + 2nF(\mathbf{k})]}$$

from which follows

{using $\sinh(x) = \tanh(x)\,[1 - \tanh^2(x)]^{-1/2}$; $\cosh(x) = [1 - \tanh^2(x)]^{-1/2}$}

$$(30.44) \qquad u(\mathbf{k})\,v(\mathbf{k}) = -\frac{nF(\mathbf{k})}{\sqrt{[\omega(\mathbf{k}) - \omega(0) + 2nF(\mathbf{k})]^2 - 4n^2F^2(\mathbf{k})}}$$

$$(30.45) \qquad u^2(\mathbf{k}) + v^2(\mathbf{k}) = \frac{\omega(\mathbf{k}) - \omega(0) + 2nF(\mathbf{k})}{\sqrt{[\omega(\mathbf{k}) - \omega(0) + 2nF(\mathbf{k})]^2 - 4n^2F^2(\mathbf{k})}}$$

By substitution into (30.39), this leads immediately to

$$(30.46) \quad \tilde{\omega}(k) = \sqrt{[\omega(\mathbf{k}) - \omega(0) + 2nF(\mathbf{k})]^2 - 4n^2F^2(\mathbf{k})}$$

$$= \sqrt{[\omega(\mathbf{k}) - \omega(0)]^2 + 4nF(\mathbf{k})\,[\omega(\mathbf{k}) - \omega(0)]}$$

Depending on the details of the interaction function $F(\mathbf{k})$, the quasi-particle energy $\tilde{\omega}$ can thus depend, in principle, on the momentum in the manner anticipated in Fig. 11.1, leading to the appearance of superfluid behavior as had been explained in Section 11.

The state of lowest energy of the system can now be described, in accordance with (30.38), as the "quasi-particle vacuum state" $|\tau = 0\rangle$, having the energy

$$(30.47) \quad \tilde{H}_0 = n\omega(0) + n^2F(0) + \tfrac{1}{2}\sum_{\mathbf{k}}{}' \,[\tilde{\omega}(\mathbf{k}) - \omega(\mathbf{k}) + \omega(0) - 2nF(\mathbf{k})]$$

which is obtained by substitution of

$$(30.48) \quad v^2(\mathbf{k}) = \tfrac{1}{2}[\cosh(2x) - 1] = \frac{\omega(\mathbf{k}) - \omega(0) + 2nF(\mathbf{k}) - \tilde{\omega}(\mathbf{k})}{2\tilde{\omega}(\mathbf{k})}$$

$$= \frac{2n^2F^2(\mathbf{k})}{\tilde{\omega}(k)\,[\omega(k) - \omega(0) + 2nF(\mathbf{k}) + \tilde{\omega}(\mathbf{k})]}$$

This result confirms the expectation that the quasi-particle vacuum is not identical with the state in which all ordinary particles occupy their ground state. In fact, $v^2(\mathbf{k})$ is identical with the average number of ordinary particles in the momentum state $\mathbf{k}$, because, from the first Eq. (30.36), it follows that

$$(30.49) \qquad \langle \tau = 0 | b^+(\mathbf{k})\,b(\mathbf{k}) | \tau = 0 \rangle = \overline{n(\mathbf{k})} = v^2(\mathbf{k})$$

which will, in general, be of a form anticipated in Fig. 30.4.

An analogous treatment is possible for a system consisting of N fermions, with similar results, the quasi-particle vacuum being in this case a state which, depending on the kind of interaction present between individual particles, may involve a spreading out of particle-occupation states near the Fermi surface. Consider, first of all, a system of non-interacting fermions, described by the Hamiltonian

$$(30.50) \qquad\qquad H_0 = \sum_{\mathbf{k}} \Omega(\mathbf{k}) \, a^+(\mathbf{k}) \, a(\mathbf{k}),$$

which conserves the particle number $N = \langle|\hat{N}|\rangle = \langle|\sum_{\mathbf{k}} a^+(\mathbf{k}) a(\mathbf{k})|\rangle$. In the state of lowest energy, all levels up to a certain $\mathbf{k}_F$ (the "Fermi surface") will be filled, and all levels above it will be empty. If one is interested only in small deviations from this ground state, one is led to consider instead a description of the system in terms of the Hamiltonian

$$(30.51) \qquad \tilde{H}_0 = H_0 - \Omega_F N = \sum_{\mathbf{k}} [\Omega(k) - \Omega_F] a^+(\mathbf{k}) \, a(\mathbf{k})$$

where $\Omega_F \equiv \Omega(\mathbf{k}_F)$ is the Fermi energy associated with $\mathbf{k}_F$, to be determined from the requirement

$$(30.52) \qquad\qquad\qquad \langle|\hat{N}|\rangle = N.$$

To find the eigenvalues of $\tilde{H}_0$ introduce new operators

$$(30.53) \qquad \tilde{a}(\mathbf{k}) = \begin{cases} a(\mathbf{k}) & \text{for} \quad \Omega(k) > \Omega_F \\ a^+(-\mathbf{k}) & \text{for} \quad \Omega(k) < \Omega_F \end{cases}$$

($\mathbf{k}$ may stand here for *both* momentum and spin labels of fermions), which satisfy, as had been shown in Section 17, the same anti-C.R.s as the $a(\mathbf{k})$ themselves. One has then

(30.54)

$$a^+(\mathbf{k}) \, a(\mathbf{k}) = \begin{cases} \tilde{a}^+(\mathbf{k}) \, \tilde{a}(\mathbf{k}) & \text{for} \quad \Omega(k) > \Omega_F \\ \tilde{a}(-\mathbf{k}) \, \tilde{a}^+(-\mathbf{k}) = 1 - \tilde{a}^+(-\mathbf{k}) \, \tilde{a}(-\mathbf{k}) & \text{for} \quad \Omega(k) < \Omega_F \end{cases}$$

so that one may write, using $\Omega(-\mathbf{k}) = \Omega(\mathbf{k})$,

$$(30.55) \qquad \tilde{H}_0 = \sum_{\substack{\mathbf{k} \\ \Omega(k) < \Omega_F}} [\Omega(k) - \Omega_F][1 - \tilde{a}^+(-\mathbf{k}) \, \tilde{a}(-\mathbf{k})]$$

$$+ \sum_{\substack{\mathbf{k} \\ \Omega(k) > \Omega_F}} [\Omega(k) - \Omega_F] \tilde{a}^+(\mathbf{k}) \, \tilde{a}(\mathbf{k})$$

$$= \epsilon_F + \sum_{\mathbf{k}} \tilde{\Omega}(\mathbf{k}) \, \tilde{a}^+(\mathbf{k}) \, \tilde{a}(\mathbf{k})$$

with

(30.56) $$\widetilde{\Omega}(\mathbf{k}) = |\Omega(k) - \Omega_F|; \qquad \epsilon_F = \sum_{\mathbf{k}} [\Omega(k) - \Omega_F].$$

One has, thus, a description in terms of *quasi particles* whose annihilation and creation operators are $\tilde{a}(k)$ and $\tilde{a}^+(k)$. For $\Omega(k) > \Omega_F$ a quasi particle of momentum $\mathbf{k}$ corresponds to an ordinary particle of momentum $\mathbf{k}$, but for $\Omega(k) < \Omega_F$ the quasi particle of momentum $\mathbf{k}$ corresponds to an ordinary hole of momentum $-\mathbf{k}$. The vacuum state $|0\rangle$ of ordinary particles is defined by

(30.57) $$a(\mathbf{k}) |0\rangle = 0$$

whereas the quasi-particle vacuum state $|\tilde{0}\rangle$ is defined by

(30.58) $$\tilde{a}(\mathbf{k}) |\tilde{0}\rangle = 0$$

which in the absence of interactions is physically identical with the ground state of the system of N ordinary fermions. Since

(30.59) $$\tilde{H}_0 |\tilde{0}\rangle = \epsilon_F |\tilde{0}\rangle$$

and

(30.60) $$\tilde{H}_0 \tilde{a}^+(\mathbf{k}) |\tilde{0}\rangle = [\epsilon_F + \widetilde{\Omega}(\mathbf{k})] \tilde{a}^+(\mathbf{k}) |\tilde{0}\rangle$$

one is justified in calling $\widetilde{\Omega}(\mathbf{k})$ the energy of a quasi particle of momentum $\mathbf{k}$. In the quasi-particle vacuum $|\tilde{0}\rangle$ the number of ordinary particles, represented by

(30.61)
$$\hat{N} = \sum_{\mathbf{k}} a^+(\mathbf{k}) a(\mathbf{k}) = \sum_{\substack{\mathbf{k} \\ \Omega(k) < \Omega_F}} [1 - \tilde{a}^+(\mathbf{k}) \tilde{a}(\mathbf{k})] + \sum_{\substack{\mathbf{k} \\ \Omega(k) > \Omega_F}} \tilde{a}^+(\mathbf{k}) \tilde{a}(\mathbf{k})$$

has, of course, the expectation value

(30.62) $$\langle \tilde{0} | \hat{N} | \tilde{0} \rangle = \sum_{\substack{\mathbf{k} \\ \Omega(k) < \Omega_F}} 1 = N,$$

whereas the number of quasi particles,

(30.63) $$\hat{\tilde{N}} = \sum_{\mathbf{k}} \tilde{a}^+(\mathbf{k}) \tilde{a}(\mathbf{k})$$

has the expectation value

(30.64) $$\langle \tilde{0} | \hat{\tilde{N}} | \tilde{0} \rangle = 0.$$

As in the case of noninteracting fermions, the Hamiltonian describing a system of weakly interacting fermions is most conveniently written so that its eigenvalues give the energy in terms of the deviation from the energy of the state in which all particle levels up to energy Ω_F are filled,

(30.65)

$$H = \sum_{\mathbf{k}} \sum_{s} [\Omega(k) - \Omega_F] a^+(\mathbf{k}, s) a(\mathbf{k}, s)$$

$$+ \sum_{\mathbf{k}'''} \sum_{\mathbf{k}''} \sum_{\mathbf{k}'} \sum_{\mathbf{k}} \sum_{s'} \sum_{s} \langle \mathbf{k}''' \mathbf{k}'' | F | \mathbf{k}' \mathbf{k} \rangle \delta_{\mathbf{k}'''+\mathbf{k}'', \mathbf{k}'+\mathbf{k}} \, a^+(\mathbf{k}''', s)$$

$$\times a^+(\mathbf{k}'', s') a(\mathbf{k}', s') a(\mathbf{k}, s).$$

The spin labels have been written out explicitly, but the interaction is taken to be spin independent.

It was first noticed by Cooper that the interaction, if attractive, can bring about correlations between pairs of fermions giving rise to a ground state in which the Fermi surface is smeared out, and that *this result cannot in principle be obtained by perturbation theory.*

Cooper's qualitative considerations can be made more precise by introducing, following Bogoliubov, quasi-particle operators

(30.66)
$$\tilde{a}(\mathbf{k}, 1) = u(\mathbf{k}) a(\mathbf{k}, 1) - v(\mathbf{k}) a^+(-\mathbf{k}, 2) \equiv \tilde{\alpha}(\mathbf{k})$$
$$\tilde{a}(\mathbf{k}, 2) = u(\mathbf{k}) a(-\mathbf{k}, 2) + v(\mathbf{k}) a^+(\mathbf{k}, 1) \equiv \tilde{\beta}(\mathbf{k}).$$

This transformation is canonical (see Appendix 5), i.e. the $\tilde{a}$ and $\tilde{a}^+$ satisfy the usual fermion anti-C.R.s

(30.67) $\{\tilde{a}(\mathbf{k}, s) \tilde{a}^+(\mathbf{k}', s')\} = \delta_{\mathbf{k}', \mathbf{k}} \delta_{s', s};$ all other $\{\ \} = 0,$

provided $u(\mathbf{k})$ and $v(\mathbf{k})$ are real numbers subject to the conditions

(30.68) $u(\mathbf{k}) = u(-\mathbf{k});$ $v(\mathbf{k}) = v(-\mathbf{k});$ $u^2(\mathbf{k}) + v^2(\mathbf{k}) = 1.$

If, in particular,

(30.69) $u(\mathbf{k}) = \begin{cases} 1 \\ 0 \end{cases};$ $v(\mathbf{k}) = \begin{cases} 0 & \text{for} \quad \Omega(k) > \Omega_F \\ 1 & \text{for} \quad \Omega(k) < \Omega_F \end{cases}$

one obtains essentially the transformation (30.53) for the case of non-interacting fermions, namely

(30.70)

$$\tilde{a}(\mathbf{k}, 1) = a(\mathbf{k}, 1); \quad \tilde{a}(\mathbf{k}, 2) = a(-\mathbf{k}, 2) \quad \text{for} \quad \Omega(k) > \Omega_F$$

$$\tilde{a}(\mathbf{k}, 1) = -a^+(-\mathbf{k}, 2); \quad \tilde{a}(\mathbf{k}, 2) = a^+(\mathbf{k}, 1) \quad \text{for} \quad \Omega(k) < \Omega_F.$$

One may thus, in this case, look upon

$\tilde{a}(\mathbf{k}, 1)$ as the annihilation operator of a $\begin{pmatrix} \text{particle } (\mathbf{k}, 1) \text{ outside} \\ \text{hole } (-\mathbf{k}, 2) \text{ inside} \end{pmatrix}$

the Fermisphere

$\tilde{a}(k, 2)$ as the annihilation operator of a $\begin{pmatrix} \text{particle } (-\mathbf{k}, 2) \text{ outside} \\ \text{hole } (\mathbf{k}, 1) \text{ inside} \end{pmatrix}$

the Fermisphere.

In the general case (30.66), application of an operator a engenders superposition of a particle and a hole.

Equations (30.66) can be solved for $a(\mathbf{k}, 1)$ and $a(-\mathbf{k}, 2)$ to give

(30.71)
$$a(\mathbf{k}, 1) = u(\mathbf{k}) \, \tilde{a}(\mathbf{k}, 1) + v(\mathbf{k}) \, \tilde{a}^+(\mathbf{k}, 2) = u(\mathbf{k}) \, \tilde{\alpha}(\mathbf{k}) + v(\mathbf{k}) \, \tilde{\beta}^+(\mathbf{k})$$

$$a(-\mathbf{k}, 2) = u(\mathbf{k}) \, \tilde{a}(\mathbf{k}, 2) - v(\mathbf{k}) \, \tilde{a}^+(\mathbf{k}, 1) = u(\mathbf{k}) \, \tilde{\beta}(\mathbf{k}) - v(\mathbf{k}) \, \tilde{\alpha}^+(\mathbf{k}).$$

One can now substitute these expressions into the Hamiltonian (30.65), making use of the symmetry properties of the matrix element

(30.72) $\quad \langle \mathbf{k}''' \mathbf{k}'' | F | \mathbf{k}' \mathbf{k} \rangle = \langle \mathbf{k}'' \mathbf{k}''' | F | \mathbf{k} \mathbf{k}' \rangle = \langle -\mathbf{k}' - \mathbf{k} | F | -\mathbf{k}''' - \mathbf{k}'' \rangle,$

where the last equality follows from time reversal invariance, and diagonalize the resulting transformed Hamiltonian by suitable choice of $u(\mathbf{k})$ and $v(\mathbf{k})$ in a manner analogous to the one employed in the corresponding problem for interacting bosons. The calculations are rather lengthy, and the reader is referred to the work of Beliaev, quoted at the end of this section, for details.

The principal result is a confirmation of Cooper's more qualitative theory mentioned earlier, and a description of the system in terms of a quasi-particle excitation spectrum which, as in the case of interacting bosons, satisfies the criterion for the existence of supertransfer phenomena established in Section 11.

It is instructive to express the ground state and the lowest excited states of the entire system in terms of the vacuum state and ordinary particle creation operators. The true vacuum state is defined by

(30.73) $\quad a(\mathbf{k}, s) \, |0\rangle = 0 \qquad \text{i.e.} \qquad a(\mathbf{k}, 1) \, |0\rangle = \alpha(\mathbf{k}) \, |0\rangle = 0;$

$$a(-\mathbf{k}, 2) \, |0\rangle = \beta(\mathbf{k}) \, |0\rangle = 0$$

while the ground state (the "quasi-particle vacuum") $|\tilde{0}\rangle$ is defined by

(30.74) $\quad \tilde{a}(\mathbf{k}, 1) \, |\tilde{0}\rangle = \tilde{\alpha}(\mathbf{k}) \, |\tilde{0}\rangle = 0; \qquad \tilde{a}(\mathbf{k}, 2) \, |\tilde{0}\rangle = \tilde{\beta}(\mathbf{k}) \, |\tilde{0}\rangle = 0.$

Equation (30.74) must follow from (30.73) upon application of the unitary operator [see Eq. (A5.32)]

$$(30.75) \qquad U_B = 1 - \sigma_0[1 - u(\mathbf{k})] - i\sigma_2 v(\mathbf{k})$$

where

$$(30.76) \quad \sigma_0 = 1 - \alpha^+\alpha - \beta^+\beta - 2\alpha^+\alpha\beta^+\beta; \qquad \sigma_2 = i(\alpha^+\beta^+ - \beta\alpha)$$

so that

$$(30.77) \quad \tilde{\alpha} = U_B \alpha U_B^+; \qquad \tilde{\beta} = U_B \beta U_B^+; \qquad |\tilde{0}\rangle = U_B|0\rangle.$$

Now application of σ_0 and σ_2 to the vacuum state gives

$$(30.78) \qquad \sigma_0|0\rangle = |0\rangle; \qquad \sigma_2|0\rangle = i\alpha^+\beta^+|0\rangle$$

so that

$$(30.79) \qquad |\tilde{0}\rangle = U_B|0\rangle = [u(\mathbf{k}) + v(\mathbf{k})\alpha^+\beta^+]|0\rangle$$

or, generally

$$(30.80) \qquad |\tilde{0}\rangle = \prod_{\mathbf{k}} [u(\mathbf{k}) + v(\mathbf{k})a^+(\mathbf{k}, 1)a^+(-\mathbf{k}, 2)]|0\rangle.$$

This expression was, in fact, the starting point of the famous BCS theory of superconductivity, in which the ground state is envisaged as containing correlated pairs of electrons in accordance with Cooper's idea.

State vectors describing the presence of various numbers of quasi particles are constructed by using the same procedure as above: Apply U_B to the corresponding ordinary particle occupation state. For example, for specific $\mathbf{k}$, one has

$$(30.81) \qquad |1_{\tilde{\alpha}(\mathbf{k})}\rangle = \tilde{\alpha}^+(\mathbf{k})|\tilde{0}\rangle = U_B\alpha^+(\mathbf{k})|0\rangle = U_B|1_{\alpha(\mathbf{k})}\rangle$$

with

$$(30.82) \qquad U_B = \prod_{\mathbf{k}'} \{1 - \sigma_0(\mathbf{k}')[1 - u(\mathbf{k}')] - i\sigma_2(\mathbf{k}')v(\mathbf{k}')\}.$$

It is well to keep in mind, however, that the particular deformation of the Fermi surface envisaged in the BCS theory is by no means the most general such deformation which may lead to a lower energy of the entire system as compared to a state with undeformed Fermi surface. In accordance with the most general transformation given in Appendix 5 one can, for example, consider nonstationary deformations which may conceivably lead to an even more effective minimization of the total energy of the system. In fact, such more sophisticated quasi-particle excitations have been considered by Landau in connection with the theory of liquid helium-3, which seems to admit the presence of ripples on the Fermi surface, known as "zero sound." In any case, the mathematical penetration of the problem posed by interacting bosons or fermions is far from complete and remains one of the most promising fields for the application of quantum-mechanical concepts.

NOTES

Einstein [1] discovered the condensation phenomenon now bearing his name. The possible connection between permutation symmetry and macroscopic properties of liquid helium II was stressed by London [2]. The significance of Einstein's characteristic temperature in terms of countability of particles and the limitation encountered by Maxwell's demon at low temperatures was pointed out by Kaempffer [3].

The treatment of the weakly interacting bosons is due to Bogoliubov [4]. See also the review article by Beliaev [5]. The treatment of the weakly interacting fermions is due to Bogoliubov *et al.* [6]. See also Bogoliubov [7] and Valatin [8].

The crucial observation that the correlation between pairs near the Fermi surface cannot in principle be obtained by perturbation theory follows from the work of Cooper [9], leading to the BSC theory of superconductivity [10].

The theory of Fermi liquids has been developed by Landau [11]. See also the review article by Abrikosov and Khalatnikov [12].

REFERENCES

[1] A. Einstein, *Berlin Ber.* **22**, 261 (1924).

[2] F. London, *Phys. Rev.* **54**, 947 (1938).

[3] F. A. Kaempffer, *Z. Physik* **125**, 395 (1948); 487 (1949).

[4] N. N. Bogoliubov, *J. Phys. USSR* **11**, 23 (1947).

[5] S. T. Beliaev *in* "The Many Body Problem" (C. de Witt, ed.), p. **343**. Wiley, New York, 1959.

[6] N. N. Bogoliubov, V. V. Tolmachev, and D. V. Shirkov, "A New Method in the Theory of Superconductivity." Consultants Bureau, New York, 1959.

[7] N. N. Bogoliubov, *Nuovo Cimento* **7**, 794 (1958).

[8] J. G. Valatin, *Nuovo Cimento* **7**, 843 (1958).

[9] L. N. Cooper, *Phys. Rev.* **104**, 1189 (1956).

[10] J. Bardeen, L. N. Cooper, and J. R. Schrieffer, *Phys. Rev.* **108**, 1175 (1957).

[11] L. Landau, *Soviet Phys. JETP* (*English Transl.*) **3**, 920 (1957).

[12] A. A. Abrikosov and I. M. Khalatnikov, *Rep. Progr. Phys.* **22**, 329 (1959).

The Eigenstates of Angular Momentum

The component operators of angular momentum, denoted J_1, J_2, J_3, satisfy the C.Rs

(A1.1) $$J_1 J_2 - J_2 J_1 = iJ_3 \qquad \text{(cyclically)}$$

and are, therefore, mutually incompatible observables. Since $J^2 = J_1^2 + J_2^2 + J_3^2$ commutes with all operators J_k, for example,

(A1.2)
$$J^2 J_3 - J_3 J^2 = J_1^2 J_3 + J_2^2 J_3 - J_3 J_1^2 - J_3 J_2^2$$
$$= J_1(J_3 J_1 - iJ_2) + J_2(J_3 J_2 + iJ_1)$$
$$- (J_1 J_3 + iJ_2)J_1 - (J_2 J_3 - iJ_1)J_2 = 0,$$

there should exist simultaneous eigenstates of J^2 and of one of the components J_k. By convention, a representation will be sought in which both J^2 and J_3 are diagonal matrices.

It is convenient to introduce the operators

(A1.3) $$J_0 = J_1 + iJ_2; \qquad J_0^+ = J_1 - iJ_2.$$

They have the properties

(A1.4) $$J_0 J_0^+ = J^2 + J_3 - J_3^2$$

(A1.5) $$J_0^+ J_0 = J^2 - J_3 - J_3^2$$

(A1.6) $$J_0^+ J_3 - J_3 J_0^+ = J_0^+$$

(A1.7) $$J_0 J_3 - J_3 J_0 = -J_0$$

as is easily verified by computation with (A1.3) and (A1.1).

Now denote the eigenstate of J_3 with eigenvalue m by $|\,,m\rangle$ the empty space in front of the label m being left open for the quantum number characterizing the value of J^2 in that state, so that

(A1.8) $$J_3 |\,,m\rangle = m |\,,m\rangle.$$

309

If one multiplies this equation from the left by J_0^+,

(A1.9) $$J_0^+ J_3| , m\rangle = m J_0^+| , m\rangle$$

and utilizes the C.R. (A1.6), one obtains the equation

(A1.10) $$J_3 J_0^+| , m\rangle = (m-1) J_0^+| , m\rangle$$

which says that if $| , m\rangle$ is an eigenstate of J_3 with eigenvalue m, then $J_0^+| , m\rangle$ is also eigenstate of J_3, but with eigenvalue $(m-1)$. In other words, the state $J_0^+| , m\rangle$ is, up to an as yet undetermined normalization constant c, identical with the state $| , m-1\rangle$,

(A1.11) $$J_0^+| , m\rangle = c| , m-1\rangle.$$

Similarly, upon multiplication of (A1.8) from the left with J_0 and utilization of the C.R. (A1.7) one obtains the equation

(A1.12) $$J_3 J_0| , m\rangle = (m+1) J_0| , m\rangle$$

which allows one to identify, up to a normalization constant d, the state $J_0| , m\rangle$ as an eigenstate of J_3 with eigenvalue $(m+1)$,

(A1.13) $$J_0| , m\rangle = d| , m+1\rangle.$$

The operator J_0 generates thus an ascending sequence of eigenstates $| , m\rangle$, $| , m+1\rangle$, ..., whereas the operator J_0^+ generates a descending sequence $| , m\rangle$, $| , m-1\rangle$, ...

Each sequence comes to an end, however, for the following reason. If $| , m\rangle$ is a simultaneous eigenstate of J_3 and J^2 then, as a consequence of (A1.4) and (A1.5), one has

(A1.14) $$J_0 J_0^+| , m\rangle = (J^2 + m - m^2)| , m\rangle$$

and

(A1.15) $$J_0^+ J_0| , m\rangle = (J^2 - m - m^2)| , m\rangle$$

where J^2 can be treated as a number. Since

(A1.16)
$$\langle , m| J_0 J_0^+| , m\rangle = \langle J_0^+(, m)| J_0^+(, m)\rangle = |c|^2 \langle , m-1| , m-1\rangle \geqslant 0$$
   and $= 0$ only if $J_0^+| , m\rangle = c| , m-1\rangle = 0$

and

(A1.17)
$$\langle , m| J_0^+ J_0| , m\rangle = \langle J_0(, m)| J_0(, m)\rangle = |d|^2 \langle , m+1| , m+1\rangle \geqslant 0$$
   and $= 0$ only if $J_0| , m\rangle = d| , m+1\rangle = 0,$

Eqs. (A1.14) and (A1.15) mean that

(A1.18) $$J^2 + m - m^2 \geqslant 0$$

(A1.19) $$J^2 - m - m^2 \geqslant 0.$$

For given J^2 this is consistent only if the descending sequence comes to an end for a certain m_{min} for which

(A1.20)
$$J^2 + m_{min} - m_{min}^2 = 0 \quad \text{i.e.} \quad m_{min} = +\tfrac{1}{2} - \sqrt{J^2 + \tfrac{1}{4}} \quad \text{and} \quad |\,, m_{min} - 1\rangle = 0$$

and if the ascending sequence comes to an end for a certain m_{max} for which

(A1.21)
$$J^2 - m_{max} - m_{max}^2 = 0 \quad \text{i.e.} \quad m_{max} = -\tfrac{1}{2} + \sqrt{J^2 + \tfrac{1}{4}} \quad \text{and} \quad |\,, m_{max} + 1\rangle = 0.$$

Since m changes by integers, the difference between m_{max} and m_{min} must also be an integer $\geqslant 0$. One may write this

(A1.22) $$m_{max} - m_{min} + 1 = 2\sqrt{J^2 + \tfrac{1}{4}} = 2j + 1$$

where j can have only the following values

(A1.23) $$j = 0, \tfrac{1}{2}, 1, \tfrac{3}{2}, \ldots .$$

Thus the possible values of J^2 are of the form

(A1.24) $$J^2 = j(j+1)$$

and for given j the number m can assume the $(2j+1)$ values between

(A1.25) $$m_{max} = +j; \qquad m_{min} = -j.$$

The simultaneous eigenstates of J^2 and J_3 will accordingly be labeled $|j, m\rangle$, satisfying

(A1.26)
$$J^2|j, m\rangle = j(j+1)|j, m\rangle; \qquad j = 0, \tfrac{1}{2}, 1, \tfrac{3}{2}, \ldots$$
$$J_3|j, m\rangle = m|j, m\rangle; \qquad m = -j, -j+1, \ldots, +j.$$

If all states $|j, m\rangle$ are normalized, then the constants c and d are obtained by squaring Eqs. (A1.11) and (A1.13),

(A1.27)
$$|c|^2 \langle j, m-1 | j, m-1\rangle = |c|^2 = \langle J_0^+(j, m) | J_0^+(j, m)\rangle = J^2 + m - m^2$$
$$= j(j+1) - m(m-1)$$

(A1.28)
$$|d|^2 \langle j, m+1 | j, m+1\rangle = |d|^2 = \langle J_0(j, m) | J_0(j, m)\rangle = J^2 - m - m^2$$
$$= j(j+1) - m(m+1)$$

so that one may write, up to an arbitrary phase factor Eqs. (A1.11) and (A1.13) explicitly

(A1.29) $J_0^+|j,m\rangle = \sqrt{j(j+1)-m(m-1)}\,|j,m-1\rangle$

(A1.30) $J_0|j,m\rangle = \sqrt{j(j+1)-m(m+1)}\,|j,m+1\rangle.$

If one represents the states $|j,m\rangle$ for given j by unit vectors with $2j+1$ components, so that with a conventional choice of phase one can write

(A1.31)

$$|j,m=j\rangle = \begin{pmatrix} 1 \\ 0 \\ 0 \\ \vdots \end{pmatrix}; \qquad |j,m=j-1\rangle = \begin{pmatrix} 0 \\ 1 \\ 0 \\ \vdots \end{pmatrix}; \qquad \text{etc.};$$

then by (A1.29) and (A1.30) the only nonvanishing matrix elements of J_0^+ and J_0 for given j are

(A1.32)

$$(J_0^+)_{m-1,\,m} = \sqrt{j(j+1)-m(m-1)}; \qquad (J_0)_{m+1,\,m} = \sqrt{j(j+1)-m(m+1)}$$

In particular, one obtains for $j=\tfrac{1}{2}$ the representations

(A1.33)

$$|\tfrac{1}{2},\tfrac{1}{2}\rangle = \begin{pmatrix} 1 \\ 0 \end{pmatrix}; \qquad |\tfrac{1}{2},-\tfrac{1}{2}\rangle = \begin{pmatrix} 0 \\ 1 \end{pmatrix}; \qquad J_0^+ = \begin{pmatrix} 0 & 0 \\ 1 & 0 \end{pmatrix}; \qquad J_0 = \begin{pmatrix} 0 & 1 \\ 0 & 0 \end{pmatrix};$$

$$J_1 = \tfrac{1}{2}(J_0^+ + J_0) = \frac{1}{2}\begin{pmatrix} 0 & 1 \\ 1 & 0 \end{pmatrix}; \qquad J_2 = (i/2)(J_0^+ - J_0) = \frac{1}{2}\begin{pmatrix} 0 & -i \\ i & 0 \end{pmatrix};$$

$$J_3 = \frac{1}{2}\begin{pmatrix} 1 & 0 \\ 0 & -1 \end{pmatrix}; \qquad J^2 = \frac{3}{4}\begin{pmatrix} 1 & 0 \\ 0 & 1 \end{pmatrix},$$

and for $j=1$ the representations

(A1.34) $|1,1\rangle = \begin{pmatrix} 1 \\ 0 \\ 0 \end{pmatrix}; \qquad |1,0\rangle = \begin{pmatrix} 0 \\ 1 \\ 0 \end{pmatrix}; \qquad |1,-1\rangle = \begin{pmatrix} 0 \\ 0 \\ 1 \end{pmatrix};$

$$J_0^+ = \begin{pmatrix} 0 & 0 & 0 \\ \sqrt{2} & 0 & 0 \\ 0 & \sqrt{2} & 0 \end{pmatrix}; \qquad J_0 = \begin{pmatrix} 0 & \sqrt{2} & 0 \\ 0 & 0 & \sqrt{2} \\ 0 & 0 & 0 \end{pmatrix};$$

$$J_1 = \frac{1}{\sqrt{2}}\begin{pmatrix} 0 & 1 & 0 \\ 1 & 0 & 1 \\ 0 & 1 & 0 \end{pmatrix}; \qquad J_2 = \frac{1}{\sqrt{2}}\begin{pmatrix} 0 & -i & 0 \\ i & 0 & -i \\ 0 & i & 0 \end{pmatrix};$$

$$J_3 = \begin{pmatrix} 1 & 0 & 0 \\ 0 & 0 & 0 \\ 0 & 0 & -1 \end{pmatrix}; \qquad J^2 = 2\begin{pmatrix} 1 & 0 & 0 \\ 0 & 1 & 0 \\ 0 & 0 & 1 \end{pmatrix}.$$

The representation (A1.33) is identical with the one obtained in Sections 1 and 2 for the spin states and operators, because one has the identities $|\frac{1}{2},\frac{1}{2}\rangle = |a_+\rangle$ and $|\frac{1}{2}, -\frac{1}{2}\rangle = |a_-\rangle$.

If one considers the special case of the *orbital* angular momentum **L** of an object, whose position and momentum are denoted by **Q** and **P**, respectively, then $\mathbf{J} = \mathbf{L}$ can be decomposed according to (12.2) in the form

(A1.35) $$\mathbf{L} = (\mathbf{Q} \times \mathbf{P})$$

Working in coordinate representation one can derive (A1.26) again, *but with the restriction that* $j = l$ *be an integer only.* Thus any angular momentum with half-odd integer value of j must at least partly be due to an intrinsic or spin angular momentum that does not permit a decomposition of the form (A1.35).

This derivation is most conveniently carried out in coordinate representation by the introduction of spherical coordinates r, ϑ, φ, such that

(A1.36)
$$Q_1 = r \sin \vartheta \cos \varphi; \qquad Q_2 = r \sin \vartheta \sin \varphi; \qquad Q_3 = r \cos \vartheta;$$

and

(A1.37)
$$(\partial/\partial Q_1) = \sin \vartheta \cos \varphi (\partial/\partial r) + \cos \vartheta \cos \varphi (1/r)(\partial/\partial \vartheta) - \sin \varphi (1/r \sin \vartheta)(\partial/\partial \varphi)$$
$$(\partial/\partial Q_2) = \sin \vartheta \sin \varphi (\partial/\partial r) + \cos \vartheta \sin \varphi (1/r)(\partial/\partial \vartheta) + \cos \varphi (1/r \sin \vartheta)(\partial/\partial \varphi)$$
$$(\partial/\partial Q_3) = \cos \vartheta (\partial/\partial r) - \sin \vartheta (1/r)(\partial/\partial \vartheta)$$

so that the angular momentum component operators are represented by

(A1.38)
$$L_3 = -i(\partial/\partial \varphi)$$
$$L_0^+ = L_1 - iL_2 = -e^{-i\varphi}[(\partial/\partial \vartheta) - i \cot \vartheta (\partial/\partial \varphi)]$$
$$L_0 = L_1 + iL_2 = e^{i\varphi}[(\partial/\partial \vartheta) + i \cot \vartheta (\partial/\partial \varphi)]$$
$$L^2 = -(1/\sin \vartheta)(\partial/\partial \vartheta)[\sin \vartheta (\partial/\partial \vartheta)] - (1/\sin^2 \vartheta)(\partial^2/\partial \varphi^2)$$

which do not contain r explicitly. It is therefore reasonable to introduce for given orbital angular momentum $j = l$ a ψ function, namely $\langle \vartheta, \varphi | l, m \rangle$, where $|\vartheta, \varphi\rangle$ is the state in which the angular momentum of amount l is with certainty aligned in direction ϑ, φ, and denote it

(A1.39) $$Y_{l,m}(\vartheta, \varphi) = \langle \vartheta, \varphi | l, m \rangle$$

so that

(A1.40) $$|\vartheta, \varphi\rangle = \sum_{l,m} |l, m\rangle \langle l, m | \vartheta, \varphi\rangle = \sum_{l,m} |l, m\rangle Y^*_{l,m}(\vartheta, \varphi)$$

and

(A1.41)

$$|l, m\rangle = \int |\vartheta, \varphi\rangle \, d\Omega \, \langle\vartheta, \varphi|l, m\rangle = \int_0^\pi \int_0^{2\pi} |\vartheta, \varphi\rangle \sin\vartheta \, d\vartheta \, d\varphi \, Y_{l, m}(\vartheta, \varphi),$$

which subject the functions $Y_{l, m}(\vartheta, \varphi)$ to the normalization condition

(A1.42)

$$\langle l', m'|l, m\rangle = \int_0^\pi \int_0^{2\pi} Y^*_{l', m'}(\vartheta, \varphi) \, Y_{l, m}(\vartheta, \varphi) \sin\vartheta \, d\vartheta \, d\varphi = \delta_{l' l} \delta_{m' m}.$$

Equations (A1.26) can now be written as eigenvalue equations for the ψ functions $Y_{l, m}(\vartheta, \varphi)$, by applying the operators (A1.38) to (A1.41) after the fashion of (8.9). One obtains

(A1.43) $-i(\partial/\partial\varphi) \, Y_{l, m}(\vartheta, \varphi) = m Y_{l, m}(\vartheta, \varphi)$

and

(A1.44)
$$\{-(1/\sin\vartheta) (\partial/\partial\vartheta) [\sin\vartheta(\partial/\partial\vartheta)] - (1/\sin^2\vartheta) (\partial^2/\partial\varphi^2)\} \, Y_{l, m}(\vartheta, \varphi)$$

$$= l(l+1) \, Y_{l, m}(\vartheta, \varphi)$$

From Eq. (A1.43) one deduces immediately that

(A1.45) $Y_{l, m}(\vartheta, \varphi) = F_{l, m}(\vartheta) \, e^{im\varphi}$

where now the as yet undetermined function $F_{l, m}(\vartheta)$ must, according to (A1.44), satisfy

(A1.46)
$$(1/\sin\vartheta) (\partial/\partial\vartheta) [\sin\vartheta(\partial F_{l, m}/\partial\vartheta)] + [l(l+1) - (m^2/\sin^2\vartheta)] F_{l, m} = 0$$

with $l \geqslant m \geqslant -l.$

If one requires of the solutions $F_{l, m}(\vartheta)$ that for all $0 \leqslant \vartheta \leqslant \pi$ they should be unique, finite, and differentiable, and also subject to normalization after (A1.42)

(A1.47) $\displaystyle\int_{\vartheta=0}^\pi \int_{\varphi=0}^{2\pi} |Y_{l, m}|^2 \sin\vartheta \, d\vartheta \, d\varphi = 2\pi \int_{\vartheta=0}^\pi |F_{l, m}(\vartheta)|^2 \sin\vartheta \, d\vartheta = 1,$

then the general solutions are

(A1.48)

$$Y_{l,m}(\vartheta, \varphi) = \frac{1}{2^l l!} \sqrt{\frac{(2l+1)(l-m)!}{4\pi(l+m)!}} \frac{e^{im\varphi}}{\sin^m \vartheta} \underbrace{\left\{ \frac{1}{\sin\vartheta} \frac{d}{d\vartheta} \left(\cdots \frac{1}{\sin\vartheta} \frac{d}{d\vartheta} (\sin^{2l}\vartheta) \right) \right\}}_{l-m \text{ differentiations}}$$

$$= \sqrt{\frac{(2l+1)(l-m)!}{4\pi(l+m)!}} P_l^m(\cos\vartheta) e^{im\varphi}$$

$$\propto \sin^m\vartheta(\cos^{l-m}\vartheta + a\cos^{l-m-2}\vartheta + \ldots) e^{im\varphi}$$

where l must be a non-negative *integer*. Non-integer values of l cannot lead to solutions satisfying all the requirements.

For $l = 0$, $l = 1$, $l = 2$ these solutions read explicitly

(A1.49)

$$Y_{0,0}(\vartheta, \varphi) = \frac{1}{\sqrt{4\pi}} \;;$$

$$Y_{1,1}(\vartheta, \varphi) = \sqrt{\frac{3}{8\pi}} \sin\vartheta \, e^{i\varphi}; \quad Y_{1,0} = \sqrt{\frac{3}{4\pi}} \cos\vartheta \,;$$

$$Y_{1,-1} = \sqrt{\frac{3}{8\pi}} \sin\vartheta \, e^{-i\varphi} \,;$$

$$Y_{2,2} = Y_{2,-2}^* = \frac{1}{4}\sqrt{\frac{15}{2\pi}} \sin^2\vartheta \, e^{2i\varphi}; \quad Y_{2,1} = Y_{2,-1}^* = \frac{1}{2}\sqrt{\frac{15}{2\pi}} \sin\vartheta \cos\vartheta \, e^{i\varphi};$$

$$Y_{2,0} = \sqrt{\frac{5}{4\pi}} \left(\frac{3}{2}\cos^2\vartheta - \frac{1}{2} \right)$$

Another phase convention used frequently in the literature consists of multiplying each $Y_{l,m}$ by $(-1)^m$ to yield

(A1.50) $$Y_l^m(\vartheta, \varphi) = (-1)^m Y_{l,m}(\vartheta, \varphi).$$

NOTES

The content of this appendix can be found in practically all texts on quantum mechanics. Monographs on the subject have been written by Edmonds [1] and Rose [2].

REFERENCES

[1] A. R. Edmonds, "Angular Momentum in Quantum Mechanics." Princeton Univ. Press, Princeton, New Jersey, 1957.
[2] M. E. Rose, "The Elementary Theory of Angular Momentum." Wiley, New York, 1957.

The Addition of Two Angular Momenta

Suppose two angular momentum operators $\mathbf{J}_1$ and $\mathbf{J}_2$ are given, each satisfying $\mathbf{J}_\alpha \times \mathbf{J}_\alpha = i\mathbf{J}_\alpha$ ($\alpha = 1, 2$), so that there exist eigenstates $|j_\alpha, m_\alpha\rangle$ of J_α^2 and $J_{\alpha 3}$ with eigenvalues

(A2.1)

$$J_\alpha^2 |j_\alpha, m_\alpha\rangle = j_\alpha(j_\alpha + 1) |j_\alpha, m_\alpha\rangle; \qquad j_\alpha = 0, \tfrac{1}{2}, 1, \tfrac{3}{2}, \ldots$$

$$J_{\alpha 3} |j_\alpha, m_\alpha\rangle = m_\alpha |j_\alpha, m_\alpha\rangle; \qquad j_\alpha \geqslant m_\alpha \geqslant -j_\alpha; \quad \Delta m_\alpha \text{ integer.}$$

Now consider an object whose total angular momentum $\mathbf{J}$ can be represented as the vector sum of the angular momentum operators $\mathbf{J}_1$ and $\mathbf{J}_2$ of the two constituents making up the object,

(A2.2)
$$\mathbf{J} = \mathbf{J}_1 + \mathbf{J}_2$$

so that $\mathbf{J}$ still satisfies the C.R.s

(A2.3)
$$\mathbf{J} \times \mathbf{J} = i\mathbf{J}.$$

There must then exist eigenstates $|j, m\rangle$ of J^2 and J_3 having the property

(A2.4)
$$J^2 |j, m\rangle = j(j+1) |j, m\rangle; \qquad j = 0, \tfrac{1}{2}, 1, \tfrac{3}{2}, \ldots$$
$$J_3 |j, m\rangle = m |j, m\rangle; \qquad j \geqslant m \geqslant -j; \qquad \Delta m \text{ integer.}$$

One tries now to represent the state $|j, m\rangle$ as linear combination of the direct products $|j_1, m_1\rangle |j_2, m_2\rangle$ in the form

(A2.5) $$|j, m\rangle = \sum_{m_1 = +j_1}^{-j_1} \sum_{m_2 = +j_2}^{-j_2} C(j, j_1, j_2; m, m_1, m_2) |j_1, m_1\rangle |j_2, m_2\rangle.$$

The coefficients C of this unitary transformation are called the "Clebsch-Gordan coefficients" or, sometimes, the "Wigner coefficients." This way of writing $|j, m\rangle$ is suggested by degeneracies of the states $|j, m\rangle$ and

$|j_1, m_1\rangle |j_2, m_2\rangle$ with respect to the quantum numbers m and m_1, m_2, respectively, i.e. the dimensionality of the space spanned by $|j, m\rangle$ is the same as the dimensionality of the product space spanned by $|j_1, m_1\rangle |j_2, m_2\rangle$.

To see this, label by convention the constituents so that $j_1 \geqslant j_2$. The total angular momentum will assume any of the values j for which $j = j_1 + j_2, j_1 + j_2 - 1, \ldots, j_1 - j_2 + 1, j_1 - j_2$. For each such value j there is a degeneracy $2j + 1$, so that the sum of all degeneracies, i.e. the degeneracy of $|j, m\rangle$ with respect to m, is

(A2.6) $[2(j_1 + j_2) + 1] + [2(j_1 + j_2 - 1) + 1] + \ldots + [2(j_1 - j_2 + 1) + 1]$

$$+ [2(j_1 - j_2) + 1]$$

$$= (2j_1 + 1)(\text{number of brackets } [\ ]) = (2j_1 + 1)(2j_2 + 1)$$

which is equal to the degeneracy of the product state $|j_1, m_1\rangle |j_2, m_2\rangle$ with respect to m_1, m_2.

If the normalizations and phases of $|j_\alpha, m_\alpha\rangle$ are given, then up to an arbitrary phase factor the coefficients C are completely determined by (A2.4) and the normalization of $|j, m\rangle$. It turns out that j_1 and j_2 add to

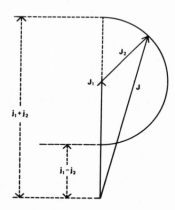

FIG. A2.1. The vector addition of two angular momenta $\mathbf{J}_1$ and $\mathbf{J}_2$.

j vectorially (see Fig. A2.1), whereas m_1 and m_2 add to m algebraically, so that

(A2.7) $m_1 = m - m_2$

and one may simplify (A2.5) by writing

(A2.8) $\quad |j,m\rangle = \sum_{m_1=+j_1}^{-j_1} C(j,j_1,j_2;m_1,m-m_1)|j_1,m_1\rangle|j_2,m-m_1\rangle.$

For the purpose of this work it is sufficient to tabulate the Clebsch-Gordan coefficients for addition of an angular momentum $j_2 = \frac{1}{2}$ to an arbitrary j_1 (Table A2.1) and of an angular momentum $j_2 = 1$ added to an arbitrary j_1 (Table A2.2).

As an example, consider an object " 1 " and an object " 2 " both having spin $\frac{1}{2}$. According to (A2.6) the degeneracy with respect to m is fourfold,

TABLE A2.1

THE FOUR NONVANISHING COEFFICIENTS $C(j,j_1,\frac{1}{2};m,m_1,m_2)$

	$m_1 = m-\frac{1}{2}$ $m_2 = \frac{1}{2}$	$m_1 = m+\frac{1}{2}$ $m_2 = -\frac{1}{2}$
$j = j_1+\frac{1}{2}$	$\left(\dfrac{j_1+m+\frac{1}{2}}{2j_1+1}\right)^{1/2}$	$\left(\dfrac{j_1-m+\frac{1}{2}}{2j_1+1}\right)^{1/2}$
$j = j_1-\frac{1}{2}$	$-\left(\dfrac{j_1-m+\frac{1}{2}}{2j_1+1}\right)^{1/2}$	$\left(\dfrac{j_1+m+\frac{1}{2}}{2j_1+1}\right)^{1/2}$

and one reads from (A2.8) and Table A2.1 the four spin states of the object composed of objects " 1 " and " 2 ":

(A2.9)

$j = 0:$ $\qquad |0,0\rangle = (1/\sqrt{2})(|\frac{1}{2},\frac{1}{2}\rangle_1|\frac{1}{2},-\frac{1}{2}\rangle_2 - |\frac{1}{2},-\frac{1}{2}\rangle_1|\frac{1}{2},\frac{1}{2}\rangle_2)$

$j = 1:$ $\quad \begin{cases} |1,1\rangle = |\frac{1}{2},\frac{1}{2}\rangle_1|\frac{1}{2},\frac{1}{2}\rangle_2 \\ |1,0\rangle = (1/\sqrt{2})(|\frac{1}{2},\frac{1}{2}\rangle_1|\frac{1}{2},-\frac{1}{2}\rangle_2 + |\frac{1}{2},-\frac{1}{2}\rangle_1|\frac{1}{2},\frac{1}{2}\rangle_2) \\ |1,-1\rangle = |\frac{1}{2},-\frac{1}{2}\rangle_1|\frac{1}{2},-\frac{1}{2}\rangle_2 \end{cases}$

Whenever the values j_1 and j_2 are equal it is important to maintain the labeling of the product states made up out of the objects " 1 " and " 2 ", as has been done here by the subscripts as in $|\ \rangle_1$ and $|\ \rangle_2$, because the transposition operator T_{12}, defined in Section 27, operates on these labels.

One recognizes by inspection the singlet state belonging to $j = 0$ as an antisymmetric eigenstate of T_{12} and the triplet of states belonging to $j = 1$ as symmetric eigenstates of T_{12}.

As a second example, the states resulting from the addition of two

TABLE A2.2

THE NINE NONVANISHING COEFFICIENTS $C(j, j_1, 1; m, m_1, m_2)$

	$m_1 = m-1$ $m_2 = 1$	$m_1 = m$ $m_2 = 0$	$m_1 = m+1$ $m_2 = -1$
$j = j_1+1$	$\left[\dfrac{(j_1+m)(j_1+m+1)}{(2j_1+1)(2j_1+2)}\right]^{1/2}$	$\left[\dfrac{(j_1-m+1)(j_1+m+1)}{(2j_1+1)(j_1+1)}\right]^{1/2}$	$\left[\dfrac{(j_1-m)(j_1-m+1)}{(2j_1+1)(2j_1+2)}\right]^{1/2}$
$j = j_1$	$-\left[\dfrac{(j_1+m)(j_1-m+1)}{2j_1(j_1+1)}\right]^{1/2}$	$\dfrac{+m}{[j_1(j_1+1)]^{1/2}}$	$\left[\dfrac{(j_1-m)(j_1+m+1)}{2j_1(j_1+1)}\right]^{1/2}$
$j = j_1-1$	$\left[\dfrac{(j_1-m)(j_1-m+1)}{2j_1(2j_1+1)}\right]^{1/2}$	$-\left[\dfrac{(j_1-m)(j_1+m)}{j_1(2j_1+1)}\right]^{1/2}$	$\left[\dfrac{(j_1+m+1)(j_1+m)}{2j_1(2j_1+1)}\right]^{1/2}$

angular momenta $j_1 = 1$ and $j_2 = 1$ will be recorded here. In this case, Table A2.2 becomes applicable to (A2.8), and one has

(A2.10)

$j = 0$:
$$|0,0\rangle = (1/\sqrt{3})(|1,1\rangle_1|1,-1\rangle_2 - |1,0\rangle_1|1,0\rangle_2 + |1,-1\rangle_1|1,1\rangle_2)$$

$j = 1$:
$$\begin{cases} |1,1\rangle = (1/\sqrt{2})(|1,1\rangle_1|1,0\rangle_2 - |1,0\rangle_1|1,1\rangle_2) \\ |1,0\rangle = (1/\sqrt{2})(|1,1\rangle_1|1,-1\rangle_2 - |1,-1\rangle_1|1,1\rangle_2) \\ |1,-1\rangle = (1/\sqrt{2})(|1,0\rangle_1|1,-1\rangle_2 - |1,-1\rangle_1|1,0\rangle_2) \end{cases}$$

$j = 2$:
$$\begin{cases} |2,2\rangle = |1,1\rangle_1|1,1\rangle_2 \\ |2,1\rangle = (1/\sqrt{2})(|1,1\rangle_1|1,0\rangle_2 + |1,0\rangle_1|1,1\rangle_2) \\ |2,0\rangle = (1/\sqrt{6})(|1,1\rangle_1|1,-1\rangle_2 + 2|1,0\rangle_1|1,0\rangle_2 + |1,-1\rangle_1|1,1\rangle_2) \\ |2,-1\rangle = (1/\sqrt{2})(|1,0\rangle_1|1,-1\rangle_2 + |1,-1\rangle_1|1,0\rangle_2) \\ |2,-2\rangle = |1,-1\rangle_1|1,-1\rangle_2 \end{cases}$$

Once again, inspection reveals the transformation properties of these spin states under transposition T_{12} of particle labels: The triplet of states belonging to $j = 1$ are antisymmetric, and the singlet state belonging to $j = 0$, as well as the quintet of states belonging to $j = 2$, are symmetric.

NOTE

The classic source for Clebsch-Gordan coefficients is the work of Condon and Shortley [1].

REFERENCE

[1] E. U. Condon and G. H. Shortley, "Theory of Atomic Spectra." Cambridge Univ. Press, London and New York, 1957.

Vector Spherical Harmonics

By taking in (A2.5) $j_1 = l$ and $j_2 = s = 1$, one obtains for each j three different angular momentum states,

$$\text{(A3.1)} \quad |j, m; l, s = 1\rangle = \sum_{m_l = +l}^{-l} \sum_{m_s = +1}^{-l} C(j, l, 1; m, m_l, m_s) |l, m_l\rangle |1, m_s\rangle$$

where $m = m_l + m_s$, corresponding to the three possibilities $j = l+1, l, l-1$ or $l = j+1, j, j-1$. The spin states belonging to $s = 1$ can be represented by ψ functions which transform as vectors, for example, the eigenstates of s_3 [see Section 18, Eq. (18.68)], namely $\boldsymbol{\xi}^1$, $\boldsymbol{\xi}^0$, $\boldsymbol{\xi}^{-1}$, whereas the orbital states can be represented by spherical harmonics which depend on $\mathbf{n}$, the direction of momentum, so that the total ψ function of an object of spin 1 and energy ω must be a linear combination of three "vector spherical harmonics"

(A3.2)
$$\mathbf{Y}_{jlm}(\mathbf{n}) = \sum_{m_l = +l}^{-l} \sum_{m_s = +1}^{-1} C(j, l, 1; m, m_l, m_s) \, Y_{l, m_l}(\mathbf{n}) \, \boldsymbol{\xi}^{m_s}; \qquad m_l = m - m_s;$$

corresponding to the three possible values of l, namely $j+1, j, j-1$, and may therefore be written

$$\text{(A3.3)} \qquad \mathbf{f}_{\omega jm}(\mathbf{n}) = \sum_{l=j+1}^{j-1} a_l(\omega) \, \mathbf{Y}_{jlm}(\mathbf{n}).$$

For photons, the expansion coefficients $a_l(\omega)$ are restricted by the condition of transversality (18.72), which reads here

$$\text{(A3.4)} \qquad \mathbf{n} \cdot \mathbf{f}_{\omega jm}(\mathbf{n}) = 0.$$

The vector $\mathbf{n}$ can be represented by its components n^μ defined by

$$\text{(A3.5)} \qquad \mathbf{n} = \sum_{\mu = +1}^{-1} n^\mu \boldsymbol{\xi}^\mu.$$

With the representations (18.68) one finds for n^μ in terms of the cartesian components $n_1 = \sin\vartheta \cos\varphi$, $n_2 = \sin\vartheta \sin\varphi$, $n_3 = \cos\vartheta$,

$$\text{(A3.6)} \quad \begin{aligned} n^{+1} &= -(1/\sqrt{2})\,(n_1 - in_2) &&= -(1/\sqrt{2})\sin\vartheta\, e^{-i\varphi} \\ n^0 &= n_3 &&= \cos\vartheta \\ n^{-1} &= (1/\sqrt{2})\,(n_1 + in_2) &&= (1/\sqrt{2})\sin\vartheta\, e^{i\varphi}. \end{aligned}$$

Now from the three vectors $\mathbf{Y}_{jlm}(\mathbf{n})$ one can construct three linearly independent combinations, labeled $\mathbf{Y}_{jm}^{\lambda}$ ($\lambda = +1, 0, -1$), of which the first two are transverse, i.e. satisfy

(A3.7) $$\mathbf{n} \cdot \mathbf{Y}_{jm}^{\lambda} = 0 \qquad \text{for} \quad \lambda = +1, 0,$$

and the last one is longitudinal, i.e. satisfies

(A3.8) $$\mathbf{n} \times \mathbf{Y}_{jm}^{-1} = 0.$$

These linear combinations are, with proper normalization,

(A3.9) $$\mathbf{Y}_{jm}^{+1} = \sqrt{\frac{j}{2j+1}}\,\mathbf{Y}_{j,j+1,m} + \sqrt{\frac{j+1}{2j+1}}\,\mathbf{Y}_{j,j-1,m}$$

(A3.10) $$\mathbf{Y}_{jm}^{0} = \mathbf{Y}_{j,j,m}$$

(A3.11) $$\mathbf{Y}_{jm}^{-1} = \sqrt{\frac{j}{2j+1}}\,\mathbf{Y}_{j,j-1,m} - \sqrt{\frac{j+1}{2j+1}}\,\mathbf{Y}_{j,j+1,m}$$

and one has, by solving these equations for $\mathbf{Y}_{j,l,m}$, the following decompositions of the vector spherical harmonics (A3.2) into longitudinal and transverse parts,

(A3.12) $$\mathbf{Y}_{j,j+1,m} = \frac{1}{\sqrt{2j+1}}\,(\sqrt{j}\,\mathbf{Y}_{jm}^{+1} - \sqrt{j+1}\,\mathbf{Y}_{jm}^{-1})$$

(A3.13) $$\mathbf{Y}_{j,j,m} = \mathbf{Y}_{jm}^{0}$$

(A3.14) $$\mathbf{Y}_{j,j-1,m} = \frac{1}{\sqrt{2j+1}}\,(\sqrt{j+1}\,\mathbf{Y}_{jm}^{+1} + \sqrt{j}\,\mathbf{Y}_{jm}^{-1}).$$

The ψ function of a transverse photon with definite angular momentum and energy will therefore be, in general, a linear superposition

(A3.15) $$\mathbf{f}_{jm} = a_1\,\mathbf{Y}_{jm}^{+1} + a_0\,\mathbf{Y}_{jm}^{0}$$

with coefficients a_1 and a_0 subject only to the normalization condition $|a_1|^2 + |a_0|^2 = 1$, and depending only on ω, $a_\lambda = a_\lambda(\omega)$.

The arbitrariness in the choice of the parameter λ is removed if one requires that a photon state have a definite parity. From the transformation property of the spherical harmonics $Y_{l,m_l}(\mathbf{n})$ under inversion of coordinates,

(A3.16) $$\Pi Y_{l,m_l}(\mathbf{n}) = (-1)^l Y_{l,m_l}(\mathbf{n}) = Y_{l,m_l}(-\mathbf{n})$$

and the vector nature of $\boldsymbol{\xi}$ in $\mathbf{n}$ space,

(A3.17) $$\Pi \boldsymbol{\xi} = -\boldsymbol{\xi}$$

follows that

(A3.18) $$\Pi \mathbf{Y}_{jlm}(\mathbf{n}) = (-1)^{l+1} \mathbf{Y}_{jlm}(\mathbf{n})$$

and the transverse and longitudinal parts transform as

(A3.19) $$\Pi \mathbf{Y}_{jm}^{\lambda} = (-1)^{j+1-\lambda} \mathbf{Y}_{jm}^{\lambda}.$$

Transverse photon states belonging to $\lambda = +1$ are called "electric multipole" states and have parity $(-1)^j$, whereas transverse photon states belonging to $\lambda = 0$ are called "magnetic multipole" states and have parity $(-1)^{j+1}$. States belonging to $j = 1, 2, 3$, etc., are called dipole, quadrupole, octopole, etc., states. This terminology derives from the asymptotic behavior of the corresponding ψ functions in coordinate space which resembles that of the classical fields emitted by the respective multipoles. For details the reader is referred to the references at the end of this Appendix.

It will be noticed that for $j = 0$ there exists only one vector spherical harmonic, namely $\mathbf{Y}_{0,1,0} = -\mathbf{Y}_{0,0}^{-1}$, which is necessarily longitudinal. This means there exists no single transverse photon state of total angular momentum $j = 0$.

As has been shown in Section 27, two-photon states of odd parity are associated with the antisymmetric spin states (A2.10) $|s = 1; m_s\rangle$ and states of even parity belong to the symmetric spin states $|s = 0, 2; m_s\rangle$. The transversality condition requires for each photon that

(I) if the spin states $|s = 1; m_s\rangle$ are represented by a vector spherical harmonic, then this must be the longitudinal $\mathbf{Y}^{-1}$, and

(II) states of even parity must be represented by linear superpositions of spin states $|s = 0; 0\rangle$ and $|s = 2; m_s\rangle$.

Because of the three-component nature of spin vectors, the ψ function associated with a two-photon state can be represented by a tensor in spin space, denoted $[f(\varkappa)]_{\alpha_1 \alpha_2}$ where α_1 and α_2 are spin labels of the individual photons, and $\varkappa$ refers to the relative momentum of the two photons, which is the only remaining variable after the center of mass momentum has been separated. In this notation, the transversality condition for each photon may be written

(A3.20) $$(\varkappa)_{\alpha_1} [f(\varkappa)]_{\alpha_1 \alpha_2} = [f(\varkappa)]_{\alpha_1 \alpha_2} (\varkappa)_{\alpha_2} = 0.$$

Now, for two-photon states of odd parity, the spin states are antisymmetric, which means $f_{\alpha_1 \alpha_2}$ is a skew tensor in three dimensions, and therefore representable in the form

(A3.21) $$f_{\alpha_1 \alpha_2}^{\text{odd}} = \begin{pmatrix} 0 & A_3 & -A_2 \\ -A_3 & 0 & A_1 \\ A_2 & -A_1 & 0 \end{pmatrix},$$

with an axial vector $\mathbf{A}$, allowing one to write the transversality condition

(A3.22) $\mathbf{n} \times \mathbf{A} = 0.$

This means the vector $\mathbf{A}$ representing the antisymmetric spin state of two photons must be longitudinal, and therefore, for given j and m, according to (A3.11)

(A3.23) $\mathbf{A} = \mathbf{Y}_{jm}^{-1} = \sqrt{\dfrac{j}{2j+1}}\,\mathbf{Y}_{j,j-1,\,m} - \sqrt{\dfrac{j+1}{2j+1}}\,\mathbf{Y}_{j,j+1,\,m}.$

This function has odd parity only when j is even. Therefore, *there exist no two-photon states of odd parity if* j *is odd*. In the language of selection rules, one can thus say that a particle of odd integer angular momentum and odd parity cannot decay into two photons under conservation of these quantities.

For two-photon states of even parity the spin states are symmetric, which means $f_{\alpha_1\alpha_2}^{\mathrm{even}}$ is a symmetric tensor in three dimensions, having six linearly independent components, corresponding to the six different values of l possible for given j and m, namely $l = j,\, j \pm 1,\, j \pm 2$ for $s = 2$ and $l = j$ for $s = 0$. Since the parity $(-1)^l$ is specified to be even, however, for even $j \geqslant 2$ only four components can be unequal to zero, namely $l = j,\, j \pm 2$ for $s = 2$ and $l = j$ for $s = 0$, whereas for odd $j \geqslant 3$ only two components can be nonvanishing, namely $l = j \pm 1$ for $s = 2$. Special cases are $j = 0$, when there exist only two components, corresponding to $l = 2$ for $s = 2$ and $l = 0$ for $s = 0$, and $j = 1$, when there is only one component, corresponding to $l = 2$ for $s = 2$. Among these, the transversality condition causes further restrictions. For even $j \geqslant 2$ there are only two transverse states and for odd $j \geqslant 3$ there is only one transverse state. In the special case $j = 0$ there is only one transverse state, and for $j = 1$ there is no transverse state.

Further details may be obtained from the references listed below.

REFERENCES

A. I. Akhiezer and V. B. Berestetskii, "Quantumelectrodynamics." USAEC transl. 1st Russian ed., Oak Ridge, Tennessee, 1957; abbreviated English version, 2nd Russian ed., Oldbourne Press, London, 1963; complete English transl., 2nd Russian ed.; Wiley, New York, 1964.

J. M. Blatt and V. F. Weisskopf, "Theoretical Nuclear Physics." Wiley New York, 1952.

M. E. Rose, "Multipole Fields." Wiley, New York, 1955.

The Invariance of Dirac's Equation under Lorentz Transformations

Dirac's equation, in momentum and energy representation

(A4.1) $$(\gamma_4\,\omega-\boldsymbol{\gamma}\mathbf{k})\,A \;=\; mA$$

can be written, with the conventions

(A4.2) $$A_\mu B_\mu = A_4 B_4 - \mathbf{AB}; \qquad \omega = k_4$$

in the form

(A4.3) $$(\gamma_\nu\,k_\nu - m)\,A(k_\nu) \;=\; 0.$$

Requiring that the matrices γ are the same in all coordinate frames, covariance of Dirac's equation means that if (A4.3) holds, then

(A4.4) $$(\gamma_\lambda\,k_\lambda' - m)\,A'(k_\lambda') \;=\; 0$$

should also hold, where the connection between $A'(k_\lambda')$ and $A(k_\nu)$ is established by a unitary operator U in spin-chirality space,

(A4.5) $$A'(k_\lambda') \;=\; U A(k_\nu); \qquad A(k_\nu) \;=\; U^{-1} A'(k_\lambda')$$

to be determined from the Lorentz transformation

(A4.6) $$k_\lambda' \;=\; L_{\lambda\nu}\,k_\nu; \qquad k_\nu \;=\; L_{\nu\lambda}^{-1}\,k_\lambda'$$

which leaves

(A4.7) $$k_\lambda'\,k_\lambda' \;=\; k_\nu\,k_\nu \;=\; m^2$$

invariant. This means that

(A4.8) $$L_{\mu\lambda}\,L_{\nu\lambda} = \delta_{\mu\nu}; \qquad L_{\mu\lambda} = L_{\lambda\mu}^{-1}; \qquad L_{\mu\lambda}\,L_{\lambda\nu}^{-1} = \delta_{\mu\nu}.$$

Note the convention

(A4.9)
$$\delta_{\mu\nu} = \left\{ \begin{array}{ll} -1 & \text{for}\quad \mu = \nu = 1,2,3 \\ +1 & \text{for}\quad \mu = \nu = 4 \\ 0 & \text{otherwise} \end{array} \right\} \qquad \text{so that} \qquad A_\mu \delta_{\mu\nu} = A_\nu.$$

Now write Eq. (A4.3) with the help of (A4.4) and (A4.6) as

(A4.10)

$$(\gamma_\nu L_{\nu\lambda}^{-1} k'_\lambda U^{-1} - m U^{-1}) A'(k'_\lambda) = 0 \quad \text{i.e.} \quad (L_{\lambda\nu} U \gamma_\nu U^{-1} k'_\lambda - m) A'(k'_\lambda) = 0.$$

This is identical with (A4.4) provided

(A4.11) $$L_{\lambda\nu} U \gamma_\nu U^{-1} = \gamma_\lambda \quad \text{or} \quad U^{-1} \gamma_\lambda U = L_{\lambda\nu} \gamma_\nu.$$

This equation is true for *all* Lorentz transformations, including the improper ones. The main interest here lies, however, in the proper Lorentz transformations, i.e. those whose determinant is $+1$ and for which $L_{44} > 0$, so that they may be thought of as evolving continuously from the identity transformation, and do not contain transformations reversing the sign of the energy.

Now consider the case of the infinitesimal transformation,

(A4.12) $$k'_\lambda = k_\lambda + \epsilon_{\lambda\nu} k_\nu \quad \text{with} \quad \epsilon_{\lambda\nu} = -\epsilon_{\nu\lambda}$$

which guarantees (A4.8). Thus

(A4.13)

$$L_{\lambda\nu} = \delta_{\lambda\nu} + \epsilon_{\lambda\nu} \quad \text{and} \quad L_{\lambda\nu} L_{\lambda\mu} = \delta_{\lambda\nu}\delta_{\lambda\mu} + \delta_{\lambda\nu}\epsilon_{\lambda\mu} + \delta_{\lambda\mu}\epsilon_{\lambda\nu}$$
$$= \delta_{\nu\mu} + \epsilon_{\nu\mu} + \epsilon_{\mu\nu} = \delta_{\nu\mu}.$$

A solution of Eq. (A4.11) will now be sought by putting

(A4.14)

$$U = I + (i/2) \epsilon_{\mu\nu} M_{\mu\nu} \quad \text{with} \quad M_{\mu\nu} = -M_{\nu\mu}$$

where the $M_{\mu\nu}$ are 4×4 matrices numbered by the pair of indices (i.e. $M_{\mu\nu}$ is *not* a matrix element). There are thus six independent matrices $M_{\mu\nu}$. Alternatively one may write

(A4.15) $$U = I + i\epsilon_\alpha S_\alpha$$

with

(A4.16)

$$\epsilon_1 = \epsilon_{23} \quad \epsilon_2 = \epsilon_{31} \quad \epsilon_3 = \epsilon_{12} \quad \epsilon_4 = \epsilon_{14} \quad \epsilon_5 = \epsilon_{24} \quad \epsilon_6 = \epsilon_{34}$$
$$S_1 = M_{23} \quad S_2 = M_{31} \quad S_3 = M_{12} \quad S_4 = M_{14} \quad S_5 = M_{24} \quad S_6 = M_{34}$$

Substituting this into Eq. (A4.11), keeping only terms linear in $\epsilon_{\mu\nu}$, one obtains

(A4.17)

$$(i/2) \epsilon_{\mu\nu}(\gamma_\lambda M_{\mu\nu} - M_{\mu\nu} \gamma_\lambda) = \epsilon_{\lambda\nu} \gamma_\nu = \tfrac{1}{2}\epsilon_{\mu\nu}(\delta_{\mu\lambda}\gamma_\nu - \delta_{\nu\lambda}\gamma_\mu).$$

It is seen that, up to terms which commute with all γ_λ, $M_{\mu\nu}$ must satisfy

(A4.18) $$i(\gamma_\lambda M_{\mu\nu} - M_{\mu\nu}\gamma_\lambda) = \delta_{\mu\lambda}\gamma_\nu - \delta_{\nu\lambda}\gamma_\mu.$$

This is obviously solved by

(A4.19)

$$M_{\mu\nu} = -(i/4)(\gamma_\mu\gamma_\nu - \gamma_\nu\gamma_\mu) = \begin{cases} -(i/2)\gamma_\mu\gamma_\nu & \text{for} \quad \mu \neq \nu \\ 0 & \text{for} \quad \mu = \nu. \end{cases}$$

Indeed,

(A4.20)

$$\tfrac{1}{2}(\gamma_\lambda\gamma_\mu\gamma_\nu - \gamma_\mu\gamma_\nu\gamma_\lambda) = \tfrac{1}{2}(\gamma_\lambda\gamma_\mu\gamma_\nu + \gamma_\mu\gamma_\lambda\gamma_\nu - 2\delta_{\nu\lambda}\gamma_\mu) = \delta_{\mu\lambda}\gamma_\nu - \delta_{\nu\lambda}\gamma_\mu.$$

The C.R.s between the $M_{\mu\nu}$ can now be established by straightforward computation:

(A4.21)

$$[M_{\kappa\lambda}, M_{\mu\nu}] = i(\delta_{\kappa\mu}M_{\lambda\nu} + \delta_{\lambda\nu}M_{\kappa\mu} - \delta_{\lambda\mu}M_{\kappa\nu} - \delta_{\kappa\nu}M_{\lambda\mu}).$$

Writing

(A4.22) $$[M_{\kappa\lambda}, M_{\mu\nu}] = \tfrac{1}{2}C_{\kappa\lambda,\,\mu\nu,\,\rho\sigma}M_{\rho\sigma}$$

(the factor $\tfrac{1}{2}$ is needed because the summation over ρ and σ counts each term twice) one has for the structure constants $C_{\kappa\lambda,\,\mu\nu,\,\rho\sigma}$

(A4.23)

$$C_{\kappa\lambda,\,\mu\nu,\,\rho\sigma} = i(\delta_{\kappa\mu}\delta_{\rho\lambda}\delta_{\sigma\nu} + \delta_{\lambda\nu}\delta_{\rho\kappa}\delta_{\sigma\mu} - \delta_{\lambda\mu}\delta_{\rho\kappa}\delta_{\sigma\nu} - \delta_{\kappa\nu}\delta_{\rho\lambda}\delta_{\sigma\mu})$$

so that

(A4.24)

$$\text{trace}\,(M_{\alpha\beta}M_{\gamma\delta}) = C_{\alpha\beta,\,\mu\nu,\,\rho\sigma}C_{\rho\sigma,\,\mu\nu,\,\gamma\delta} = 4(\delta_{\alpha\gamma}\delta_{\beta\delta} - \delta_{\alpha\delta}\delta_{\beta\gamma}).$$

REFERENCE

W. Pauli, Die Allgemeinen Prinzipien der Wellenmechanik, *in* "Encyclopedia of Physics" (S. Flügge, ed.), Vol. 5, Pt. I, Section 19. Springer, Berlin, 1958.

The Most General Canonical Transformation of a Pair of Fermion Operators

Any unitary operator U is of the form

(A5.1) $$U = e^{iS}; \qquad S = S^+$$

so that any operator A transforms as

(A5.2) $$\tilde{A} = e^{iS} A e^{-iS} = A + i[SA] + (i^2/2!)[S[SA]] + \ldots$$

Consider now two fermion operators α and β, which together with their hermitean adjoints satisfy the anti-C.R.s

(A5.3) $$\{\alpha^+\alpha\} = I; \qquad \{\beta^+\beta\} = I; \qquad \text{all other } \{ \ \} = 0.$$

For fermions which are completely specified by a momentum label $\mathbf{k}$ and a spin label s, the operators α and β may, for example, be identified with annihilation operators $a(\mathbf{k}_1,s_1)$ and $a(\mathbf{k}_2,s_2)$ provided the two sets of labels $(\mathbf{k}_1,s_1)$ and $(\mathbf{k}_2,s_2)$ are not identical. The anticommutation properties of α and β allow one to form not more than eight linearly independent products of even order in the operators $\alpha,\alpha^+,\beta,\beta^+$, namely

(A5.4)
$$I; \quad \mu^+ = \alpha^+\beta^+; \quad \mu = \beta\alpha; \quad \nu^+ = \alpha^+\beta; \quad \nu = \beta^+\alpha; \quad n_\alpha = \alpha^+\alpha;$$

$$n_\beta = \beta^+\beta; \quad \rho = \alpha^+\beta^+\alpha\beta = -n_\alpha n_\beta = -\mu^+\mu = \nu^+\nu - n_\beta.$$

These operators commute with any other such set made up out of operators α', β' belonging to a different set of fermion labels $(\mathbf{k}',s')$. Therefore, the most general hermitean operator S which will guarantee that the various operators U, U' belonging to different sets $(\mathbf{k},s)$, $(\mathbf{k}',s')$ commute, will be of the form

(A5.5)
$$S = c_0 + c_1\mu^+ + c_2\mu + c_3\nu^+ + c_4\nu + c_5 n_\alpha + c_6 n_\beta + c_7\rho$$

where the coefficients c_i must still satisfy certain reality conditions so that $S^+ = S$ as required by (A5.1).

This representation can be simplified if new linear combinations are introduced, a simplification to which one is led by the following *theorem of Koppe and Mühlschlegel*:

Let A be an operator such that

(A5.6) $A^2 = 0$ and $AA^+A = A$

then the operators

(A5.7)

$$\Sigma_1 = A^+ + A; \quad \Sigma_2 = i(A^+ - A); \quad \Sigma_3 = AA^+ - A^+A; \quad \Sigma_0 = A^+A + AA^+$$

satisfy the relations

(A5.8)

$$\Sigma_i \Sigma_j = i\Sigma_k \quad (i,j,k \text{ cycl.}); \quad \Sigma_i^2 = \Sigma_0; \quad \Sigma_i \Sigma_0 = \Sigma_0 \Sigma_i = \Sigma_i; \quad \Sigma_0^2 = \Sigma_0.$$

The operators (A5.7) are thus *isomorphic* to the Pauli matrices and the unit matrix. Note, however, that Σ_0 need not be identical with the unit operator I.

The premises of this theorem are satisfied for $A = \mu$ and $A = \nu$. Therefore the following sets of operators satisfy *separately* the relations (A5.8):

(A5.9) $\sigma_1 = \mu^+ + \mu;$

$\sigma_2 = i(u^+ - \mu);$

$\sigma_3 = \mu\mu^+ - \mu^+\mu = I - n_\alpha - n_\beta;$

$\sigma_0 = \mu^+\mu + \mu\mu^+ = I - n_\alpha - n_\beta + 2n_\alpha n_\beta$

(A5.10) $\tau_1 = \nu^+ + \nu;$

$\tau_2 = i(\nu^+ - \nu);$

$\tau_3 = \nu\nu^+ - \nu^+\nu = n_\beta - n_\alpha;$

$\tau_0 = \nu^+\nu + \nu\nu^+ = n_\alpha + n_\beta - 2n_\alpha n_\beta.$

One has, in addition, the relations

(A5.11) $\sigma_0 + \tau_0 = I; \qquad \sigma_0\tau_0 = \tau_0\sigma_0 = 0$

so that, because of (A5.8) *all products between operators σ and operators τ vanish.*

The most general unitary operator can therefore be written

(A5.12) $U = e^{it_0 \tau_0} e^{it\tau} e^{is_0 \sigma_0} e^{is\sigma}$

where $(t_0, \mathbf{t}, s_0, \mathbf{s})$ are eight *real* coefficients.

The C.R.s between the operators σ, τ, and the various product operators made up out of $\alpha, \beta, \alpha^+, \beta^+$ allow one to sum the series (A5.2), representing the various transformations, in closed form. By expending some labor one finds, for example, the following general transformation formulae:

(A5.13)

$$\tilde{\alpha} = A_s A_t (D + 2iCn_\beta)\,\alpha - A_s B_t (D + 2iCn_\alpha)\,\beta$$
$$- B_s A_t (D^* - 2iCn_\alpha)\,\beta^+ - B_s B_t (D^* - 2iCn_\beta)\,\alpha^+$$

(A5.14)

$$\tilde{\beta} = A_s B_t^* (D + 2iCn_\beta)\,\alpha + A_s A_t^* (D + 2iCn_\alpha)\,\beta$$
$$- B_s B_t^* (D^* - 2iCn_\alpha)\,\beta^+ + B_s A_t^* (D^* - 2iCn_\beta)\,\alpha^+$$

where

(A5.15) $A_s = \cos|s| + (is_3/|s|)\sin|s| = A_{-s}^*$

(A5.16) $B_s = [i(s_1 + is_2)/|s|]\sin|s| = -B_{-s}$

(A5.17) $C = C(s_0 - t_0) = \sin(t_0 - s_0) = -C(t_0 - s_0)$

(A5.18) $D = D(s_0 - t_0) = e^{i(s_0 - t_0)} = D^*(t_0 - s_0).$

The transformation of Bogoliubov [Section 30, Eqs. (30.66)] is contained in (A5.13) and (A5.14) as a special case. By putting $\mathbf{t} = 0$, $t_0 = 0$, $s_0 = 0$, $s_1 = s_3 = 0$ and letting only $s_2 \neq 0$ one has $A_s = \cos(s_2)$, $A_t = 1$, $B_s = -\sin(s_2)$, $B_t = 0$, $C = 0$, $D = 1$, so that

(A5.19) $\tilde{\alpha} = \cos(s_2)\,\alpha + \sin(s_2)\,\beta^+$

(A5.20) $\tilde{\beta} = \cos(s_2)\,\beta - \sin(s_2)\,\alpha^+$

This transformation coincides with Bogoliubov's transformation if one makes the identifications

(A5.21)

$\alpha = a(\mathbf{k}, 1); \qquad \beta = a(-\mathbf{k}, 2); \qquad u(\mathbf{k}) = \cos(s_2); \qquad v(\mathbf{k}) = -\sin(s_2)$

with the understanding that s_2 depends on $|\mathbf{k}|$ only so that the relations (30.68) are satisfied.

Another special case contained in the transformations (A5.13) and (A5.14) is the operator (27.49) representing the transposition of particle labels. The conditions reducing these transformations to

(A5.22) $\tilde{\alpha} = \beta$ and $\tilde{\beta} = \alpha$

are obviously

(A5.23)

$C = 0, \quad D = 1, \quad B_s = 0, \quad A_t = 0, \quad -A_s B_t = A_s B_t^* = 1,$

$$\text{or} \quad A_s = B_t = i.$$

They can be satisfied by putting,

(A5.24)
$$s_0 = t_0 = \pi, \quad s_1 = s_2 = 0, \quad s_3 = \pi/2, \quad t_1 = \pi/2, \quad t_2 = t_3 = 0.$$

Then [using (A5.9)–(A5.11)]

(A5.25)
$$U = e^{i\pi\tau_0} e^{i(\pi/2)\tau_1} e^{i\pi\sigma_0} e^{i(\pi/2)\sigma_3} = e^{i\pi(\tau_0+\sigma_0)} e^{i(\pi/2)\tau_1} e^{i(\pi/2)\sigma_3}$$
$$= -i\exp[i(\pi/2)(\alpha^+\beta+\beta^+\alpha)]\exp[i(\pi/2)(I-n_\alpha-n_\beta)]$$

coincides with the operator T_{12} (27.50) if one makes the identifications

(A5.26) $$a(1) = \alpha \quad \text{and} \quad a(2) = \beta.$$

It is often convenient to have an explicit expression for the operator U, which can be evaluated by using

(A5.27)
$$(\mathbf{s\sigma})^2 = s^2\sigma_0; \quad \sigma_0\mathbf{\sigma} = \mathbf{\sigma}; \quad \sigma_0^2 = \sigma_0; \quad (\mathbf{t\tau})^2 = t^2\tau_0; \quad \tau_0\mathbf{\tau} = \mathbf{\tau}; \quad \tau_0^2 = \tau_0.$$

One has

(A5.28)
$$e^{i\mathbf{s\sigma}} = 1+(i^2/2!)s^2\sigma_0+(i^4/4!)s^4\sigma_0+\ldots+i(\mathbf{s\sigma})+(i^3/3!)s^2(\mathbf{s\sigma})+\ldots$$
$$= 1-\sigma_0(1-\cos|\mathbf{s}|)+i[(\mathbf{s\sigma})/|\mathbf{s}|]\sin|\mathbf{s}|$$

and

(A5.29)
$$e^{is_0\sigma_0} = 1+is_0\sigma_0+(i^2/2!)s_0^2\sigma_0+\ldots = 1-\sigma_0(1-e^{is_0})$$

and completely analogous expressions for $e^{i\mathbf{t\tau}}$ and $e^{it_0\tau_0}$, so that

(A5.30)
$$U = \{1-\tau_0(1-e^{it_0})\}\{1-\tau_0(1-\cos|\mathbf{t}|)+i[(\mathbf{t\tau})/|\mathbf{t}|]\sin|\mathbf{t}|\}$$
$$\times \{1-\sigma_0(1-e^{is_0})\}\{1-\sigma_0(1-\cos|\mathbf{s}|)+i[(\mathbf{s\sigma})/|\mathbf{s}|]\sin|\mathbf{s}|\}.$$

Bogoliubov's transformation is obtained from this as the special case

(A5.31) $$-\sin(s_2) = v(k); \quad \cos(s_2) = u(k)$$

with all other s, t vanishing:

(A5.32) $$U_B = 1-\sigma_0[1-u(k)]-i\sigma_2 v(k).$$

REFERENCE

H. Koppe and B. Mühlschlegel, *Z. Physik* **151**, 613 (1958).

The Delta Function and its Application to Phase Space Considerations

For a scalar variable α the delta function $\delta(\alpha)$ is defined as the integral

$$(A6.1) \qquad \delta(\alpha) = \frac{1}{2\pi} \int_{-\infty}^{+\infty} e^{i\alpha t} \, dt.$$

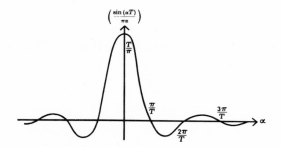

FIG. A6.1. The function $[\sin^2(\alpha T)]/\pi\alpha$ for large T.

For the purpose of visualization (see Fig. A6.1), this may be thought of as the limit

$$(A6.2) \qquad \delta(\alpha) = \lim_{T \to \infty} \frac{1}{2\pi} \int_{-T}^{+T} e^{i\alpha t} \, dt = \lim_{T \to \infty} \frac{\sin(\alpha T)}{\pi\alpha}.$$

It has the properties

$$(A6.3) \quad g(\alpha) = \int_{-\infty}^{+\infty} g(\alpha')\,\delta(\alpha-\alpha')\,d\alpha' \qquad \text{and} \qquad \int_{-\infty}^{+\infty} \delta(\alpha)\,d\alpha = 1.$$

335

Alternatively, one may think of $\delta(\alpha)$ as the limit (see Fig. A6.2)

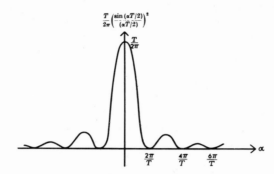

$$\frac{T}{2\pi}\left(\frac{\sin(\alpha T/2)}{(\alpha T/2)}\right)^2$$

FIG. A6.2. The function $[T/2\pi]\,(\sin{}^2(\alpha T/2)/(\alpha T/2))^2$ for large T.

$$(A6.4)\qquad \delta(\alpha) = \lim_{T\to\infty}\frac{1}{2\pi T}\left|\int_{-T/2}^{+T/2} e^{i\alpha t}\,dt\right|^2 = \lim_{T\to\infty}\frac{T}{2\pi}\left(\frac{\sin(\alpha T/2)}{(\alpha T/2)}\right)^2.$$

Comparing (A6.4) with (A6.2) one finds the relation

$$(A6.5)\qquad [\delta(\alpha)]^2 = \frac{1}{2\pi}\delta(\alpha)\int_{-\infty}^{+\infty} dt.$$

For a four-vector P one defines similarly

$$(A6.6)\qquad \delta(P) = \frac{1}{(2\pi)^4}\int e^{iPx}\,d^4x = \delta(\mathbf{P})\,\delta(P_0)$$

and has

$$(A6.7)\qquad [\delta(P)]^2 = \frac{1}{(2\pi)^4}\delta(P)\int d^4x.$$

Of great usefulness for applications of the δ function is the following general formula governing the δ function having a function $f(\alpha)$ as its argument:

$$(A6.8)\qquad \delta[f(\alpha)] = \frac{\sum_r [\delta(\alpha - \alpha_r)]}{|df/d\alpha|},$$

where α_r are the simple roots of $f(\alpha) = 0$ in the region under consideration.

To prove this divide the region of integration into small intervals so that in each interval there is only one root of $f(\alpha) = 0$. For the root α_r, in particular, let this interval be $\alpha_r - \epsilon_r < \alpha < \alpha_r + \epsilon_r$. One has then in this interval the expansion [using $f(\alpha_r) = 0$ by definition, and $f'(\alpha_r) \neq 0$ if α_r is a simple root]

(A6.9) $$f(\alpha) = f(\alpha_r + \epsilon) = f'(\alpha_r)\,\epsilon + (1/2!)f''(\alpha_r)\,\epsilon^2 + \dots$$

and thus

(A6.10) $$\int_{\alpha_r - \epsilon_r}^{\alpha_r + \epsilon} g(\alpha)\,\delta[f(\alpha)]\,d\alpha = \int_{-f'(\alpha_r)\epsilon_r + \dots}^{+f'(\alpha_r)\epsilon_r + \dots} g(\alpha)\,\frac{\delta(f)\,df}{df/d\alpha}$$

$$= \begin{cases} \dfrac{g(\alpha_r)}{f'(\alpha_r)} & \text{for} \quad f'(\alpha_r) > 0 \\[2ex] \dfrac{-g(\alpha_r)}{f'(\alpha_r)} & \text{for} \quad f'(\alpha_r) < 0 \end{cases}$$

$$= \frac{g(\alpha_r)}{|f'(\alpha_r)|}$$

from which the formula (A6.8) follows by summation. Equation (A6.10) is based on the understanding that ϵ_r is chosen such that the signs of the integration limits are determined by the term linear in ϵ_r.

If one takes, for example, $f(\alpha) = \alpha^2 - c^2$ so that there are two roots $\alpha = \pm c$, one has the special case

(A6.11) $$\delta(\alpha^2 - c^2) = \frac{1}{2|c|}[\delta(\alpha - c) + \delta(\alpha + c)].$$

As an application consider the Compton effect and describe it in a laboratory frame of reference in which, initially, the electron is at rest, $\mathbf{k} = 0$, $\Omega = m$, and one photon of energy $\omega = |\varkappa|$ is present. With a normalization of states to one particle in a volume V, $w_{\tau'\tau}V$ is the transition probability per scattering center and unit time, where $w_{\tau'\tau}$ is defined by Eq. (23.22). The incident intensity I in photons per unit area and time is equal to the number of incident photons per unit volume (namely, $1/V$ with the normalization adopted above) times the velocity of light (namely 1 in natural units), giving $I = 1/V$. Thus the scattering cross section per scattering center into a fixed final state τ' is

$$(w_{\tau'\tau}V/I) = w_{\tau'\tau}V^2.$$

In any actual experiment one rarely discriminates a specific final state τ', but rather a set of final states with final momenta of electron and photon ranging between $\mathbf{k}'$ and $\mathbf{k}' + d\mathbf{k}'$, and $\varkappa'$ and $\varkappa' + d\varkappa'$, respectively.

The number of such states, for definite spin and definite polarization, is*

(A6.12) $$dN = [V^2/(2\pi)^6] d\mathbf{k}' \, d\mathbf{\varkappa}'.$$

The cross section per scattering center and unit time into this set of final states is therefore

(A6.13) $$d\sigma = w_{\tau' \tau} V^2 dN.$$

If the detection apparatus responds only to photons scattered into a solid angle dS oriented in direction ϑ, φ, one has to integrate over $d\mathbf{k}'$ and $d\omega'$, keeping ϑ and φ constant. Since $d\mathbf{\varkappa}' = \omega'^2 d\omega' \, dS$ one obtains for the cross section per scattering center encountered by photons scattered into dS

(A6.14) $$\sigma \, dS = [V^4 dS/(2\pi)^6] \iint w_{\tau' \tau} d\mathbf{k}' \, \omega'^2 \, d\omega'.$$

Since $w_{\tau' \tau}$ contains a δ function $\delta(\mathbf{P}' - \mathbf{P}) = \delta(\mathbf{k}' + \mathbf{\varkappa}' - \mathbf{\varkappa})$, the integration over $d\mathbf{k}'$ can be carried out immediately, resulting in replacement of $\mathbf{k}'$ everywhere by

(A6.15) $$\mathbf{k}' = \mathbf{\varkappa} - \mathbf{\varkappa}'$$

in accordance with conservation of momentum. *As a consequence of this, the energies Ω, Ω', ω, ω', appearing in the remaining δ function*

$$\delta(P_0' - P_0) = \delta(\Omega' + \omega' - \Omega - \omega)$$

become interdependent, and this has to be kept in mind when the integration over $d\omega'$ is carried out. Writing

(A6.16) $$(\Omega' + \omega' - \Omega - \omega) = f(\omega')$$

and using Eq. (A6.8), one may write (A6.14) as

(A6.17) $$\sigma \, dS = \frac{V^4 dS}{(2\pi)^{10}} \int \frac{|\langle \tau' | R | \tau \rangle|^2 \, \omega'^2 \, \delta(\omega' - \omega_r') \, d\omega'}{|[df(\omega')/d\omega']|}$$

$$= \frac{V^4 dS}{(2\pi)^{10}} \left(\frac{|\langle \tau' | R | \tau \rangle|^2 \, \omega'^2}{|[df(\omega')/d\omega']|} \right)_{\omega' = \omega_r'}$$

The "resonance value" ω_r' of ω' follows from

(A6.18)
$$[f(\omega')]_{\omega' = \omega_r'} = [\Omega' + \omega_r' - \Omega - \omega] = [\sqrt{m^2 + k'^2} + \omega_r' - m - \omega]$$

$$= [\sqrt{m^2 + \omega^2 + \omega_r'^2 - 2\omega \omega_r' \cos \vartheta} + \omega_r' - m - \omega] = 0,$$

---

* If V is a cube whose sides are of length L, then possible values of $\mathbf{k}$ are $k_i = n_i(2\pi/L)$ with n_i integer. The number of integers whose corresponding vectors $\mathbf{k}$ lie between $\mathbf{k}$ and $\mathbf{k} + d\mathbf{k}$ is equal to $(L/2\pi)^3 dk_1 dk_2 dk_3 = [V/(2\pi)^3] d\mathbf{k}$. Thus any sum $(1/V) \sum_{\mathbf{k}} \ldots$ can, in the limit $V \to \infty$, be replaced by an integral $[1/(2\pi)^3] \int \ldots d\mathbf{k}$.

where ϑ is the angle of scattering between $\varkappa$ and $\varkappa'$, so that $\mathbf{k}'^2 = (\varkappa - \varkappa')^2$ $= \kappa^2 + \kappa'^2 - 2\kappa\kappa' \cos\vartheta = \omega^2 + \omega'^2 - 2\omega\omega' \cos\vartheta$. The result is

(A6.19)
$$\omega'_r = \frac{m\omega}{m + \omega(1 - \cos\vartheta)}$$

and one has also

(A6.20)

$$|[df(\omega')/d\omega']|_{\omega'=\omega_r} = \left(\frac{\omega' - \omega\cos\vartheta}{\sqrt{m^2 + \omega^2 + \omega'^2 - 2\omega\omega'\cos\vartheta}} + 1\right)_{\omega'=\omega_{r'}} = \frac{m\omega}{\Omega'_r\omega'_r}$$

where Ω'_r is the energy of the scattered electron at resonance,

(A6.21) $\quad \Omega'_r = [\sqrt{m^2 + k'^2}]_{\omega'=\omega_{r'}} = \sqrt{m^2 + \omega^2 + \omega_r'^2 - 2\omega\omega'_r\cos\vartheta}.$

The expression for the differential cross section (A6.17) can therefore be cast in the form

(A6.22) $\qquad \sigma\, dS = [V^4\, dS/(2\pi)^{10}]\,|\langle \tau'|R|\tau\rangle|^2\,[\omega_r'^3\,\Omega'_r/m\omega]$

with the understanding that the matrix element has to be evaluated with the resonance value (A6.19) and (A6.21) for the energies in the final state and under observation of conservation of momentum (A6.15).

Finally, if the target and the incident photon beam are unpolarized, and if one does not observe the polarizations of either scattered electrons or scattered photons, the differential cross section must be averaged over initial spins and polarizations, and summed over final spins and polarizations, giving (since only transverse photons are involved)

(A6.23) $\qquad \langle \sigma\, dS\rangle_{\text{Av}} = \frac{1}{4}\sum_{r=1}^{2}\sum_{S=1}^{2}\sum_{r'=1}^{2}\sum_{S'=1}^{2}\sigma\, dS.$

NOTES

The δ function was introduced into physics by Dirac [1]. For a more detailed description of the δ function see Iwanenko and Sokolow [2].

A complete treatment of the Compton effect is contained, for example, in Mandl [3].

REFERENCES

[1] P. A. M. Dirac, "Quantum Mechanics," 4th ed., Univ. Press, Oxford, London and New York, 1958. §15.
[2] D. Iwanenko and A. Sokolow, "Klassische Feldtheorie," Chapter 1. Akademie Verlag, Berlin, 1953.
[3] F. Mandl, "Introduction to Quantum Field Theory," Chapter 15. Wiley (Interscience), New York, 1959.

If Galileo Had Known Quantum Mechanics

Galileo would presumably have been keenly interested in the transformation properties of the time-dependent ψ function $\Psi(\mathbf{q},t)$ of a free particle of mass m, satisfying

$$(A7.1) \qquad i(\partial\Psi/\partial t) - (1/2m)\nabla^2\Psi = 0; \qquad \boldsymbol{\nabla} = (\partial/\partial\mathbf{q}),$$

under the general transformation bearing his name,

$$(A7.2) \qquad \mathbf{q} \to \mathbf{q}' = R\mathbf{q} + \mathbf{V}t + \mathbf{a}; \qquad t \to t' = t + b,$$

where R is a constant orthogonal matrix representing a spatial rotation, $\mathbf{V}$ a constant vector representing a pure Galileo transformation, $\mathbf{a}$ a constant vector representing a displacement in space, and b a constant representing a displacement in time.

Denoting the set of numbers $(R, \mathbf{V}, \mathbf{a}, b)$ by G, the transformation (A7.2) should be representable by a unitary operator $U(G)$ so that, up to some phase factor, $|\mathbf{q}(t)\rangle' = U(G)|\mathbf{q}(t)\rangle = |\mathbf{q}'(t')\rangle$ is the transformed state. Accordingly, the state can be characterized in the transformed frame by a ψ function $\Psi'(\mathbf{q},t)$ which differs from the untransformed ψ function taken at the transformed point at most by a phase factor,

$$(A7.3) \qquad \Psi'(\mathbf{q},t) = e^{if(\mathbf{q}',t')}\Psi(\mathbf{q}',t').$$

Invariance under Galileo transformations means Ψ' and Ψ must satisfy the same Schroedinger equation. Thus

$$(A7.4) \qquad i(\partial\Psi'/\partial t) - (1/2m)\nabla^2\Psi' = 0.$$

This equation imposes conditions on the phase function f. Using the relations, following from (A7.2) and the orthogonality of R,

$$(A7.5)$$
$$(\partial/\partial t) = (\partial/\partial t') + \mathbf{V}\cdot\boldsymbol{\nabla}'; \qquad \boldsymbol{\nabla} = R\boldsymbol{\nabla}' \qquad \text{so that} \qquad \nabla^2 = \nabla'^2$$

one can write (A7.4) upon substitution of (A7.3)

$$(A7.6) \quad [-(\partial f/\partial t') - \mathbf{V}\cdot\boldsymbol{\nabla}'f + (1/2m)(\boldsymbol{\nabla}'f)^2 - (i/2m)\nabla'^2 f]e^{if}\Psi$$
$$+ i[\mathbf{V} - (1/m)\boldsymbol{\nabla}'f]e^{if}\boldsymbol{\nabla}'\Psi$$
$$+ [i(\partial\Psi/\partial t') - (1/2m)\nabla'^2\Psi]e^{if} = 0.$$

341

The last term vanishes on account of the Schroedinger equation (A7.1) (which must hold for any $\mathbf{q}$ and t), and since Ψ and $\mathbf{\nabla}' \Psi$ are linearly independent, one has the two conditions

(A7.7) $$\mathbf{\nabla}' f = m\mathbf{V}$$

(A7.8) $$(\partial f/\partial t') = -\mathbf{V} \cdot \mathbf{\nabla}' f + (1/2m)(\mathbf{\nabla}' f)^2 - (i/2m)\mathbf{\nabla}'^2 f.$$

Substituting (A7.7) into (A7.8) and keeping in mind that $\mathbf{V}$ is a constant vector, one finds

(A7.9) $$(\partial f/\partial t') = -\tfrac{1}{2}mV^2.$$

Equations (A7.7) and (A7.9) can be integrated immediately, yielding

(A7.10)
$$f(\mathbf{q}', t') = m\mathbf{V} \cdot \mathbf{q}' - \tfrac{1}{2}mV^2 t' + C \qquad (C \text{ a constant}).$$

The remarkable conclusion to be drawn from this result is that *the phase factor* f *cannot, in general, be eliminated by judicious choice of the integration constant* C. This has a profound consequence, first noticed by Bargmann, namely:

It is impossible to have in nonrelativistic quantum mechanics states which are linear superpositions of states describing particles of different masses. This means one cannot grasp in nonrelativistic quantum mechanics states with a mass spectrum, or states describing unstable elementary particles.

To see this, consider a linear superposition

(A7.11) $$\Psi = \Psi_1 + \Psi_2$$

where Ψ_1 and Ψ_2 transform according to (the constant phase C has been put equal to zero for simplicity's sake)

(A7.12)
$$\Psi'_\alpha(\mathbf{q}, t) = \{\exp[im_\alpha(\mathbf{V} \cdot \mathbf{q}' - \tfrac{1}{2}V^2 t')]\}\,\Psi_\alpha(\mathbf{q}', t'); \quad (\alpha = 1, 2).$$

Now perform the following sequence of transformations, amounting to the identity,

(A7.13)
$$G_I = G_4 G_3 G_2 G_1 = (I, -\mathbf{V}, 0, 0)(I, 0, -\mathbf{a}, 0)(I, \mathbf{V}, 0, 0)(I, 0, \mathbf{a}, 0)$$
$$= (I, 0, 0, 0)$$

corresponding to a sequence of coordinates and velocities

(A7.14)

$\mathbf{q}_4 = \mathbf{q};$	$\mathbf{q}_3 = \mathbf{q} + \mathbf{V}t;$	$\mathbf{q}_2 = \mathbf{q} + \mathbf{V}t + \mathbf{a};$	$\mathbf{q}_1 = \mathbf{q} + \mathbf{a}$
$t_4 = t;$	$t_3 = t;$	$t_2 = t;$	$t_1 = t$
$\mathbf{V}_4 = -\mathbf{V};$	$\mathbf{V}_3 = 0;$	$\mathbf{V}_2 = \mathbf{V};$	$\mathbf{V}_1 = 0$

giving rise, for each m_α, to a sum of phases

(A7.15)
$$m_\alpha \sum_{j=1}^{4} (\mathbf{V}_j \cdot \mathbf{q}_j - \tfrac{1}{2} V_j^2 t_j) = m_\alpha [\mathbf{V} \cdot (\mathbf{q}_2 - \mathbf{q}_4) - V^2 t] = m_\alpha \mathbf{V} \cdot \mathbf{a}$$

so that the transformed ψ function becomes

(A7.16)
$$\Psi^T = U(G_I) \, \Psi = [\exp{(im_1 \mathbf{V} \cdot \mathbf{a})}] \, \Psi_1 + [\exp{(im_2 \mathbf{V} \cdot \mathbf{a})}] \, \Psi_2.$$

This means, a transformation amounting to the identity can affect the norm of the superposition (A7.11). In other words, the relative phase of two ψ functions describing particles of different mass is completely arbitrary if one demands Galileo invariance. To avoid inconsistency one must conclude that a superposition of the type (A7.11) is without meaning, and that there can exist no operators which sponsor transitions between states characterized by different masses m_1 and m_2. *This amounts to existence of a superselection rule which guarantees the strict conservation of mass in nonrelativistic quantum mechanics.*

It should be noted that this conclusion is not valid in relativistic quantum mechanics. Consider, for example, the equation governing the ψ function of a spinless particle

(A7.17)
$$\Box^2 \psi - m^2 \psi = 0$$

which is invariant under the inhomogeneous Lorentz transformations

(A7.18)
$$x \rightarrow x' = Lx + u$$

where L is a constant Lorentz matrix and u a constant four vector. Demanding that

(A7.19)
$$\psi'(x) = e^{ig(x')} \psi(x')$$

satisfy the same equation

(A7.20)
$$\Box^2 \psi' - m^2 \psi' = 0$$

leads, because of

(A7.21)
$$\Box = L\Box' \qquad \text{so that} \qquad \Box^2 = \Box'^2,$$

to the equation

(A7.22)
$$[i\Box'^2 g - \Box' g \Box' g] e^{ig} \psi + 2i\Box' g \Box' \psi e^{ig} + [\Box'^2 \psi - m^2 \psi] e^{ig} = 0.$$

The last term vanishes on account of (A7.17), and the remaining conditions are

(A7.23)
$$\Box' g = 0; \qquad i\Box'^2 g - \Box' g \Box' g = 0,$$

which require

(A7.24)
$$g = \text{constant.}$$

It is perhaps instructive to set down here the alternative treatment of the Galileo transformation in momentum-energy representation, in which a free particle of momentum $\mathbf{k}$ and energy ω is characterized by the ψ function

$$(A7.25) \qquad \Phi(\mathbf{k}, \omega) = \int e^{i[\omega t - \mathbf{k} \cdot \mathbf{q}]} \Psi(\mathbf{q}, t) \, d\mathbf{q} \, dt$$

satisfying the Schroedinger equation

$$(A7.26) \qquad [\omega - (k^2/2m)] \Phi(\mathbf{k}, \omega) = 0.$$

The transformed function is then given by

$$(A7.27) \qquad \Phi'(\mathbf{k}, \omega) = \int e^{i[\omega t - \mathbf{k} \cdot \mathbf{q}]} \Psi'(\mathbf{q}, t) \, d\mathbf{q} \, dt.$$

Using

$$(A7.28) \qquad \mathbf{q} = R^{-1}(\mathbf{q}' - \mathbf{V}t' + \mathbf{V}b - \mathbf{a}); \qquad t = t' - b,$$

and the transformation formulae (A7.3) and (A7.10), the integrand can be expressed in terms of the transformed coordinates $\mathbf{q}', t'$,

$$(A7.29)$$
$$\Phi'(\mathbf{k}, \omega) = \int \exp \{i[\omega(t' - b) - \mathbf{k} \cdot R^{-1}(\mathbf{q}' - \mathbf{V}t' + \mathbf{V}b - \mathbf{a})$$
$$+ m\mathbf{V} \cdot \mathbf{q}' - \tfrac{1}{2} m V^2 t' + C]\} \Psi(\mathbf{q}', t') \, d\mathbf{q}' \, dt'.$$

Extracting the terms not containing $\mathbf{q}', t'$ from under the integral, dropping the primes on the integration variables, and using the orthogonality of R by writing $\mathbf{k} \cdot R^{-1} \mathbf{a} = R\mathbf{k} \cdot \mathbf{a}$, etc., one obtains

$$(A7.30)$$
$$\Phi'(\mathbf{k}, \omega) = \exp [i(- \omega b + R\mathbf{k} \cdot \mathbf{a} - bR\mathbf{k} \cdot \mathbf{V} + C)] \int \exp [i(\omega' t - \mathbf{k}' \cdot \mathbf{q})]$$
$$\Psi(\mathbf{q}, t) \, d\mathbf{q} \, dt$$

with

$$(A7.31) \qquad \omega' = \omega + \mathbf{V} \cdot R\mathbf{k} + \tfrac{1}{2} m V^2; \qquad \mathbf{k}' = R\mathbf{k} + m\mathbf{V}.$$

The quantities ω' and $\mathbf{k}'$ can be called the transformed energy and momentum, because they satisfy

$$(A7.32) \qquad \omega'^2 - (k'^2/2m) = \omega^2 - (k^2/2m).$$

Expressing ω and $\mathbf{k}$ in terms of ω' and $\mathbf{k}'$,

$$(A7.33) \qquad \omega = \omega' - \mathbf{V} \cdot \mathbf{k}' + \tfrac{1}{2} m V^2; \qquad \mathbf{k} = R^{-1}(\mathbf{k}' - m\mathbf{V}),$$

one can write the transformation formula (A7.30)

$$(A7.34)$$
$$\Phi'(\mathbf{k}, \omega) = \exp \{i[(b/2) m V^2 - m\mathbf{V} \cdot \mathbf{a} + C]\} \exp [i(- b\omega' + \mathbf{a} \cdot \mathbf{k}')] \Phi(\mathbf{k}', \omega').$$

As in case of the coordinate representation, the phase factors cannot be eliminated entirely by judicious choice of the constant C. The sequence of transformations (A7.13), corresponding to

(A7.35)

$$\mathbf{k}_4 = \mathbf{k}; \qquad \mathbf{k}_3 = \mathbf{k} + m\mathbf{V}; \qquad \mathbf{k}_2 = \mathbf{k} + m\mathbf{V}; \qquad \mathbf{k}_1 = \mathbf{k}$$
$$\mathbf{a}_4 = 0; \qquad \mathbf{a}_3 = -\mathbf{a}; \qquad \mathbf{a}_2 = 0; \qquad \mathbf{a}_1 = \mathbf{a}$$
$$b_4 = 0; \qquad b_3 = 0; \qquad b_2 = 0; \qquad b_1 = 0$$

give once again, for each m_α, rise to a change in phase

(A7.36)
$$\sum_j \mathbf{a}_j \cdot \mathbf{k}_j = \mathbf{a}_1 \cdot \mathbf{k}_1 + \mathbf{a}_3 \cdot \mathbf{k}_3 = -m_\alpha \mathbf{a} \cdot \mathbf{V}$$

leading again to Bargmann's superselection rule.

A representation for the generator $\mathbf{u}$ of the pure Galileo transformation characterized by the parameter $\mathbf{V}$, so that in momentum-energy representation

(A7.37) $$|\mathbf{k}, \omega\rangle^T = |\mathbf{k}', \omega'\rangle = U(\mathbf{V})|\mathbf{k}, \omega\rangle = e^{i\mathbf{V}\cdot\mathbf{u}}|\mathbf{k}, \omega\rangle,$$

can be obtained by demanding, in accordance with (A7.31),

(A7.38)
$$\mathbf{k}' = \langle\mathbf{k}', \omega'|\mathbf{P}|\mathbf{k}', \omega'\rangle = \langle\mathbf{k}, \omega|\mathbf{P}|\mathbf{k}, \omega\rangle + m\mathbf{V} = \mathbf{k} + m\mathbf{V}$$

and

(A7.39)
$$\omega' = \langle\mathbf{k}', \omega'|H|\mathbf{k}', \omega'\rangle = \langle\mathbf{k}, \omega|H|\mathbf{k}, \omega\rangle + \langle\mathbf{k}, \omega|\mathbf{V}\cdot\mathbf{P}|\mathbf{k}, \omega\rangle + \tfrac{1}{2}mV^2$$
$$= \omega + \mathbf{V}\cdot\mathbf{k} + \tfrac{1}{2}mV^2.$$

Using the expansions

$$e^{-i\mathbf{V}\cdot\mathbf{u}}\,\mathbf{P}\,e^{i\mathbf{V}\cdot\mathbf{u}} = \mathbf{P} - i(\mathbf{V}\cdot[\mathbf{u}]\,\mathbf{P}] + \ldots$$

(A7.40)

$$e^{-i\mathbf{V}\cdot\mathbf{u}}\,H\,e^{i\mathbf{V}\cdot\mathbf{u}} = H - i(\mathbf{V}\cdot[\mathbf{u}]\,H] + (i^2/2!)\,(\mathbf{V}\cdot[\mathbf{u}]\,(\mathbf{V}\cdot[\mathbf{u}]\,H]] - + \ldots$$

one has then the requirements

(A7.41) $$-i(\mathbf{V}\cdot[\mathbf{u}]\,\mathbf{P}] = m\mathbf{V}$$

(A7.42)
$$-i(\mathbf{V}\cdot[\mathbf{u}]\,H] = \mathbf{V}\cdot\mathbf{P}; \qquad -(i/2)\,(\mathbf{V}\cdot[\mathbf{u}]\,(\mathbf{V}\cdot\mathbf{P})] = \tfrac{1}{2}mV^2.$$

The second equation (A7.42) is identically satisfied if (A7.41) holds, and for the unknown operator $\mathbf{u}$ there remain the C.R.s

(A7.43) $$[u_j P_k] = im\,\delta_{jk}; \qquad [\mathbf{u}H] = i\mathbf{P},$$

which can be solved by putting

(A7.44) $$\mathbf{u} = im(\partial/\partial\mathbf{P}) + i\mathbf{P}(\partial/\partial H).$$

The general Galileo transformation is thus associated with the 10 operators

$\mathbf{J} = \mathbf{L} + \mathbf{S} = (\mathbf{Q} \times \mathbf{P}) + \mathbf{S} = -i(\mathbf{P} \times \partial/\partial\mathbf{P}) + \mathbf{S}$, of total angular momentum, generating the rotations R,

$\mathbf{u} = im(\partial/\partial\mathbf{P}) + i\mathbf{P}(\partial/\partial H)$, which can be identified as coordinate operator $\mathbf{Q} = m\mathbf{u}$, generating the pure Galileo transformations $\mathbf{V}$,

$\mathbf{P}$, of linear momentum, generating the displacements in space $\mathbf{a}$, and

H, of energy, generating the displacements in time b.

They satisfy the C.R.s

(A7.45)

$$[J_1 J_2] = iJ_3 \quad \text{(cycl.)}; \qquad [J_1 u_2] = iu_3 \quad \text{(cycl.)};$$
$$[u_j u_k] = 0;$$

$$[J_1 P_2] = iP_3 \quad \text{(cycl.)}; \qquad [J_k H] = 0$$
$$[u_j P_k] = im\,\delta_{jk}; \qquad\qquad [u_k H] = iP_k$$
$$[P_j P_k] = 0; \qquad\qquad\qquad [P_k H] = 0$$
$$[HH] = 0$$

It is interesting to note that the algebra engendered by the Galileo transformation admits the two invariants

(A7.46)

$$P^2 - 2mH = 2mE \quad \text{and} \quad [m\mathbf{J} - (\mathbf{u} \times \mathbf{P})]^2 = m^2 S^2 = m^2 s(s+1),$$

where E and S can be interpreted as the *intrinsic* energy and *intrinsic* angular momentum of the particle under consideration. At this point there becomes apparent the feasibility of a nonrelativistic quantum mechanics of particles of mass $m = 0$, which are characterized by two invariants, P^2 and $(\mathbf{u} \times \mathbf{P})^2$, with $\mathbf{u} = i\mathbf{P}(\partial/\partial H)$. Further details of this intriguing aspect of invariance under Galileo transformations can be found in the references listed below.

REFERENCES

V. Bargmann, *Ann. Math.* **59**, 1 (1954).

A. S. Wightman, *Nuovo Cimento Suppl.* **14**, 86 (1959).

M. Hamermesh, *Ann. Phys. N.Y.* **9**, 518 (1960).

J.-M. Levy-Leblond, *J. Math. Phys.* **4**, 776 (1963).

Author Index

A

Abrikosov, A. A., 231, 307
Adamskii, V. B., 192
Aharonov, Y., 175
Akhiezer, A. I. ,150, 250, 326
Albertson, J., 42
Ambler, E., 97
Archibald, W. J., 150

B

Bardeen, J., 307
Bargmann, V., 346
Becker, R., 168
Beliaev, S. T., 307
Berestetskii, V. B., 150, 250, 326
Blatt, J. M., 326
Bogoliubov, N. N., 150, 216, 231, 307
Bohm, D., 175
Born, M., 54, 86
Brenig, W., 202
Brueckner, K., 288

C

Chambers, R. G., 175
Chraplyvy, Z. V., 168
Condon, E. U., 321
Cooper, L. N., 307

D

Darwin, C. G., 168
De Broglie, L., 64
De Sobrino, L., 140
D'Espagnat, B., 288
Dirac, P. A. M., 9, 23, 30, 118, 129,
    149, 168, 175, 250, 339
Dyson, F. J., 231

E

Edmonds, A. R., 315
Ehrenfest, P., 64
Einstein, A., 307
Ekstein, H., 202

F

Fano, U., 30
Feldman, D., 202
Fermi, E., 9, 186, 202, 288
Feynman, R. P., 75, 113, 168, 202, 216
Fierz, M., 118
Furry, W., 262

G

Gell-Mann, M., 186, 231, 262, 288
Gerlach, W., 9
Glashow, S. L., 186
Goldhaber, M., 288
Goldstein, H., 86

H

Haag, R., 202
Hamermesh, M., 346
Hayward, R. W., 97
Heisenberg, W., 54, 69, 86, 250
Henley, E. M., 262
Hoppes, D. D., 97
Horn, D., 129
Hudson, P. R., 97

I

Iwanenko, D., 339

J

Jacobsohn, B. A., 262
Jauch, J. M. 149, 202, 216
Jona-Lasinio, G., 118
Jordan, P., 23, 54, 86, 129, 149

K

Kaempffer, F. A., 118, 307
Källén, G., 150, 216, 231
Khalatnikov, I. M., 231, 307
Klein, O., 149
Konopinski, E., 129

347

Subject Index

A

Absolute zero of temperature, 72
Action integral, 189
Addition of,
  angular momenta, 139, 243, 245,
    265, 274, 317–321
  intensities, 41
  isospins, 265, 273
  probability amplitudes, 40, 223
  velocities, 61, 62
Affinities, 189
Amplification, by stimulated emission, 75
Amplitudes,
  addition of probability, 40
  probability, 6, 19, 275
  transition, 255
Analytic properties of true propagators, 228–230
Angular momentum,
  eigenstates, 309–315
  intrinsic, see Spin
  operators, 16, 77, 103, 134, 273,
    309–315
  orbital, 77, 93, 238, 313
Annihilation operator, 119, 131, 195,
  234, 273
Anticommutation relations, 120, 152
Antifermion state, 112
Antimatter, 99
Antisymmetric state, 242, 266
Antiunitary operator, 20, 85, 102,
  111, 125, 163, 256, 278, 280, 287
Antiworld, 89
Approximation procedure by Landau,
  226
Arbitrary,
  coupling parameter, 178
  phase of state vector, 7, 14, 25, 96,
    105, 169
Attributes,
  additive, 111
  compatible, 2, 8
  complete set of compatible, 2, 17,
    31, 248

Attributes—*continued*
  continuous, 55
  dichotomic, 5, 7, 8
  incompatible, 2, 29
  intrinsic, 106, 126, 129, 263, 269
  obscure, 286
  of strongly interacting particles,
    263–287
Average value of,
  attribute, 11, 12, 19, 28
  density matrix, 26
  operator, 37
  polarization, 29

B

Balance,
  detailed, 100, 234, 257–261
  overall, 261
Barycentric energy, 296
Baryonic charge, 183, 184
Baryon number, 112, 129, 277–280,
  286, 287
Beta decay of Cobalt, 89
Boson,
  number, 131
  states, 117, 131–149
Branching ratio, 276
Broken symmetry, 269

C

Canonical,
  transformation, 299, 331–334
  variables, 43, 66
Charge,
  conjugation, 90, 161 (*see also* Par-
    ticle conjugation)
  conservation, 239, 272, 278
  independence, 271, 278
  renormalization, 227
  state,
    of nucleon, 8, 264
    of pion, 9, 273
  unrenormalized, 228

Operator—*continued*
  projection, 26, 202, 271
  quasi-particle, 299, 304
  rotation, 237
  scattering, 197, 202, 203, 255
  spin, 14
  time-odd, 105
  time-ordering, 197, 200
  transposition, 241, 245–249, 252
  unitary, 20, 21
  unpaired, 208
  vector potential, 144, 234, 252
Optical transitions, 239, 254
Optimum,
  knowledge, 7
  state, 3, 66
Orbital angular momentum, 77, 93, 238
Orthogonal,
  complements, 108
  systems of eigenvectors, 12
Orthonormality, 7
Overall balance, 261

P

Pairing, chronological, 205
Parity,
  conservation of, 87–96, 234
  definite, 92, 94, 135, 237
  intrinsic, 234, 236–240
  nonconservation of, 91
  relative, 239
  space, 92
  vacuum state, 96
Partially gauge invariant theories, 184
Particle,
  concept, 1, 115–117
  conjugation, 111, 126, 127, 136, 234, 251–254, 279, 287
Pauli matrices, 76, 139, 265
Permutation symmetry, 241–249
Perturbation theory, 203–215, 221, 227, 230
Phase,
  arbitrary, of state vector, 7, 14, 25, 96, 105, 169
  change in, 170
  convention, 124, 156, 279
  nonintegrability of, 170
  randomly distributed, 41

Phase—*continued*
  space, 29, 198, 266
  transformation, 173, 177, 239
Phonon, 73, 115, 290
Photon,
  lines, 210
  –photon interaction, 245
  propagator, 209, 219, 224, 228, 230
  stability of, 185
  state, 234
  theory of, 133
Polarization,
  average, 29
  circular, 141
  correlation of, 236
  degree of, 29, 145
  elliptical, 145
  events, chain of, 185
  longitudinal, 147
  operator, 139, 245
  photon, 133, 138, 245
  space, 135, 138, 146, 244
  state, 244
  time-like, 147
  transverse, 140
  vacuum, 185
  vector, 27, 50
Position variable, 59
Positron,
  lines, 209
  theory of, 151–168
Positronium, 239, 251–254
Potential in quantum mechanics, 172
Precession of Polarization vector, 50
Probability,
  amplitudes, 6, 19, 235
    -addition, of 40, 223
  definition, 37
  interpretation, 6, 7
  transition, 198
Product,
  chronological, 203
  normal, 204
  time-ordered, 203
Projection,
  operator, 26, 202, 271
  state vector, 6, 12, 19
Propagator,
  concept, 203–215
  hierarchy, 217–231
  true, 218

Date Due

MAY 7 1996			